Global Information Society Watch

2018

APC
IDRC | CRDI
Canada
International Development Research Centre
Centre de recherches pour le développement international

Global Information Society Watch 2018
Community Networks

Operational team
Roxana Bassi (APC)
Valeria Betancourt (APC)
Kathleen Diga (APC)
Alan Finlay (APC)
Michael Jensen (APC)
Carlos Rey-Moreno (APC)

APC project coordination team
Namita Aavriti (APC)
Roxana Bassi (APC)
Valeria Betancourt (APC)
Kathleen Diga (APC)
Anriette Esterhuysen (APC)
Flavia Fascendini (APC)
Alan Finlay (APC)
Chat Garcia Ramilo (APC)
Michael Jensen (APC)
Carlos Rey-Moreno (APC)

GISWatch 2018 advisory committee
Carlos Baca (REDES)
Luca Belli (FGV)
Jane Coffin (ISOC)
Kazanka Comfort (Fantsuam Foundation)
Stéphane Couture (York University)
Alison Gillwald (Research ICT Africa)
Michuki Mwangi (ISOC)
Leandro Navarro (PANGEA)
Dorothy Okello (WOUGNET/Makerere University)
Nico Pace (AlterMundi)
Steve Song (Village Telco/Rhizomatica)
Ritu Srivastava (DEF)

Project coordinator
Kathleen Diga / Roxana Bassi (APC)

Editor
Alan Finlay

Assistant editor and proofreading
Lori Nordstrom (APC)

Graphic design
Monocromo
info@monocromo.com.uy
Phone: +598 2400 1685

Cover illustration
Matías Bervejillo

This work was carried out with the aid of a grant from the International Development Research Centre (IDRC), Ottawa, Canada, as part of the APC project "Community access networks: How to connect the next billion to the Internet". More information at: https://www.apc.org/en/project/local-access-networks-can-unconnected-connect-themselves
The views expressed herein do not necessarily represent those of IDRC or its Board of Governors.

IDRC | CRDI Canada
International Development Research Centre
Centre de recherches pour le développement international

Financial support provided by
Sida

This edition of GISWatch came into being alongside a brand new baby boy. Welcome to the world, Ronan Diga!

Published by APC
2018

Printed in USA

Global Information Society Watch 2018 - print
ISBN 978-92-95113-05-3
APC-201810-CIPP-R-EN-P-295

Table of contents

Preface

Chat Garcia Ramilo
EXECUTIVE DIRECTOR, APC

This year's Global Information Society Watch (GISWatch) focuses on community networks. Community networks are "communication networks built, owned, operated, and used by citizens in a participatory and open manner."[1] This is a starting point. As the 43 country reports gathered here show, in practice, "community networks" can be hybrid systems, with different political and practical objectives. The country reports cover a diverse range of countries such as Georgia, Nepal, South Africa, India, Argentina, Honduras, Portugal, Germany and the Democratic Republic of Congo. Key ideas like participatory governance systems and community ownership and skills transfer, and the "do-it-yourself" spirit that drives community networks, give community networks across the globe a shared purpose and implementation methodology.

The country reports are framed by eight thematic reports. Some – for example, those by Steve Song, Mike Jensen and Nic Bidwell – draw on research conducted under the two-year "Community access networks: How to connect the next billion to the Internet" project implemented by the Association for Progressive Communications (APC) and Rhizomatica in collaboration with the Internet Society and funded by the International Development Research Centre (IDRC). The thematic reports deal with issues critical to the emerging global community network movement of which the Community access networks project is a part. The themes include the need for telecommunication regulation institutions to take into account the steep reductions in costs that wireless technologies have effected and to redesign regulation to further community networks; the need to increase awareness of "community stories" and the power structures embedded in those stories; the need to foster the transformation of local social structures and power relationships to enhance the agency of women and give them real power; the need to increase meaningful local content that is conducive to social change; and the need to explore ways for community networks to achieve financial sustainability.

This GISWatch edition was supported by a group of experts whose contributions, as members of the advisory committee, are gratefully acknowledged. We are pleased to present this edition to raise awareness of the tremendous potential of community networks to help achieve universal access.

1 See the introduction to this year's edition of GISWatch, "The rise and fall and rise of community networks".

Introduction
The rise and fall and rise of community networks

Steve Song, Carlos Rey-Moreno, Anriette Esterhuysen, Mike Jensen and Leandro Navarro
Fellow in Residence, Mozilla Foundation; Association for Progressive Communications (APC); Pangea.org
https://manypossibilities.net; https://www.apc.org; https://pangea.org

Community networks[1] pre-date the commercial internet. They have their roots in the early email and electronic bulletin board systems (BBSes) that emerged in the mid-1980s. These systems grew into networks, which were adopted by enthusiasts as technologies that could be easily built with dial-up modems and the newly emerging, low-cost personal computers. These networks were also adopted by social activists who immediately saw their potential for improved organising, knowledge sharing, and awareness raising.

The Association for Progressive Communications (APC), a global network of social activist organisations, has its roots in connecting community groups around the world with email and news[2] at a time when email was limited to a tiny fraction of society using standalone BBSes or computers in academic departments. When it became possible to network these systems together – using FidoNet[3] or UUCP,[4] for example – their affordability and accessibility helped to spread their use to social and political movements in communities around the world, particularly among those who did not have other reliable ways of communicating internationally.[5]

In this early "pre-internet" phase, APC facilitated the use of email and maintained hundreds of private discussion forums by and for non-governmental organisations, United Nations agencies, trade unions, universities, journalists and activists. These forums were batch-replicated among APC member or partner organisations around the world and made available locally, initially through dial-up modems and later also through public networks.[6] APC was, as a result, a global computer communications and information network maintained by many local member "networks" – organisations offering local access to this global resource network, often run as self-sustaining cooperatives or collectives.

With the growth of the commercial internet in the 1990s and the birth of the World Wide Web,[7] FidoNet and UUCP began to give way to commercial internet service providers (ISPs) who offered the entire internet as opposed to just email and newsgroups.[8] The users of these early internet services mainly relied on dial-up modems operating over copper phone lines. Unlike the store-and-forward nature of FidoNet/UUCP, dial-up internet required continuous use of a phone line. While this service spread rapidly around the world, it was limited to those who had their own phone lines with stable connections, and who could afford the monthly subscription fee (and, outside of North America which had free local calls, large phone bills). Not surprisingly, people in developing countries and the poor

1 Communication networks that are built, owned, operated and used by citizens in a participatory and open manner.

2 News at the time was exchanged through "newsgroups", primarily through "Usenet newsgroups" which allowed users to share news articles as well as discuss the content with others. Some newsgroups were used purely for debate and discussion. Usenet was developed over a decade before the public internet and the World Wide Web. APC provided access to both the public Usenet newsgroup, as well as to APC-run newsgroups. https://en.wikipedia.org/wiki/Usenet

3 https://en.wikipedia.org/wiki/FidoNet and, for a good introduction by Randy Bush, see https://www.fidonet.org/inet92_Randy_Bush.txt

4 https://en.wikipedia.org/wiki/UUCP

5 Murphy, B. (2001). Mike Jensen and the code that stitched together the APC: The pre-internet days and early efforts at linking APC nodes. *APC Annual Report 2000*. https://www.apc.org/about/history/mike-jensen-pre-internet-days

6 APC nodes were themselves interconnected by a wide range of technologies, from LEO satellites, to international X.25 packet lines, as well as local university links and international dial-up connections using the latest high-powered modems (such as the Trailblazers).

7 The first APC website in the internet archive: https://web.archive.org/web/19961028120226/http://www.apc.org:80

8 Surman, M. (2001). Where do we go from here? APC after the internet explosion. *APC Annual Report 2000*. https://www.apc.org/about/history/apc-after-internet-explosion

everywhere were the most constrained in their ability to access the internet. A solution that emerged around that time to directly address affordability, particularly in Europe and North America, which had good telecommunications infrastructure, was a type of community network called a Free-net.[9] Free-nets offered no-charge dial-up access and public terminals which allowed ordinary citizens to participate in discussion forums about local city topics and problems. Many of these Free-nets evolved into community ISPs.

But it was another low-cost, commodity technology that really enabled the independent growth of community networks. In 2003, it was discovered that Linksys, a manufacturer of Wi-Fi access points, had used software licensed under the GNU General Public License on the firmware of their access points.[10] Under such a licence, anyone using or changing the software must release it into the public domain on the same terms as the freely available original source. Linksys was compelled to release the source code for its flagship Wi-Fi access point, the WRT54G, to the community. This triggered a wave of tinkering and innovation with these devices, which in turn led to several important innovations.

Wi-Fi hackers discovered that the access points could be connected as peers to create a mesh network, allowing them to extend connectivity by placing them in proximity to each other. They further discovered that the antennas could be replaced with homemade directional "can-tennas" which could direct connectivity over several kilometres. The source code evolved into open-source operating systems for network devices, such as OpenWrt.[11] This gave birth to the community wireless movement which thrived in cities and universities around the world.

Although these networks were largely limited to reasonable proximity to an existing internet connection, they had a profound effect on affordable access as a single dial-up (ADSL) internet connection could be shared with an entire community. Examples such as Free2Air[12] in Europe and others in the global North provided the basis to learn and document experiences[13] that were then shared[14] and piloted in the global South.[15] Additional momentum was gained via the International Summit for Community Wireless Networks (IS4CWN) events held from 2004 to 2013, as well as the BattleMesh,[16] an annual event organised by developers of open source, ad hoc network routing protocols.

Towards the end of the 2000s, things changed. The rise of mobile data networks, first with 3G and then LTE services, provided an alternative to community wireless networks that was reasonably affordable and was often more reliable. Also, as demand for broadband grew, community networks often struggled to keep pace with demand for capacity. As a result, many of these networks either disappeared or shifted their focus to content hosting and services. Many commercial wireless ISPs suffered as well.

Some community networks like guifi.net[17] in the Iberian peninsula evolved to embrace fibre optic infrastructure as well as Wi-Fi and developed the practice of their community network through the exploration of the principles of common pool resources as developed by Elinor Ostrom.[18] The community networks that survived often relied on the extraordinary efforts of a few talented volunteers and a commitment to build and rely on their own cooperative networking and access infrastructure.

Fast-forward to the latter half of this decade and new trends have emerged with implications for community networks. Perhaps most significantly, the value of being connected has risen to the point where access to affordable communication has begun to rival access to other basic services in terms of personal priorities. More than a decade ago, researchers established that simple proximity to a communication network was directly correlated with a reduction in the probability of dying from malaria.[19] Today, with smartphones delivering powerful generic services like group and personal messaging and more specific apps aimed at critical sectors such as education, agriculture and others, communication networks are approaching the status of essential infrastructure for people's livelihoods. Affordable access to communication has gone from luxury to necessity no matter where you live or what your income.

And yet, over half of the world's population still does not have access to the internet. Traditional solutions are showing signs of having reached their

9 https://en.wikipedia.org/wiki/Free-net

10 Miklas, A. (2003, 7 June). Linksys WRT54G and the GPL. *LKML.ORG*. https://lkml.org/lkml/2003/6/7/164

11 https://en.wikipedia.org/wiki/OpenWrt

12 https://wiki.p2pfoundation.net/Free2Air

13 wndw.net

14 https://www.apc.org/en/project/wireless-lac-tricalcar and https://www.apc.org/en/wireless

15 www.fmfi.org.za/wiki/index.php/First_Mile_First_Inch_Home_Page

16 https://battlemesh.org

17 See the Catalonia country report in this edition of GISWatch.

18 https://en.wikipedia.org/wiki/Elinor_Ostrom

19 Mozumder, P., & Marathe, A. (2007). Role of information and communication networks in malaria survival. *Malaria Journal, 6*, 136. https://malariajournal.biomedcentral.com/articles/10.1186/1475-2875-6-136

limits. Mobile subscriber growth is slowing as the current economics of mobile network operators struggle to find viability in markets with subsistence-level incomes and/or in sparsely populated regions. It is also noteworthy that the same situation is being mirrored in the number of internet users, whose annual growth has slowed from 12% in 2016 to only 7% in 2017. Varied attempts to address this problem, through universal service strategies/ funds, private sector initiatives or philanthropy, have met with limited success.

This presents a conundrum for policy makers and regulators where value continues to accrue to those with affordable access to communication infrastructure while the unconnected fall further and further behind by simply staying in the same place. Those who most desperately need support are cut off from access to opportunity, to social and healthcare safety nets, to education, to information that can improve lives, and to platforms to demand change. It is ironic, or perhaps tragic, that the voices of the unconnected are not heard on this issue for the very reason that they are unconnected. And the problem extends beyond the unconnected. There are also the underserved. Lack of choice in access alternatives often results in a cost of access that is unaffordable for a significant percentage of the population (especially in rural areas) and/or in low quality or speed of service. In a context where government shutdowns are becoming a trend, and data privacy is becoming a growing concern to many, this lack of alternatives also compromises the freedom of expression of many users.

These unattended needs represent a challenge and an opportunity for community networks. If community networks were able to thrive and provide services effectively in the pre-internet era, might it be possible that they can do so again? There are a number of factors that suggest that the telecommunications infrastructure landscape has shifted yet again.

The spread of fibre optic infrastructure, both undersea and terrestrial, is changing the access market. Fibre optic networks are the deep water ports of the internet. While there is no question that fibre optic networks are increasing the ability of existing operators to deliver broadband, those same networks are opening up possibilities for new players who can now deliver more targeted, localised, affordable solutions to unserved and underserved populations. Where open access[20] policies exist, the spread of undersea and terrestrial fibre optic networks has democratised access to broadband.

Changes in last-mile technology are also opening up new possibilities. The spread of Wi-Fi as an access technology is empowering commercial, government and community access initiatives to offer local services. Dynamic spectrum technologies such as television white space (TVWS) also show promise as alternative access technologies.

Finally, the meteoric growth of access combined with mass manufacturing has brought down the cost and complexity of access technologies to the point where they are within the reach of small-scale operators. For example, low-cost solar-powered open source GSM base stations can be deployed for a fraction of the cost compared to the proprietary equipment used by existing mobile network operators.

All of these changes in the infrastructure and the technologies available are now being exploited in many imaginative forms by communities around the world to meet their communication needs. Those needs vary, and relate to issues such as a lack of services, the affordability or quality of access to voice and data services, or the lack of locally relevant content and services, often ignored by mainstream providers. But a community's communication needs go beyond just technical issues. In places where both commercial and community providers exist, users may choose to access communications via a community network because of trust, because of its commitment to local development, because it is customer friendly, or it preserves and defends their privacy better than other options available.

Yet while there are many good examples of community network success stories across the world, community networks are not yet the norm that they might become.

There are several reasons for this.

First there is a lack of awareness of opportunity. The more advanced community networks like B4RN[21] in the United Kingdom and guifi.net in the Iberian peninsula are offering broadband services that the incumbents cannot match on either speed or price, yet neither their performance nor the innovative commons-based business models they operate on are well known. Similarly, in the state of Oaxaca in Mexico, a non-profit, Rhizomatica,[22] is helping communities build their own GSM base stations and services. But their similarly remarkable achievement is also not as well known as it should be. More needs to be done to spread the word on

20 ITU. (2011). Open access regulation in the digital economy. https://www.itu.int/net/itunews/issues/2011/07/43.aspx

21 https://b4rn.org.uk

22 See the Mexico country report in this edition of GISWatch.

how these organisations are taking community networks from proven feasibility to proven scalability.

A second limiting factor is the current state of policy and regulation for community networks. Community networks have largely succeeded in spite of existing regulation rather than because of it. Regulatory frameworks were designed with large, relatively slow-moving, monolithic operators in mind. Changes in access policy and regulation are required, in particular with regard to spectrum management, in order to encourage communities to address their own access challenges. As the International Telecommunication Union (ITU) has recommended, it is "important that administrations, in their radio-spectrum planning and licensing activities, consider mechanisms to facilitate the deployment of broadband services in rural and remote areas by small and non-profit community operators."[23]

Another factor is related to the lack of technical and financial support to backstop those who may see the opportunities of a community network but lack either the technical expertise or the seed funding to get started. Universal, affordable access to communication will only be achieved when communities are empowered to solve their own local access challenges, instead of just waiting to be connected.

Finally, communities are composed of people, with their own background, social dynamics, and history. Community networks, like any other other collective initiative, also have to deal with the different sensitivities of everyone in the community to avoid clubs that perpetuate existing inequalities, with regard to gender, economic resources, or technical skills, amongst other areas. This is not always possible, and tensions and issues need to be resolved to enable everyone in the community to enjoy the benefits of the network.

Still, as the 43 country reports in this year's Global Information Society Watch show, many collectives around the world have managed to overcome these challenges. And, as in the pre-internet days, they are collaborating among themselves, exchanging information and learning from each other, and taking collective action at the local, national, regional and global levels to consolidate their work, and encourage more and more people to join what has become a global movement.

23 Recommendation ITU-D 19. See: International Telecommunication Union. (2017). *World Telecommunication Development Conference (WTDC-17): Final Report*, p. 634. https://www.itu.int/en/ITU-D/Conferences/WTDC/WTDC17/Documents/WTDC17_final_report_en.pdf

Thematic reports

At the limits of the internet: Technology options for community networks[1]

Leandro Navarro, Leonardo Maccari and Renato Lo Cigno
Pangea, Universitat Politècnica de Catalunya, Ninux, University of Trento
https://pangea.org, https://upc.edu, ninux.org, https://www.unitn.it

Introduction

Community networks are kinds of networking infrastructures built to provide local and global connectivity to interconnect people and devices and transfer messages and content. As with other critical infrastructures, the challenge is to reach everyone and anything requiring connectivity. This relies on a combination of technologies to optimise affordability, complexity, quality and performance.

This report discusses how connectivity works in general, the specifics of network access and backhaul technology, and the software considerations when setting up a community network. It then offers a list of key software and other resources useful to community networks.

Connectivity

Connectivity – the ability to connect or communicate with others – comes in units of links that are part of the internet. Links bring access to people or devices in a given location or as they move from location to location. Interconnected links spread over geographic areas to provide coverage forming access networks, or autonomous systems. Regions which are at the edges of the internet are connected through routers and long-distance links, also known as the backhaul network, which connects access networks to the core of the internet. In some cases, access providers use internet exchange points (IXPs), where autonomous systems meet and exchange or trade internet traffic to reach local content or transit providers. Transit providers allow customer networks to cross or "transit" the provider's network, usually to reach the rest of the internet. This can take the form of offering backhaul connectivity to networks. They do not offer connectivity to individual end-users. Technology in the form of standards, hardware and software artifacts, and their complexity, restrictions, performance, cost and evolution, determine the availability of connectivity or the lack of it.

The *electromagnetic spectrum* refers to the range of all frequencies of electromagnetic radiation. The internet relies on devices that generate, carry and read information encoded in this radiation in the form of waves in cables, in the air or even as light waves in optic fibres. Simply speaking, waves are oscillations, and at each oscillation some information is moved. The bandwidth of a signal refers to how wide the frequency range of the oscillation is. A wider frequency range, a *broad band*, results in higher data transfer speeds. Therefore, broadband is an old term to refer to connectivity from the perspective of the allocated spectrum bandwidth.[2] By being an always-on and faster internet access, fit for a wide range of uses, *broadband* is differentiated from slow and fragile access through telephone lines (dial-up, or *narrowband*). The term is carefully defined and politically loaded in the telecom regulation in every region[3] as it affects criteria for digital inclusion policies and public subsidies or investment.

Generally speaking, transmitting two different wireless signals at the same frequency creates interference, and may make it impossible to receive any information. For these reasons the public electromagnetic spectrum is divided in intervals (bands) and each band may be *licensed* or *unlicensed*. There are international agreements and regulations regarding this public spectrum,[4] but to use a licensed band one must obtain an authorisation from a local regulator (a public body that has the task of assigning bands of the public spectrum commons to operators, generally at a cost). Unlicensed bands[5] are free for use, which means that they must use

1 This work has been partially funded by the European Commission, H2020-ICT-2015 Programme, Grant Number 688768 "netCommons" (Network Infrastructure as Commons).

2 Analogous to comparing traffic characteristics of vehicles (packets) on roads (links) referring to the thickness of their asphalt (signal to noise ratio) or width of lanes (narrow or broad).

3 https://en.wikipedia.org/wiki/Broadband; regional definitions – EU: https://ec.europa.eu/digital-single-market/en/broadband-glossary and UK: https://www.ispreview.co.uk/broadband.shtml

4 https://en.wikipedia.org/wiki/Radio_spectrum; https://en.wikipedia.org/wiki/ITU_Radio_Regulations

5 https://en.wikipedia.org/wiki/ISM_band

FIGURE 1.

Spectrum properties. Frequency ranges (Hz), wavelength (metres), radiation. Wi-Fi is on the range of 109 Hz and light on 1014

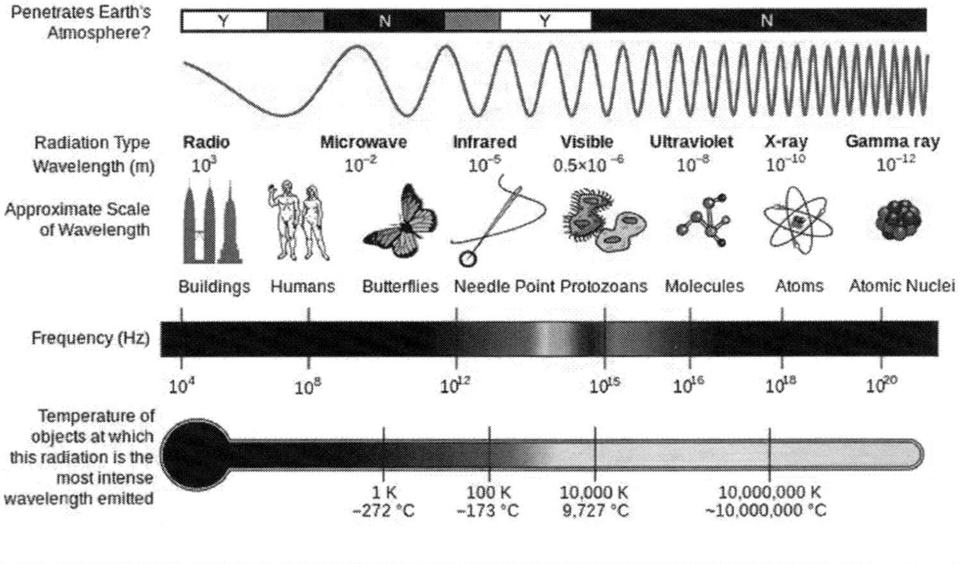

Source: https://commons.wikimedia.org/wiki/File:EM_Spectrum_Properties_edit.svg

technical means to survive congestion, and in general are considered less reliable. Cables and fibres create their own "private spectrums" that are separate from the public spectrum and do not interfere with each other.

More speed allows for more communication to happen simultaneously. Adequate connectivity allows running one or multiple apps with no visible degradation, which means not limited by data rate or fluctuations (congestion) at any part of the network path. Using the analogy of roads, if broadband is the asphalt, lanes carry data packets (cars) and roundabouts correspond to routers that do packet switching and queueing. Quality also includes reliability, the quality of being trustworthy or performing consistently well, and latency, the time interval between input and response, which depends on the length of the network path, given the unavoidable physical limits of propagation of electromagnetic signals.[6]

Access

Access generally refers to the first network link between a hardware device (also called a "terminal") and the network that reaches each user or server device. Mobile phones, and computers in general, link human beings to the internet, but many types of hardware serve as terminal nodes, such as servers, printers, cameras, and environmental sensors. These terminals are typically connected through cables, or without (wireless), with the first option offering a wider bandwidth at the cost of a higher price (cables need ducts to be deployed and reach our houses), and the second option being generally poorer in performance, but cheaper and supporting mobility.

Wired access usually reuses an existing wired cable such as a telephone copper line (dial-up, DSL) or TV distribution cable. The evolution towards more data traffic has led to the replacement of copper network segments with faster fibre[7] – for example, fibre to the cabinet in the neighborhood (FTTC) or to the premises/building (FTTP) extended by reusing existing copper cables to each premises, and full fibre access networks to each unit, business or home (FTTH). As fibre cables and fibre network devices get standardised and become easy and cheap to deploy, community networks have adopted the FTTH access model, with examples of full fibre in the Broadband for the Rural North (B4RN)[8] community network, and, mixed with some participants using wireless while others use full fibre, in the case of guifi.net.[9] These give typical access rates of around 1-10 Gbps.

6 Less than 300 metres per microsecond (light), comparable to the execution time of a CPU instruction.

7 https://en.wikipedia.org/wiki/Fiber_to_the_x

8 https://b4rn.org.uk

9 https://guifi.net/en/node/38392

Wireless cellular access networks: The road to 5G

Mobile devices get connected through some form of GSM (Global System for Mobile communications) standards or evolutions of it, nowadays described in terms of technology generations (from 1G to 5G). Each connectivity provider sets up base stations with omnidirectional antennas that cover a certain area (a "cell") and pays a fee to have a licence for the exclusive use of a part of the electromagnetic spectrum to serve its customers. The more spectrum allocated to a given mobile provider, the more customers will be able to communicate at the same time in the same cell. That leads to a competitive privatisation of the public spectrum, typically at country level, for commercial usage: optimised to maximise the profit, and therefore prioritising and optimising the most profitable market of dense (urban) population with higher disposable income.

The allocation of public spectrum is a good source of income for governments (through spectrum auctions) and a good source of business for mobile operators, but leaves less spectrum available for other public, community or private uses. In underserved areas, there is the tragedy of lock-out of allocated spectrum, kept idle by the licensee due to lack of profitability and preventing its usage by anyone else.

The evolution of mobile technology has brought faster data rates with more efficient data encodings, better support for mobility, roaming, internet packets, and different coverages for dense (a few metres with pico/femto cells) or sparse macrocell deployments up to a 100 km radius.

The future generation of mobile connectivity (5G) deserves special attention, as it is intended to be not only a technological update but a leap forward. The goal of 5G is to provide a 1,000-fold increase of the aggregate network capacity, with up to 10,000 connected devices per base station. Apparently, 5G will change the way we access the internet. This requires a large spatial densification of the base stations, and therefore a huge investment to install and connect them through high-capacity links (typically fibre or high-capacity directive radio links). Operators will generate revenues selling new kinds of applications that take advantage of the breakthrough in terms of bandwidth and communication delay. The need for densification shows the focus on more speed in smaller areas, as opposed to a focus on increased coverage. This will increase the cost per user and the profit of the service provider in a given location.

It is worth noting that more than eight years after the roll-out of 4G,[10] only 29% of the five billion worldwide mobile subscribers use it. Since 5G calls for an even larger capital expenditure for new infrastructure, it is legitimate to ask: who will mostly benefit from 5G? Will it ever reach the populations in the developing world, especially in rural areas? Or – most likely – will it simply widen the divide between those who are already well connected and can afford better connectivity at a higher price, and those who are still unconnected today, roughly 50% of the world population?

At the other end of the technological spectrum, we find community networks that work with mobile phones as user devices that are cheaper, simpler and lighter than larger devices. These cellular community networks rely on low-cost cellular access points (base stations). Hardware for these base stations is becoming available at lower prices (in the range of a few thousand USD and 50 W of power consumption) and runs open source software. Alongside technical advancements, proposals are being made on innovative ways to access spectrum, such as progressive regulation for spectrum access to promote social rights for communities or secondary spectrum access for digital inclusion.[11]

Wi-Fi access networks

The term *Wi-Fi* refers to a family of technologies for wireless radio components (technically belonging to the IEEE 802.11 standards, with multiple revisions and updates: 11b/g/a/n/ac/ax etc.) that have reached ubiquitous diffusion. Contrary to cellular access networks, Wi-Fi uses unlicensed spectrum, which cuts the deployment costs but also increases the risks of congestion.

At a very low price, a Wi-Fi *access point* can be used for an access link to another Wi-Fi device or to the internet (coverage of 100 metres or less) or, using directional antennas, Wi-Fi can provide high-performance point-to-point links (between only two devices, separated even by tens of kilometres).

Interconnected with access points through wired or wireless point-to-point links, Wi-Fi can expand the coverage of access networks and also create a backhaul network. The multipoint links, with sector antennas, result in a mesh network.

A *mesh network* is a network topology in which each node is capable of relaying data for any user of the network, not just the node owner. In mesh networks, all nodes cooperate in the distribution of data throughout the network to the mutual benefit of its participants. With each participating node, the reach, throughput and resilience of the network expands.

10 https://www.gsma.com/mobileeconomy/#techmigration

11 https://www.rhizomatica.org/blog

FIGURE 2.

Alternatives for Wi-Fi wireless radio links

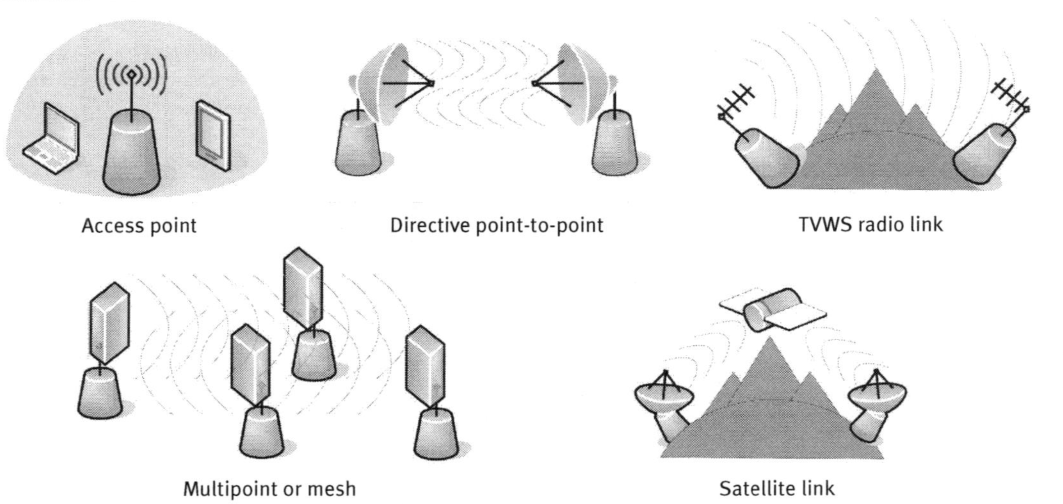

Source: Authors

Mesh networks are able to adapt to changes: when a node joins or leaves the network, the others automatically reconfigure to guarantee connectivity to the modified network. In some sense, they can grow "organically" with the growth of the community of people that use/manage them. A key ingredient of mesh networks is the routing protocol that can automatically select routes to enable multi-hop communication between any two nodes on the network. Combined with access points for user devices and internet gateways to reach the internet, mesh networks allow access networks to transform as new participants join, new areas are reached, and more capacity is added to links and internet gateways.

Many community networks based on mesh networks exist today, often led by volunteers who are able to set up networks to give coverage to large areas at a fraction of the cost it would require with cellular technology or cables.

The key observation is that if the price to bootstrap a network is lowered while capacity and cost grow incrementally with the growth of the community, this technology makes it possible to create networks that gradually expand with little planning or human coordination and give the time for the community to face the technical and organisational issues that come up along the way. Without a large initial capital expenditure for a spectrum licence and expensive infrastructure, it is much easier and less risky to create low-cost, bottom-up network infrastructures owned and managed by initially small communities of participants.

Other wireless access technologies

Another opportunity for long-distance communication is the use of the "white spaces" of TV spectrum, so-called TVWS, which are lower frequencies than Wi-Fi, and which were allocated to analog TV broadcasting (UHF and VHF) but are not used anymore. The standards for these radios are IEEE 802.11af and IEEE 802.22, also referred to as "White-Fi" and "Super Wi-Fi". The antennas look like TV antennas (both for access points and users) and have very good coverage, in the range of a radius of tens of kilometres without the need for line of sight.

What is known as the "internet of things" (IoT) allows us to connect "slow" devices, such as sensors, using very-long-range transmissions (more than 10 km in rural areas) with low power consumption and very slow data rates. One popular example of this is LoRa/LoRaWAN.[12]

On the higher part of the spectrum, beyond microwaves, we find millimetre waves, in the range of 30-300 GHz, with one licence-free ISM band[13] at 60 GHz. The IEEE 802.11ad standard, also known as wireless gigabit or WiGig, promised very directive in-room or open space multi-Gbps communication

12 The Things Network follows a model comparable to a federation of community networks. See: https://www.thethingsnetwork.org

13 The industrial, scientific and medical (ISM) radio bands are radio bands (portions of the radio spectrum) reserved internationally for the use of radio frequency energy for industrial, scientific and medical purposes other than telecommunications. https://en.wikipedia.org/wiki/ISM_band

FIGURE 3.

Coverage of different technologies, applicable to licensed and unlicensed bands

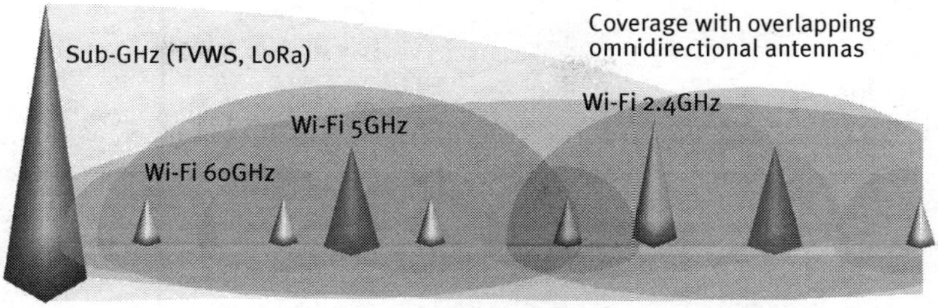

Areas can be covered by multiple antennas in different bands and coverage with omnidirectional antennas or point-to-point links:
- Sub-GHz (TVWS, LoRa ~ 10-30 Km omni)
- Wi-Fi 2.4GHz ~ 100 m omni + few walls, ~ 100 Km p2p
- Wi-Fi 5GHz~10 m omni, 0-1 walls, ~100 Km p2p
- Wi-Fi 60GHz, few metres omni, 0.1-1 Km p2p

Higher frequencies => faster speeds but less coverage
Results heavily influenced by power, obstacles, interference

Source: Authors

in the range of a few to perhaps a few hundred metres. However, this specification has not succeeded in the market, needing expensive and niche devices. Instead there are alternative wireless gigabit proprietary products in the 24 GHz ISM band.[14] IEEE 802.11ay targets even higher speeds of up to 20 Gbps with a final specification expected by 2019 and products a few months later.

Satellite access[15] only appears as a competitive solution for low population density areas, in the range of less than a few tens of inhabitants per square kilometre. Beyond that it becomes too costly compared to the alternatives. The added latency comes from propagation delay considering the radius of the orbits, 1,000 km high or 12 ms for low earth orbits (LEO), 10,000 km or 120 ms for medium earth orbits (MEO), and 36,000 km or 480 ms for geostationary orbit (GEO). The satellite latency is a reason in favour of high-altitude planes or balloon networks that operate in the stratosphere, at altitudes around 20 km[16] with less than 1 ms latency.

The service cost for satellite is determined by the number of subscribers in the coverage area, the cost of the satellite in orbit, and the base stations on the supplier side. The frequency bands used for internet traffic are C: 4-8 GHz, Ku: 12-18 GHz, Ka: 26-40 GHz with antennas for users of 2.5 metres in the C band, 1 metre in Ku band or less than that in Ka band. The achievable data rates, assuming the total capacity of the satellite is not saturated, can be up to 16 Mbps for 99.995% of the time for the C band, 64 Mbps for 99.9% of the time for the Ku band, and 512 Mbps for 99.7% of the time for the Ka band in a temperate climate, as rain has more of a fading (degradation) effect on higher frequencies. Satellite internet access has unique characteristics in coverage, but is expensive given the cost of build, launch, capacity and latency. Although an expensive access technology for any individual user, some community networks in remote environments may benefit from sharing the cost and capacity of satellite connectivity as one of the sources of connectivity in the backhaul to reach the internet, but ideally not the only one.

Beyond technological details and choices, both Wi-Fi and mobile technologies evolve side by side: while mobile operators evolve towards LTE and 5G, successive generations of Wi-Fi technology also offer faster and cheaper devices capable of serving more users (e.g. MIMO) with faster data rates (new modulation schemes) in the range of gigabits per second but covering smaller areas. Who will win out between mobile operators or Wi-Fi device vendors? Probably both will coexist and complement each other, but definitely one is based on a "centralised" operator model with its own reserved radio spectrum, and the other is "self-provided" or "decentralised" using shared and unlicensed radio spectrum.

14 See, for example: https://www.ubnt.com/airfiber/airfiber24-hd

15 https://youtu.be/YDedVZo4aqk?t=8s

16 https://en.wikipedia.org/wiki/High-altitude_balloon; https://en.wikipedia.org/wiki/Loon_(company)

Backhaul

Beyond access networks, network interconnection relies on long distance links that carry aggregate traffic, IP packets, from/to the internet. These links are provided by internet service providers (ISPs), which can be retail providers (one community network sharing one or several retail internet connections: fibre, DSL or satellite), wholesale internet transit providers, with points of presence reachable over fibre or high-speed point-to-point radio links, or IXPs, with the presence of multiple network providers (internet carriers) and content providers (content distribution networks or CDNs). The interconnection fees in these IXPs may depend on the symmetry of the traffic (cheaper or even free for a community network with a balanced mix of content that is generated by the network and users or readers of content on the network, while more expensive for networks that only have users or readers of content, the latter also called "eyeball networks"). Community networks can even formally or informally become IXPs in regions without any. As mentioned before, satellite can be one ingredient of the backhaul for community networks in remote areas, but ideally not the only one.

A growing development to facilitate connectivity is the availability of open access optical regional networks (or fibre-equivalent radio links) that provide wholesale/volume connectivity to reach interconnection points, carriers or build access networks. These shared infrastructures,[17] developed cooperatively or competitively, benefit nearly everyone locally, and therefore may be supported by large users such as governments, education institutions or the private sector, and may create economies of scale of competitive dark, active fibre or ethernet circuits to facilitate regional connectivity. Availability and cost efficiency in regional connectivity increase the opportunities to provide more services to more people. Recommendations regarding functional separation can keep incumbents from unhealthy competition and overbuilding, and facilitate community networks to scale up their deployments while reducing the cost. Community networks are effective in aggregating traffic from different stakeholders and sharing internet access, which directly translates into a significant reduction of cost for internet connectivity. This is the case with several community networks that share and rent wholesale open-access fibre for regional connectivity.[18]

Another barrier for the deployment of backhaul cables is the occupation of public space (through rights of way) by "private" infrastructures for private use. Municipalities are in charge of regulating this. Beyond laws to facilitate deployments, the guifi.net Foundation has developed the *universal deployment model*,[19] a template for a municipal ordinance to help promote the development of commons infrastructures. This template helps municipalities to avoid any discrimination and facilitate infrastructure deployments that are mutually beneficial for governments and private and community use. The principle is that any cable for private use to be deployed on public land is required to assign fibres for public (municipal) and shared/commons usage. This results in a public and community infrastructure at minimal cost (the private actor takes on the installation and maintenance costs in exchange). Therefore, the universal deployment model simultaneously allows for the three uses described, which results in infrastructure to expand community networks. The model can be extended from municipal land nationally, regionally and internationally (overseas), and even govern the use of undersea fibres.[20]

Hardware and software

The behaviour of the building blocks that produce connectivity is controlled by software. There is the software needed to run the network: routing protocols, authentication systems, and wireless/wired drivers for link adapters. There is also the software needed to monitor and manage the network, or network management and planning tools. We do not have a wide range of these in open source, or not as stable as proprietary products. The typical discussion in community networks is on the tradeoffs with regard to openness when choosing between proprietary and open source solutions (efficacy vs lock-in in integrated components).

Hardware is an area where openness is lacking. Mostly anything directly related to it is still kept

17 For more information, see the APC project "Infrastructure Sharing for Supporting Better Broadband and Universal Access": https://www.apc.org/en/infrastructuresharing

18 See, for example, Waites, W., et al. (2016). *RemIX: A Distributed Internet Exchange for Remote and Rural Networks*. https://arxiv.org/abs/1603.08978 and Baig, R., et al. (2016). *Making Community Networks Economically Sustainable: The guifi.net experience*. people.ac.upc.edu/leandro/pubs/baig-sigcomm.pdf

19 Most recent version in English (outdated): people.ac.upc.edu/leandro/docs/ordinancePEIT-rev14-en.pdf; updated version in Catalan: https://fundacio.guifi.net/web/ content/2322?unique=cef4bebe39b45ba50ed5ebb5e2a63ecaf07e6cb4&download=true

20 Navarro, L. (2018). Network deployments for universal connectivity. Presentation at IETF 102, 14-20 July, Montreal, Canada. people.ac.upc.edu/leandro/docs/ietf-102-universal.pdf

closed source, with protected intellectual property and product secrets held by the industry for hardware such as radio boards, radio firmware, device drivers, and programming interfaces. Notable exceptions of open hardware are the Mesh Potato,[21] LibreRouter[22] and Turris Omnia[23] routers and the software-defined radios for GSM such as UmTRX[24] or the USRP[25] family.

Proprietary hardware often requires the use of proprietary software, more expensive and potentially less secure due to the lack of public scrutiny. It is also less adaptable because it lacks the possibility of contributing bug fixes or alternative implementations. Most community networks rely on proprietary hardware and software black boxes, or a mix of open and closed source, for the previous reasons. Fortunately, there are a range of standards and public specifications which allow interoperable interconnection of components from different sources. The community of open source developers has made and is working on an impressive list of key solutions for community networks (see our list below for details).

Open specifications and standards,[26] in comparison to proprietary specifications, are key to promoting software and hardware alternatives, reducing cost and promoting specialised and optimised components that are interoperable. Public research helps to address the needs of the population, exploring challenges with high societal impact, in contrast to research in industry, which is typically focused on the development of competitive advantages and economic benefits that benefit private profits and shareholders first. In fact, Elinor Ostrom identified this requirement in the task of designing sustainable, complex human-resource systems: "Building respectful collaborations between local users, public officials, and scientific experts is a vital requisite of adaptive governance."[27]

Software and other resources for community networks

The following is a set of typical software and related resources used in community networks.

21 https://villagetelco.org/mesh-potato

22 https://librerouter.org

23 https://en.wikipedia.org/wiki/Turris_Omnia

24 https://umtrx.org

25 https://en.wikipedia.org/wiki/Universal_Software_Radio_Peripheral

26 Such as those promoted by organisations such as the Institute of Electrical and Electronics Engineers (IEEE) or the Internet Engineering Task Force (IETF).

27 Ostrom, E. (2008). The Challenge of Common-Pool Resources. *Environment: Science and Policy for Sustainable Development, 50*(4), 8-21. https://doi.org/10.3200/ENVT.50.4.8-21

Wi-Fi access points

- **OpenWISP:** A software platform that can be used to implement a complete Wi-Fi service, including managing access point devices, captive portals, user credentials, accounting data and monitoring.[28]

Cellular mobile access

- **Osmocom:** An umbrella project focused on open source mobile communications; includes software and tools implementing a variety of mobile communication standards, including GSM, DECT, TETRA and others.[29]
- **OpenBTS:** BTS stands for base transceiver station. OpenBTS is an open source software-based GSM access point, allowing standard GSM-compatible mobile phones to be used as SIP endpoints in VoIP networks.[30]

Firmware for routers

- **OpenWrt:** An open source project for an embedded operating system based on Linux, primarily used on embedded devices to route network traffic.[31]
- **Quick Mesh Project (qMp):** A system for easily deploying Mesh/MANET networks using Wi-Fi technology. qMp has been designed to be used in any scenario, such as free community networks, corporate networks, large social events, quick network deployments, etc. The qMp firmware, based on OpenWrt, works on many embedded Wi-Fi network devices.[32]
- **LibreMesh:** LibreMesh is an initiative of community network members from different continents to unite efforts in developing tools to facilitate the deployment of free networks for any community in the world. The main tool is the LibreMesh firmware, based on OpenWrt, which standardises the creation of Wi-Fi communities and provides roaming to existing ones. This project was initiated to merge a number of pre-existing firmware projects: AlterMesh (from AlterMundi, Argentina), qMp (from guifi.net, Catalonia) and eigenNet (from eigenLab, Ninux, Italy).[33]

28 http://openwisp.org

29 https://osmocom.org

30 https://en.wikipedia.org/wiki/OpenBTS

31 https://en.wikipedia.org/wiki/OpenWrt

32 https://qmp.cat

33 https://libremesh.org

Routing protocols

There are two families of protocols: distance-vector routing protocols, based on the Bellman-Ford algorithm, share only aggregated information about the path metrics, whereas link-state routing protocols, based on the Dijkstra algorithm, share the whole view of the network, and the metric of every single link is known by every router. Therefore, in link-state routing, every router has a global map of the network, whereas distance-vector routing only takes into account vectors (arrays) of distances to the other routers in the network. The most popular distance-vector routing protocols in community networks are Babel and BMX, while Batman-adv and OLSR are widely used for link-state routing.[34]

- **Babel:** A distance-vector routing protocol for internet protocol packet-switched networks that is designed to be robust and efficient on both wireless mesh networks and wired networks. In June 2016, an Internet Engineering Task Force (IETF)[35] working group was created whose main goal is to produce a standard version of Babel.[36]
- **BMX6/7:** BMX6 (or the new version BMX7) is an IPv6 native dynamic routing protocol which offers very advanced features and a small network overhead (thanks to the distance-vector strategy and a set of optimisations). Version 7 has, in addition, security routing extensions.[37]
- **Batman-adv:** B.A.T.M.A.N. (better approach to mobile ad-hoc networking) Advanced, known as Batman-adv, is a link-state mesh routing protocol which runs in the kernel space. Even if the network topology is made by multiple nodes and multiple hops, it abstracts it to a single layer 2 collision domain, so from the user perspective the entire mesh will look like a single LAN. This architecture is very interesting for roaming purposes given that network connections are not lost even when moving and changing access points.[38]
- **OLSR:** The Optimised Link State Routing protocol is an IP routing protocol optimised for mobile ad hoc networks. OLSR is a proactive link-state routing protocol, which uses control messages[39] to discover and then disseminate link-state information throughout an ad hoc mobile network. Individual nodes use this topology information to compute next hop destinations for all nodes in the network using shortest hop forwarding paths.[40]

Network management

- **Prometheus:** A metrics collection and monitoring system that is particularly well suited to community networks, with data exporters for network nodes, including network traffic and BMX6/7 routing metadata.[41]

Network description

- **netJSON:** A data interchange format based on JavaScript Object Notation (JSON) designed to describe the basic building blocks of layer 2 and layer 3 networks. It defines several types of JSON objects and the manner in which they are combined to represent a network: network configuration of devices, monitoring data, routing information, and network topology.[42]

Applications

- **FreeSWITCH:** A free and open source application server for real-time communication, WebRTC, telecommunications, video and VoIP.[43]

Fibre planning

- **FiberFy:** An application for those who develop fibre networks. It allows implementers to plan deployments and maintenance, define coverage areas, prepare projects and budgets, etc. It allows the sharing of information among actors who can intervene in a fibre deployment: suppliers, local administrations, and retailers.[44]

Governance and economics

See guifi.net's governance principles[45] and economic compensation system,[46] or the community shares for investment in B4RN.[47]

34 http://dsg.ac.upc.edu/eval-mesh-routing-wcn

35 https://www.ietf.org

36 https://www.irif.fr/~jch//software/babel

37 http://bmx6.net

38 https://www.open-mesh.org

39 https://en.wikipedia.org/wiki/Optimized_Link_State_Routing_Protocol#Messages

40 https://en.wikipedia.org/wiki/Optimized_Link_State_Routing_Protocol

41 https://prometheus.io

42 http://netjson.org

43 https://en.wikipedia.org/wiki/FreeSWITCH

44 https://github.com/guifi; https://guifi.net/ca/node/107850

45 Baig, R., Roca, R., Freitag, F., & Navarro, L. (2015). guifi.net, a crowdsourced network infrastructure held in common. *Computer Networks, 90*. people.ac.upc.edu/leandro/pubs/crowds-guifi-en.pdf

46 Baig, R., Dalmau, L., Roca, R., Navarro, L., Freitag, F., & Sathiaseelan, A. (2016). Making Community Networks Economically Sustainable: The guifi.net experience. Paper presented at the 2016 Workshop on Global Access to the Internet for All (GAIA), 22-26 August, Florianopolis, Brazil. people.ac.upc.edu/leandro/pubs/baig-sigcomm.pdf

47 https://b4rn.org.uk/b4rn-community/investors

Legal framework for community networks in Latin America

Erick Huerta
Rhizomatica
https://www.rhizomatica.org

Introduction

Although community networks are not a recent phenomenon in Latin America,[1] they have very little regulatory oversight in the region, given that most of the legislation has focused on addressing the behaviour of large service providers and the markets in which they operate.

The development of regulations that facilitate the coverage of areas not yet served by internet providers necessarily involves the creation of a regulatory framework that provides certainty and access to the infrastructure required by community networks to function. This is because up until now, community networks have been the only sustainable model for connectivity in underserved areas in the region.

This report provides a starting point for the creation of a regulatory framework for community networks. We start from the basis that the best regulation is that which only appears where it is necessary, since over-regulation can constitute an obstacle to the growth of any industry and to the achievement of the objectives it intends to serve.[2] While new regulations are necessary, the report shows that the existing legal framework in the region can be drawn on when it comes to issues such as spectrum allocation, essential infrastructure or, where appropriate, the licences that community networks require. Although this report is based on Latin American examples and experience, its logic can probably be applied in any country.

Legal nature of community networks

Considering their network architectures, business models, operating and organisational models and purposes, community networks have a specific legal character that finds its place in existing categories of regulation, regardless of whether or not there is a specific category called "community network" in the legislation of a given country.

The legal nature of a network allows us to establish the parameters with respect to which it must be regulated, whether or not it needs to have a licence and, if applicable, the characteristics that such a licence should have.

In order to establish the legal nature of a network, it is necessary to understand its architecture, its form of organisation and its purposes. This allows us to consider the legal categories that already exist and that are applicable to it.

In general, community networks can be grouped for legal purposes into three categories: those that can be categorised as self-provisioning networks, those providing services, and mixed or hybrid systems. There may be subdivisions of these categories, but while these may be useful in establishing regulatory particularities,[3] they are not essential when defining the legal nature of community networks.

The subcategories are also defined according to criteria that are important for each country. For example, for one country it may be relevant to establish a distinction between state-owned and commercial networks, while for another this distinction may not be necessary. Because of these particularities, we only discuss the three categories mentioned above in this report.

Self-provisioning networks

This type of network is made up of communities or organised groups that decide to share a telecommunications service through their own network;

1 See Galperin, H., & Girard, B. (2007). Microtelcos in Latin America and the Caribbean. In H. Galperin & J. Mariscal (Eds.), *Digital Poverty: Latin American and Caribbean Perspectives*. IDRC. https://www.idrc.ca/en/book/digital-poverty-latin-american-and-caribbean-perspectives

2 The Telecommunications Regulation Handbook published by the International Telecommunication Union (ITU) in 2011 clearly defines the objectives of telecommunications regulation, and explains that it is not about regulating just to regulate, but to meet four basic objectives: to increase access to technology and services, avoid market failure, foster effective competition and protect consumer interests. The Handbook is available at: https://www.itu.int/pub/D-PREF-TRH.1-2011

3 An interesting study that describes the different organisational models of different community networks is Navarro L. et al. (2017). *Report on the Governance Instruments and their Application to CNs (v2)*, produced as part of the EU-funded netCommons project. https://www.netcommons.eu/sites/default/files/d1.4_cn-governance_v1.0-2017-12-30.pdf

they have a non-profit purpose and they build or share common infrastructure.

This type of network can use free or licensed spectrum; its interconnection is generally through another network, through which they connect with the internet.[4] Legally they might form a non-profit association or consumer cooperative, just to name a few of the possible governance models.

Examples of these networks are the QuintanaLibre network started by AlterMundi[5] in Argentina and Telecomunicaciones Indígenas Comunitarias (TIC A.C.)[6] in Mexico. QuintanaLibre is a community mesh network that provides internet services in remote areas of the Córdoba region and TIC A.C. is a cellular telephony community network operated by indigenous communities in the state of Oaxaca, Mexico.

Because of their characteristics, the networks are private networks, since they only serve their members and do not normally manage direct interconnection links.[7] In a way we can say that they are similar to a switch in an office building. Basically, they receive services from one or more provider, and redistribute them inside their network, sharing the costs. In the case of AlterMundi, these are internet services, and in the case of TIC A.C., cellphone services.

Most countries make allowances for private networks that do not require licensing in their legislation, as long as they are fixed networks or use free-use spectrum for their wireless links.

When this type of network uses licensed spectrum, it is necessary to request a licence or permit, depending on the existing regulations. At this point it becomes essential to distinguish these networks not only in terms of their legal nature, but in terms of their purpose; otherwise there would be no difference between how we treat a private network of a commercial company, a public-private network, and a private network set up by a marginalised community and serving an area that no other service provider is interested in.

In these cases it is necessary to establish a specific modality that recognises the purposes of the network and even the type of community that requests it. In some countries this is already the case: in Mexico a social sector licence exists with two variations, one for "community" groups and one for "indigenous communities". Both of them can be granted access to spectrum without being subject to an auction.

The lack of recognition of the social purpose of a community network, as happens in countries that have the auction as the only model to assign licences, significantly limits the possibilities of access to the spectrum for community networks. This can constitute a barrier to competition and, at the same time, deprive communities of several human rights.

In summary, we can say that if a community network meets the following characteristics, it does not require a licence (depending on the specific laws of each country, of course):

- It operates as a self-provisioning network.
- In the case of telephone networks, it does not have direct interconnection, and only redistributes an access service.
- It uses free-use spectrum.

However, if a network complies with a) and b) but uses licensed spectrum, then it will require a licence. In these cases it is necessary for the country to have specific legislation in place considering the purposes of the network and the areas in which it intends to operate. This could even mean obtaining a different licensing scheme for primary use (where protection from interference is guaranteed) or secondary use (where protection from interference is not guaranteed), depending on whether or not the spectrum has been allocated to another provider.

Networks that provide services to third parties

There are networks that have a telecommunications infrastructure constituted as a common good, but can provide services to third parties that are not necessarily owners/members of the network. This small difference makes them providers of telecommunications services and depending on the legislation, and whether or not they use licensed spectrum, they require some type of licence.

4 When referring to "interconnection" it is necessary to distinguish between telephone networks and internet networks. In the former, when there is interconnection with other networks, it is necessary to provide services to users not belonging to the network, which is why these are no longer self-provisioning. In the case of internet networks, the connection is to a network of networks, and the interconnection does not change the nature of the service.

5 AlterMundi is a civil association based in Argentina that works to promote a new paradigm based on freedom through peer collaboration. AlterMundi explores different manifestations of this collaboration from a technological perspective and with an emphasis on community wireless networks in rural areas and small towns. Through projects such as LibreMesh and LibreRouter, they contribute to a model based on accessible technologies that can be handled by people without prior technological knowledge. See https://www.altermundi.net and the Argentina country report in this edition of GISWatch.

6 Telecomunicaciones Indígenas Comunitarias A. C. is a federated network of cellular telephone networks of indigenous communities. It has published a Manual of Community Cellular Telephony that describes its operations (available at: https://www.redesac.org.mx/telefoniacomunitaria). See https://www.tic-ac.org and the Mexico country report in this edition of GISWatch.

7 https://en.wikipedia.org/wiki/Interconnection

Networks that are self-service but perform direct interconnection in telephony also fall into this category, and will require numbering resources, quality of service agreements, and all other obligations that arise from the interconnection of these types of networks.

Examples of this category of community networks are B4RN[8] in the United Kingdom or the telecommunications cooperatives of Argentina. These networks, although they are constituted as community networks (i.e. the infrastructure belongs to a specific community), can provide services to non-members of the network, and therefore we can say that they are public telecommunications networks that provide services to the general public.

It is their purposes which define the particular characteristics of this type of network, not their architecture. That is, their specific legal treatment derives from their form of economic or social organisation, rather than their networking architecture. This way of organising can give them special tax or legal treatment, for example, by being a non-profit association or cooperative.

Although these types of networks require a licence even if they do not use spectrum, in some countries they can benefit from a simplified licensing model or from spectrum reserved for them.

This distinction is normal in the case of radio broadcasting, and a similar principle should apply in telecommunications. For example, if a country recognises special legislation for community broadcasting that is derived from its form of governance and its purposes, when the means of transmission is changed, but not the form of organisation or purposes, the same principles should apply. This is because there is a general principle of law that says: where there is the same reason, there must be the same provision.

If we also take into account that technological convergence allows the provision of different communication services, by establishing an artificial distinction for a certain type of community media, based only on the kind of technology it uses, this could turn into a barrier to the exercise of the right to freedom of expression or a barrier to entry to markets.[9]

There may also be networks with a commercial purpose, but aimed at a specific market segment that is not serviced. In these cases the licensing model can be simplified to facilitate their attention to this segment, such as the simplified licence in Brazil for operators that serve localities of less than 5,000 inhabitants.

Countries such as Mexico and Argentina have a special regime for community or social operators and, in the case of Mexico, spectrum specifically reserved for these purposes. In the case of Mexico, the spectrum segment in the GSM band assigned to social uses is not exclusive, since it can be granted at the same time for social and commercial use – the rural and remote areas are far enough away from commercially viable areas so as not to cause interference.[10]

In summary, we can say that community networks start to look more like commercial telecommunications networks, if they provide services to third parties or, in the case of telephony, if they perform direct telecommunications interconnection.[11]

In these cases, the licensing model may consider the purpose or scheme of a specific organisation and create a specific licence; but unlike the first category, these networks usually require a licence.

Mixed networks

In these cases the network constitutes a separate infrastructure from the services that are provided and has a different legal status compared to a model where there is a service operator: the network becomes a separate entity that is not owned by any operator.

This type of network consists of the aggregation of user nodes into a network, where the users contribute their local infrastructure to create a common infrastructure. There are, in this model, multiple pieces of a network that are added together to form a single one. This is something similar to what is known in civil law as servitude: where private property or goods are also used for the benefit of others, as is the case with the right of way, where the good (in this case a piece of land) remains the property of the owner but he or she must allow passage and not hinder the passage of others across that land.

8 https://b4rn.org.uk

9 "Regulation inspired by the principle of neutrality must avoid discriminatory effects among other technologies at the same time as favouring the development of ICTs. Broadly speaking, the regulatory principle of technological neutrality is based on four commitments: non-discrimination, sustainability, efficiency and consumer certainty." Cullell March, C. (2011). El principi de neutralitat tecnològica i de serveis a la UE: la liberalització de l espectre radioelèctric. *IDP. Revista de Internet, Derecho y Política, 11.* https://idp.uoc.edu/articles/abstract/10.7238/idp.voi11.1017

10 Programa Anual de Bandas de Frecuencia 2016, Instituto Federal de Telecomunicaciones, Mexico.

11 To determine if there is interconnection, it is advisable to observe whether the network requires interconnection agreements for the provision of its services, and if the interconnection is made using its own resources or that of another operator.

Its existence in the law is old and still enforced, as is the case with transhumant grazing routes.[12]

This licensing model can exist without a specific entity that owns the network, because it is enough to have a governance agreement for the network. These are interconnection agreements between different owners of nodes or network segments, which can be operators, users, universities, community networks, municipalities, governments, etc.[13] None of them owns the network and therefore nobody operates it in its entirety.

In these cases it depends on the legal nature of each node owner whether the network needs a licence or not. If a user is a telecommunications service provider, it will require a licence, but if it is a private network, it will not.

Conclusions

To determine the need for a licence for a community network, you have to consider the network architecture in the first place and the infrastructure it uses (free spectrum, licensed or shared infrastructure). If the network is private and uses free spectrum, it probably does not require a licence.

If it is a public or private network that uses spectrum, its purpose and form of social/legal organisation must be analysed to see if it fits a specific type of licensing scheme.

12 See, for example, the chapter on "Ownership, Tenure Regime and Use" in the *White Paper on Transhumance in Spain* by the Ministry of Agriculture, Food and Environment of Spain. It shows how the property regime of transhumant lands is varied, but mainly public, and how the use of the land affects its legal status. https://www.mapa.gob.es/es/desarrollo-rural/publicaciones/publicaciones-de-desarrollo-rural/LIBRO%20BLANCO%202013_tcm30-131212.pdf

13 See, for example, the commons approach of guifinet: https://guifi.net; for a more in-depth study, see Navarro et al. (2016). A Commons-Oriented Framework for Community Networks. In L. Belli (Ed.), *Community Connectivity: Building the Internet From Scratch. Annual Report of the UN IGF Dynamic Coalition on Community Connectivity*. https://www.intgovforum.org/multilingual/index.php?q=filedepot_download/4391/1163; see also the Catalonia report in this edition of GISWatch.

Community networks and telecommunications regulation

Steve Song
Fellow in Residence, Mozilla Foundation
https://manypossibilities.net

Regulation should not just be about large operators

The world of cables and radio waves that make up the underlying physical communication infrastructure on which the internet is built is fundamentally different from the digital world of the internet where permissionless innovation rules. Where affordable access to the internet exists, the barriers to manifesting a work digitally are extremely low. Although there are signs that this may be changing, the internet remains a realm largely free of regulation. Digital producers require no licence or certification to create, just the willingness to invest the time and effort in the production.

Telecommunications infrastructure comes from a very different history. As centrally controlled, top-down networks based on a command-and-control philosophy, their underlying conception is radically different from the more organic, bottom-up network of networks that is the internet. There are reasons for this. At the time the first large-scale telecommunication networks were being developed, their construction was an effort so extraordinary and expensive that they were typically only undertaken by national governments. Following the tradition of postal services, countries undertook the provision of telecommunication infrastructure as a public good.

This began to change in the early 1990s as, around the world, governments began to embrace privatisation as a means of addressing inefficiencies in state-run infrastructure monopolies as well as a means of generating revenue for the exchequer. In most countries, privatisation was accompanied by a process of market liberalisation allowing for competition for the first time. Part of the privatisation and liberalisation involved the establishment of regulatory frameworks and organisations to ensure that the public good was still being served and that the resulting privatised and liberalised market was fair, open and competitive.

It is not surprising that these regulatory bodies were designed to deal with large-scale national companies, because it required millions (even billions) of dollars of investment to build a national communication network including the international connectivity, national backhaul (long-distance, high-capacity infrastructure) and last-mile infrastructure.

Because of this, most telecommunication regulatory frameworks are designed with these large corporations in mind, with implications for organisational capacity. It is implied, for example, within most regulatory processes and requirements that the applicant has the time and resources that the legal department of a large telecommunications corporation might possess to fulfil detailed applications and reporting requirements as well as draft submissions and commentary on new proposed regulations. With the new norm of auctioning high-demand spectrum, it is assumed that any telecommunications organisation should have the millions of dollars required to bid on spectrum licences.

An exception to the above model has emerged, however, with the growth of the use of Wi-Fi technologies. Wi-Fi equipment operates in the licence-exempt frequency bands which are regulated through technological constraints rather than the requirement of a user licence. The licence-exempt nature of Wi-Fi has created a very low market barrier for both manufacturing and deployment of this technology. Wi-Fi has proven successful as both an access technology and a backhaul technology, making it suitable for a wide range of deployments. It is estimated that the Wi-Fi market will be worth USD 15.6 billion by 2022.[1] Wi-Fi has allowed people to build out broadband networks in a manner that was not foreseen by regulators. It has allowed for independent initiatives to establish connectivity in places that were either unserved or where access was deemed expensive by citizens. It has enabled the rise of small-scale operators both in the form of non-profit community-owned networks and commercial wireless internet services providers (ISPs).

1 MarketsandMarkets. (2018, 23 March). Wi-Fi Market worth 15.60 Billion USD by 2022. https://www.marketsandmarkets.com/PressReleases/global-wi-fi.asp

The disaggregation of the telecommunications network supply chain has also enabled the use of Wi-Fi for internet access at the community level. Historically, telecommunication operators managed the entire communication network infrastructure, from international backhaul links through undersea cable to national fibre optic and microwave backhaul networks to the last mile connecting consumers. Increasingly, international, national and metropolitan network infrastructures are available as wholesale services to any operator, lowering the bar to market entry for smaller operators who can focus on last-mile delivery.

Finally, communication technology in general has come down dramatically in cost. Not only has Wi-Fi technology become extremely affordable, but all kinds of communication technologies, from point-to-point microwave to GSM to LTE base stations have dropped in price; even fibre optics now have prices that are within the reach of the community network and small-scale operator.

All of these changes represent tremendous potential for community networks.

Nevertheless, most regulatory frameworks have yet to catch up with these changes and are not structured in such as way as to enable communities to easily take advantage of them. Most community networks happen in spite of existing regulatory frameworks, not because of them.

Regulation needs to evolve

Most regulatory and policy frameworks focus on the provision of broadband access primarily by a limited number of national mobile operators. This presents barriers to other models of access that can complement the existing players, whose business models are less able to cost-effectively serve remote and sparsely populated areas.

Regulators need to recognise community networks and small-scale operators as an essential part of their regulatory strategy, representing an important complementary approach to access delivery that can address geographic or sectoral gaps in service delivery. This is not a case of replacing one approach with another but of recognising that countries do not have one single economy. French historian Fernand Braudel[2] argued that economies can be understood at three different levels: ordinary economic life at the base, where local efforts are consumed locally; then the market economy of cities, markets and trade, currencies, transport systems, etc.; and at the top, capitalism, where competition for control of entire trade networks or even entire economies exists. Regulation should acknowledge the existence of these levels of economy and their value in delivering affordable access.

Metaphorically, we might think of regulation as trying to fill a glass jar with stones. Current regulation accommodates only one fist-sized type of stones. When we attempt to fill the jar, we can fit three or four stones in at best. The jar may look full, but if we were to fill the remaining space with water, it would fill more than half the volume in the jar. What is needed is regulation that enables smaller stones and even tiny pebbles so that we might fill the jar.

This approach acknowledges and continues to value larger operators but recognises that smaller-scale operators and even subsistence operators have an important role to play as well. It is important to note that the small-scale operators and community networks may not be designed to scale to the size of large operators, but rather to serve the geographic, economic and sector niche they were designed for.

In order for this to happen, there are a host of enabling regulations that are needed.

Licensing

Many countries have yet to move to a modern unified regulatory regime based on technological neutrality and simple authorisations to permit service provision. National licences are often the only type available and may come with substantial administrative reporting requirements and fees. Although a few countries like Brazil and India have adopted tiered licensing systems which provide licences at the regional or municipal level, the requirements for these are still bureaucratic, and the technical and financial requirements are beyond the means of most potential operators. In countries like New Zealand and the United States, no specific licence is required to become an operator below a certain level of operation. Awareness raising about existing good practices and capacity-building work among policy makers and regulators are needed to address this situation.

Access to radio spectrum

While licence-exempt Wi-Fi has grown exponentially in deployment and application, demand for exclusive-use licensed spectrum has also grown. Operators are now expected to pay millions of dollars at auction for an exclusive-use spectrum licence. This creates an insuperable barrier for all but the largest companies to gain access to this spectrum, and even those that do gain access may be

2 See: dannyreviews.com/h/Civilization_and_Capitalism.html

obliged to pass on the cost of that spectrum to consumers. There is a need to build on the success of licence-exempt spectrum by exploring new frequencies that might be similarly regulated. There is also a need to find a middle ground between licence-exempt and national exclusive-use licences. This may include new approaches such as dynamic spectrum regulation needed for TV white space (TVWS) spectrum or Citizens Broadband Radio Service (CBRS).[3] It might also include alternative licensing for rural, underserved regions. There is scope for a range of creative alternatives. Finally, new technologies such as radio devices which operate over a much wider range of spectrum bands and use spectrum sensing suggest that we may be on the cusp of a paradigm shift in spectrum management.

Access to passive infrastructure and backhaul

As demand for broadband grows, especially with the rise in streaming media content through services like YouTube and Netflix, access to affordable, high-capacity backhaul services becomes one of the most critical limiting factors in the delivery of affordable access. Open access policy and regulation for backhaul networks are essential to ensure equitable access. Perhaps more importantly, pricing on these networks needs to reflect the national strategic assets that networks are. Like roads or railways, broadband backhaul networks should be designed and priced to maximise traffic in order to realise the full potential for positive externalities that these networks represent. Similarly, passive infrastructure, such as the towers of mobile operators and the masts and poles of public broadcasters and energy distribution grids, should be considered from the point of view of enabling all kinds of operators.

Transparency

Even if fibre is available nearby, it is often very difficult for a new operator to know where the nearest point of presence is, so it can design and cost the network accordingly. It is also difficult to know who has been assigned licences to radio frequencies that might be unoccupied or unused in rural areas. Similarly, access to information on tower locations is needed so that both governments and other actors can identify the connectivity gaps and adopt the best approach to close them. Information on the deployment of fibre, towers and spectrum infrastructure should be a matter of public record. This is essential both from the point of view of transparency, where millions of dollars are changing hands, but also from the point of view of enabling the identification of market gaps and possible solutions.

Associated taxation

Finally, there are many taxes that add to the burden of starting and operating networks. In some countries, import taxes are up to 40% of the total cost of the equipment. Other taxes include fees per mast and device installed and contributions to universal service funds, among others. These added costs must be recovered from end-users, which further limits the service's affordability.

Conclusion

The very low barriers to digital production on the internet have enabled an explosion of creativity in content and services, which is steadily increasing the value of being connected. Those without affordable access to the internet are increasingly socially and economically left behind. In order to ensure that everyone has affordable access to communication, we need to unleash the same kind of energy that spurred the growth of internet content and services. Lowering barriers to the establishment and operation of community networks will exploit the pent-up demand (and creativity) of the underserved, allowing them to implement low-cost, local connectivity solutions that can sustainably serve their constituencies. Regulators must recognise that community networks have an essential complementary role to play in the delivery of affordable access for all.

3 CBRS is a regulatory framework under development in the United States which applies similar dynamic spectrum allocation principles to TVWS but for the delivery of LTE services in the 3.5 GHz frequency band. It is designed to enable both large and small-scale operators.

Towards financial sustainability in community-based networks

Mike Jensen
Association for Progressive Communications (APC)
https://www.apc.org

Introduction

Community networks are increasingly being seen as a means to help address the need for affordable connectivity where traditional commercial networks are not present or are too expensive to use. According to a 2018 report by the GSMA, these areas represent a substantial portion of the planet – up to 40% of the world's population will still not have internet access by 2025, while 30% will not even have voice connectivity.[1] Considering that after more than 20 years of deployments in developing countries, mobile network operators have been unable to respond to demand for even basic voice connectivity, this is clearly not a simple problem to solve.

Due to their generally small size, there are limited economies of scale in community networks, which often means more costly services to operate the network, resulting in higher per-user overhead costs than in larger networks. Since community-based networks often operate or plan to set up in remote, sparsely populated areas, costs are higher than in urban areas for providing internet connectivity and energy, as well as for transport and sourcing of the business and technical skills, which are usually scarce in these areas. And although there may be many important social and economic benefits that can be derived from a community network, it is often difficult to translate these benefits into the cash needed to pay for the network and its operations.

On the positive side, in contrast to traditional commercial operators, community-based networks are able to start at a very small scale and have a more diverse range of models for achieving financial sustainability. In addition, they are less likely to need an expensive marketing and public relations budget. While some community networks may operate much like a traditional commercial network (where users pay a monthly fee to cover all costs), others may draw to varying levels on volunteer labour, donations of equipment, donated upstream bandwidth and the use of high sites to erect towers and antennas, or subsidies from government and commercial sources.

Primarily focusing on remote or rural areas where connectivity is not available, this report looks at the different aspects that may be considered in maximising the potential for small-scale networks to achieve financial sustainability by leveraging opportunities to minimise costs and access start-up funds.

Starting small

At the outset, it should be noted that many community-based networks have started on an informal basis from very small beginnings, which require almost no initial external financial support. Considering that the high cost of internet access is a major barrier to increased connectivity, it is not surprising that the most common example is the Wi-Fi broadband network, where the cost of a link to the internet is shared among a number of users via Wi-Fi. Households and offices do this routinely, but this can easily extend to providing links to neighbours. If the users are close enough and they install their own routers, the only cost is for each user to pay their share of the monthly fee for the upstream connection to the internet, and perhaps add a small contribution for router power consumption at the location of the shared upstream connection.

Bandwidth costs and network scaling

Ensuring the lowest possible cost for upstream connectivity to the internet is often a top priority with community networks, as this usually has the single largest impact on overall operating costs, and ultimately, on the financial sustainability of the network.

Some communities have been able to negotiate with their upstream internet provider to reduce the fees for the bandwidth leased – often a university, a government infrastructure provider or perhaps a

1 GSMA. (2018). *The Mobile Economy 2018*. https://www.gsma.com/mobileeconomy

sympathetic local internet service provider (ISP). Even if a discount cannot be arranged from an ISP, and there are no other nearby supporting organisations with capacity to spare, commercial ISPs still usually charge less per Mbps for higher capacity commitments. This means the larger the initial network deployment (in terms of numbers of users), the lower the monthly cost per user. And if cheaper additional bandwidth is available to respond to demand as the network grows, the cost savings can be passed on to the users. This lower cost of participation further adds to the network effect in attracting new members.

It is also worth taking into consideration that as the network grows, bandwidth costs per user are further reduced, because usage is more evenly spread over time with a larger user base. So, for example, doubling the number of users does not require doubling the upstream capacity in order for each user to have the same network experience. As a result, even if extra capacity costs the same on a per-Mbps basis, the cost of upstream bandwidth per user reduces as the number of users grows. If this can be translated into reduced charges for cost recovery from users, this will further incentivise participation in the network.

In networks providing voice services, the economies of scale are smaller, because each voice channel requires symmetric, dedicated capacity with low latency and high quality of service. As a result, service fee increases are more linearly linked to traffic increases. Balancing the number of channels required in peak and off-peak periods can involve compromises and requires experience. In addition, there may be recurring costs associated with allocation of numbering resources. Furthermore, unless there is a favourable regulatory regime, small voice networks can struggle to meet the minimum interconnection requirements of the larger national operators, let alone gain any volume discounts from them.

Once a broadband network has grown to a sufficient size, upstream bandwidth costs are often significantly reduced by installing a caching server on the network. The server stores copies of content requested by users, thereby reducing duplication of traffic on the link whenever that content is requested again by the same or another user. Pre-fetching content and refreshing mirror servers (such as software and operating system updates, Wikipedia, etc.) during off-peak periods can further optimise the use of the link for peak traffic during the day. Some community networks have also taken additional steps to manage their expensive upstream capacity by setting up their routers to filter access to high-bandwidth websites, especially during peak periods.

Exploiting the availability of the excess internet capacity of nearby larger institutions during off-peak periods has also proved an effective strategy in cutting upstream bandwidth operating costs. For example, AlterMundi in Argentina has an agreement with the National University of Córdoba[2] for free capacity of up to about 10 Mbps during the day, but in off-peak periods the community network can access as much capacity as is available to the university, in practice about 250 Mbps. Members of the network adapt their usage accordingly, knowing that access at peak time is likely to be less efficient.

Where low-cost off-peak capacity is not available from a larger nearby institution, some community networks, such as Pamoja in the Democratic Republic of Congo (DRC), have leveraged the peak/off-peak dynamic by leasing capacity during office hours to local businesses at commercial rates (and higher service levels), while making the service available at a much lower cost (or free) to the public during the rest of the day/night. This strategy can also be adopted more generally by charging all users a differential rate for peak versus off-peak usage, or even making usage free during off-peak periods.

Finally, it should also be noted that many community networks have not aimed to provide access to the upstream internet, focusing instead on linking the community directly with each other and to locally hosted servers and content. Naturally these networks are unburdened with upstream connectivity costs, although in some cases it is assumed that the participants have their own internet connections (mainly in urban environments). In others, the networks are "islands" completely unconnected with the "rest" of the internet, such as Mesh Bukavu,[3] which hosts a large amount of content online locally.

Gaining independence

If the community network's upstream connection is provided on a purely commercial basis by a single operator, the network is essentially reselling the service in smaller chunks on their behalf and absorbing the cost of collecting the fees. In this situation the community network is also dependent on the prices for capacity charged by the operator, and must pay for all the upstream traffic, even when it is destined for other local networks nearby.

2 See the Argentina country report in this edition of GISWatch.

3 See the DRC country report in this edition of GISWatch.

If other ISPs are present, then it makes sense for community networks to establish additional direct links to at least one of the other operators as well. Although this requires greater technical knowledge and a more capable router (to be able to route traffic efficiently between multiple networks), this not only gives the network a better negotiating position on the price of upstream capacity, but also makes the network faster for the users of the interconnected networks, and reduces capacity needs on the original link. Ideally, if it is possible to establish a link to a local internet exchange point (IXP), then more networks can be reached directly, and it should normally be possible to "peer away" even more traffic, further reducing the costs for transit capacity purchased from upstream providers.

As importantly, a single link to one upstream provider also creates a vulnerable point of failure, while a network with multiple upstream connections will be more reliable, because one of the providers can go down and the network will continue to function.

Reliability quickly becomes an important concern once an affordable service has proven itself and as the community becomes more dependant on it, especially for economic activities, such as remote work. Long periods of downtime can quickly sap confidence in the network, and generally chill the level of use when connectivity returns. If there is only one source of cost-effective capacity in the area, it can still make sense to set up a 3/4G backup link for urgent traffic if a mobile network is accessible (perhaps through a long distance Yagi antenna).[4] Alternatively, a tower or high site, or even a satellite link, may be needed to reach other more distant options for obtaining backup connectivity.

Minimising the cost of additional resource needs

At the next level up in terms of costs and infrastructure required for the network, there may be a need for a tower and/or network equipment to provide mobile voice and data services, or to relay the signal to a distant community, a larger institution or a sub-set of users. In some cases, this might also involve site rental for a high site on which to locate the relay equipment, and tower insurance. Along with upstream capacity, these costs are usually the other major cost component of a community network, especially if a large tower is needed, and/or solar power and protection against lightning is required.

Sometimes the owner of the high site will accept the provision of free connectivity in return for installation of a tower on the location. If there is already a telecom tower of some form on the site or nearby, it may be cheaper to lease space on the tower than construct another. However, this may require some hard negotiating or bringing in the telecom regulator to ensure that infrastructure-sharing regulations (which should include price caps for space rental) are being adhered to. These regulations are unfortunately not widely adopted in many developing countries – for example, Airtel still charges USD 1,300/month for space on one of its towers in Rwanda, a country which prides itself on having one of the more advanced regulatory environments on the African continent.

Tower costs can often be reduced by having them locally constructed, and by mounting shorter towers on existing tall buildings, or even trees, if available. In addition, use of non-line-of-sight frequencies (most often those lower than 800 Mhz) means that towers do not have to be high enough to reach over trees, buildings and other obstacles, which considerably reduces tower deployment costs. This was noted in the Gram Marg network in India,[5] where the initial TV white space (TVWS) deployment used relatively low towers. When the network had to switch to line-of-sight 5 Ghz links due to regulatory issues, the towers required needed to be much higher. As a result, although 5 Ghz radio equipment for the links is much cheaper than the TVWS equipment, the overall deployment costs were significantly higher because the tower costs were a much larger component of the total cost.

Similarly, with mobile voice and data services, choice of lower frequencies (e.g. 700 Mhz rather than 1800 Mhz) means that towers can be shorter and/or farther apart because lower frequencies travel farther, providing greater coverage. Voice (2G) deployments in the lower frequency bands can also take advantage of the much longer distance that these signals cover relative to 3G/4G data connections.

Fortunately, the cost of equipment for generating electricity from solar power continues to drop, but the batteries, electronics and solar panels for off-grid sites can often still cost as much as the tower itself, especially when the power system needs to support mobile networks, for which the base stations consume significantly more energy than Wi-Fi. However, for off-grid locations, it should be noted that energy needs in a mobile network are concentrated at the tower and overall energy consumption

4 https://en.wikipedia.org/wiki/Yagi%E2%80%93Uda_antenna

5 See the India country report by Gram Marg in this edition of GISWatch.

in a mobile network is lower, because the end-user devices use less power than the user's router and access equipment in a Wi-Fi network.

In general, because of the reliability concerns described above, it is important to dimension the power system sufficiently to ensure that occasional long periods of cloudy weather do not cause a system outage. In addition, availability of backup equipment, ideally stored on-site, for quick replacement of broken parts also needs to be considered, as well as the lightning protection and security for the tower equipment if necessary – some community networks need to employ full-time patrols to guard against theft.

The community network may be required to pay licence, spectrum, business and other fees, for which there is often no way to reduce costs, except by spreading them across a larger user base. However, as the importance of these networks is being increasingly recognised, it is hoped that more countries will follow the example of Mexico in recognising the social purpose of these networks,[6] and making appropriate dispensations to support them by providing access to licensed spectrum and limiting bureaucratic burdens and unnecessary fees and taxes.

Aside from spectrum and licence fees, import duties should not be ignored, as these can often double the cost of the network equipment, and also often add significantly to the cost of end-user access devices. If waivers on import duties cannot be obtained from the government for community networks, it may be possible to avoid some of these taxes through partnerships with charities which have special status, or through informal import channels.

Use of open source hardware and software also helps to bring down equipment costs and provides many other advantages. This is already a relatively common strategy among community networks where proprietary Wi-Fi hardware is often modified with open source routing software (e.g. Open-Wrt). This trend is similarly found now for mobile network infrastructure thanks to projects like Osmocom for 2G/GSM and OpenAirInterface and NextEPC for 4G/LTE.

There are now also an increasing number of open hardware platforms, in particular the much-anticipated LibreRouter initiative by AlterMundi, a number of 2G base stations such as those from Fairwaves and Sysmocom, and the OpenCellular LTE base station currently under development. These new devices generally offer cost advantages over the traditional equipment commonly being used – in particular, the presence of three radios in the LibreRouter increases the available capacity on the mesh, and obviates the need for duplicate devices when acting as a relay or mesh node, while simultaneously performing hotspot functions to provide end-user access.

Buying network equipment in bulk or organising group purchases with other community networks can also help to bring down equipment costs. Community network collaboration is particularly important for helping reduce prices in small community-driven hardware projects such as the LibreRouter, which does not benefit from the same economies of scale as consumer devices mass produced by the large companies operating in this market.

In relation to the administrative and human resource aspects of a community network, the involvement of community members is usually essential to minimising costs of deployment and operations. While technical and business skills often need to be initially sourced from outside the community, with fairly minimal training, local volunteers can be used for many tasks, such as erecting towers and installing equipment on roofs, or even day-to-day technical and administrative tasks (troubleshooting, adding users, collecting fees, etc.).

Nevertheless, once the network grows beyond a certain size, the most cost-effective solution is likely to involve part-time or even permanent staff from within the community. In some cases, especially where there are multiple similar networks operating in the country with licence compliance needs and shared use of other resources (such as higher level technical expertise, a satellite link or DID numbers),[7] it can make sense to establish a national or regional organisation that can take on the burden of many of these common administrative tasks. This has been done in Mexico with TIC A.C. to support its member villages operating community networks.

Fundraising

In some locations, the members of the community may be able to fundraise internally to cover the costs of the network, especially if there are some potential businesses or other organisational users willing to contribute. In cases where telecom

6 See the discussion on TIC A.C. in the Mexico country report in this edition of GISWatch.

7 https://en.wikipedia.org/wiki/Direct_inward_dial

infrastructure is managed as a common-pool resource, finance is crowdsourced by those benefiting from the infrastructure. See for example the country report on guifi.net in Catalonia, which won a European Commission Broadband Award with this approach.

However, in most rural areas in the developing world, the resident population is unlikely to be able to provide the needed resources, and external fundraising will be required. In choosing targets for fundraising, it is worth noting that there are three intrinsic difficulties in raising funds for networks focused on remote and rural areas from traditional lenders, investors and soft funders (banks, venture funds, development institutions, etc.):

- Scale: Rural networks are likely to have far fewer customers than urban networks, rendering them less attractive to traditional investors or lenders, be they commercial or soft (development) funders. This is because the overheads for due diligence and administering smaller disbursements are not so different from those for larger-scale projects, resulting in a relatively high cost of finance, especially if they are in remote and isolated locations which may be unfamiliar to the funder. Also, many of these networks may be purely focused on provision of connectivity in a particular location, and may have little or no interest in scaling and replicating in ways that would create the larger projects that are more attractive to traditional funders.
- Real and perceived levels of risk: There may be higher actual or perceived levels of risk by potential funders because the initiatives are based on novel business models, may be run by people with limited management skills, or use new technologies in unfamiliar contexts. These initiatives may also lack land collateral or other asset sureties needed to provide guarantees for loans. Even if collateral is available, in many developing countries the cost of commercial bank loan finance is exceedingly high to reflect the high level of perceived risk, so this option is unlikely to be cost effective for a community network.
- Low surplus revenues: Networks serving remote and rural areas usually operate in locations with low income levels, and where operating costs are substantially higher in comparison to urban areas. Therefore, the ability to service a loan or provide a return on an investment may be quite limited. Furthermore, there are many networks which a) do not aim to make a profit and/or b) try to ensure that fees for service are as low as possible. This may disincentivise traditional investors in the telecommunication sector looking for higher returns.

Given these considerations, community networking initiatives are likely to find raising the needed startup funds from commercial or other traditional lenders difficult. Even soft loans from development funds are still currently more focused on large-scale national initiatives, and as conservative lenders or grant makers, they need to be convinced of the potential for the novel strategies and innovative business models of community networks. Ideally, local intermediaries acting for many networks could play a key role in this area, as they may be more familiar with the landscape and can better evaluate potential initiatives, aggregate needs, as well as manage the disbursement of funds received from large funding sources.

To meet the funding gap, a variety of other fundraising strategies can be considered:

- Universal service funds: National governments usually have universal service funds to support the provision of access in rural and underserved areas. Many of these have already accumulated large amounts of unspent funds, partly because of the limited capacity of regulators to evaluate and disburse funds, and also because of the paucity of effective projects to support. Given the recent response of regulators and policy makers who have been sensitised to the potential of community networks, it would appear that this avenue of support is likely to become increasingly fertile in future.
- Grants and awards from Regional Internet Registries (RIRs), ccTLD operators, the Internet Society, APC and other international NGOs and commercial tech organisations such as Facebook, Microsoft and Mozilla: While the funds available from these organisations are relatively small, these institutions have been the most common source of financial support for community networks to date.
- Provision of in-kind services: These can reduce the startup and operating costs of the network by tapping into the corporate social responsibility (CSR) programmes of businesses, forming partnerships with local and international NGOs operating in the area and local government offices. Examples include donation of equipment, skills/training, tech volunteers and bandwidth.
- Cross-subsidisation: As discussed earlier, in some cases, community networks may be financially sustainable by charging businesses a

monthly fee and giving discounts to the general public. Funds for the cross-subsidy can also come from other services provided, which may be unrelated to the provision of connectivity to the end-user – for example, hosting remote sensing equipment (weather, air quality, etc.) for a government or research agency, as is being experimented with at TakNet/Net2Home in Thailand.[8]

- External crowdsourced funding: Crowdsourcing funds from outside the community offers significant though untested potential. However, there may be interest from the diaspora and people in developed countries who have visited the area as volunteers or tourists, among others, in funding a local initiative.

Conclusion

This report aims to familiarise the reader with the most common strategies for minimising and sharing costs in community networks, and in raising the necessary financial and other resources to help support their long-term financial sustainability. Given the relatively short time frame and difficult conditions in which community-based networks have emerged, the extent to which these strategies will help ensure a place for community networks in meeting the needs of connecting the unconnected is still unclear. However, given the diversity of strategies that have already emerged and the level of interest in supporting community networking initiatives, the prognosis is good.

8 See the Thailand country report in this edition of GISWatch.

Community networks as a key enabler of sustainable access: A review

Michael J. Oghia
https://linkedin.com/in/mikeoghia

One of the most significant problems vexing the information society is the lack of a holistic perspective when it comes to technical and policy development. Take, for example, the issue of access to information. Often it is considered solely from a rights-based perspective – i.e. that access to information is a right that is often hindered or impeded in some way, such as by a governance or policy decision. Likewise, from technical protocol and standards development to content-related issues like hate speech, actors tend to organise among stakeholder groups and conduct their operations in silos. Although the multistakeholder model is championed as a way to alleviate this tendency, there has yet to be a silver bullet that fully addresses the lack of holistic vision that is necessary to govern an inherently collaborative and global resource such as the internet while also addressing its fundamental challenges.

If we revisit the example of access to information, we can see how this has played out with one particular and lingering problem: sustainability. Democracy is built on the ability to access information, which is why access to information is such an important pillar of the United Nations Sustainable Development Agenda[1] – along with the fact that access to information facilitates myriad social, educational and economic gains as well. At the same time, however, how can individuals – particularly those four billion or so people who are not connected to the internet – be expected to access information digitally when they not only face significant barriers to connectivity, such as poor or non-existent infrastructure and/or the lack of user capabilities, but the electrical backbone necessary to even *power* internet infrastructure is often lacking?

In a chapter I wrote titled "Community networks as a key enabler of sustainable access", which was published in the Dynamic Coalition on Community Connectivity (DC3) 2017 report[2] for the Internet Governance Forum (IGF), I described how connecting another billion people to the internet will require more than an internet-connected device; such an endeavour requires significant long-term vision, investment in both technology and human capacity building, as well as communities committed to ensuring their access is useful, meaningful and sustainable. For this to occur, however, such communities must be invested in the process of connectivity – from energy access, to network set-up and maintenance – as well as leading this process based on their own needs, context, and developmental challenges.

Community networks are vital to catalysing this investment – not in terms of financial investment, but in terms of community development. A key shift in thinking is necessary for this to happen, however, in part because

1 https://www.un.org/sustainabledevelopment/development-agenda

2 Oghia, M. J. (2017). Community networks as a key enabler of sustainable access. In L. Belli (Ed.), *Community Networks: The Internet by the People, for the People. Official Outcome of the UN IGF Dynamic Coalition on Community Connectivity*. https://bibliotecadigital.fgv.br/dspace/bitstream/handle/10438/19401/Community networks - the Internet by the people%2C for the people.pdf?sequence=1&isAllowed=y

the challenge of generating reliable energy to power infrastructure continues to pose a significant barrier to lowering costs and the ability to scale. I argued that one way this can change is by replacing the focus that sustainable development places on the role and proliferation of information and communications technologies (ICTs) with the concept of "sustainable access" – a term I coined that broadly refers to the ability for any user to connect to the internet and then stay connected over time. Sustainable access encompasses various aspects of the relationship between technology, society and the environment – everything from infrastructure, energy and the availability of radio spectrum, to the recyclability of ICTs, how internet-connected devices are manufactured, and even space junk. The concept is meant to address a larger gap in current practices *vis-à-vis* development and ICTs – i.e. that facilitating access to the internet and expanding connectivity in general must be seen as a holistic, interconnected process involving multiple stakeholders.

The core thesis of this perspective is that internet technologies are largely unsustainable at present. This relates to overall lack of design consideration of ICTs for sustainability (such as recycling or energy scaling), but also reflects serious challenges such as the exponential growth of data use and generation. Because of this, we cannot legitimately discuss internet access without addressing sustainability – even though, conversely, ICT sustainability is largely viewed as a future concern, not a present one, and is therefore largely overlooked.

This is not the case for community networks, however, which generally operate in rural or remote areas that often do not have access to electrical grids, depending on the region. This point is emphasised in the chapter I wrote by outlining the role that community networks can play in catalysing sustainable access, and focusing on efforts and initiatives used by such networks to electrify their infrastructure and ensure their energy sustainability. It also addresses how energy, the subsequent costs of infrastructure (both initial investments and upgrades), and the inability to recycle equipment or use it over the long term can significantly hinder the sustainability and growth of a community network – as well as its ability to scale.

The chapter concludes by stressing how if we truly want digital technology and the myriad emerging technological innovations that are beginning to scale to become ubiquitous, sustainability must be addressed more prominently as a core component and within the design of ICTs. We cannot disregard or downplay sustainability with the hope that the inherent problems with our digitised world disappear – time will only exacerbate them. On the contrary, it is clear that there are unexplored and underemphasised synergies and areas of collaboration between the energy and ICT sectors, which undoubtedly include the internet governance community, that could better address sustainability as a whole. Therefore, since sustainability and access are intrinsically connected, the role of community networks in ushering in the next phase of the internet's development should not be underestimated. Instead, it will benefit anyone seeking to make the internet more sustainable to offer more financial, technical, policy, legal and regulatory support to community networks, and ensure that such initiatives are viable, sustainable and successful.

Community networks: Stories and power

Nicola J. Bidwell
International University of Management, Namibia
www.ium.edu.na

Introduction

In this report I consider some of the meanings embedded in community networks, and the way they work together with power relations. A critical awareness of the interplay of meanings and power can inspire us to create new meanings that might better contribute to achieving aspirations, such as promoting the agency of all community network users. I draw on my preliminary analysis of data generated in the past eight months about the social and gender impacts of community networks in the global South, and reflections on conversations within the Community Access Networks project, a research study into community networks globally that is led by the Association for Progressive Communications (APC) and funded by the International Development Research Centre (IDRC).[1] I refer to only a few of the many power relations at work in my research about community networks and focus on just four sets of meanings.

During my research so far at five community networks in Asia, Latin America and Africa, I generated data about people's everyday practices and opinions in relation to their network in focus groups and individual interviews. Adapting my methods to suit each community network, I was privileged to interview 119 men and 103 women, individually or in groups, some repeatedly and extensively. This included community network initiators, champions, members, users and non-users.

While all networks are rural, their socioeconomic and political contexts vary widely. They are located in countries that have, according to the UNDP's 2015 measures, gross national income per capita ranging from USD 1,670 to USD 20,945, and income distributions from 0.41 to 0.63 on the Gini index, where 0 represents total equality and 1 total inequality. Indices for the population's health, longevity, living standards and knowledge also vary amongst the community networks; for instance, one is located in a country with very high human development, another in a country with high, two in countries with medium and one with low human development. These countries also differ in gender equality, one high, two medium-to-high, and two countries with low gender equality.

The intentions and the geographic scale of the operations of the initiatives I studied vary as widely as their socioeconomic and political contexts. Some initiatives prioritise ethics about human rights to communicate or net neutrality, some are driven by research about technical solutions to provide "first-mile" internet access to rural people, and others seek to integrate information and communications technologies (ICTs) into local culture to address significant humanitarian challenges. Some networks connect directly to individual people or homes, others connect via local not-for-profit or government centres; some are groups of local networks distributed over distances of 300 km, and others are geographically localised. However, all initiatives self-identify as community networks, and all aim to improve access to low-cost telecommunications for people in rural areas.

Power relations and narratives

Power relations between people enable one person, or group of people, to have more influence over another person or group. This influence operates through direct and indirect relationships between people and arises due to differences in socially agreed political or legal authority, or capability in certain domains, or economic status, or race, age or sexual orientation. In some community networks, people explicitly referred to differences that are institutionalised according to formal categories such as refugees, internally displaced persons, indigenous people, tribe, caste and "other backward" classes. For instance, members said "You don't come from our background" to a woman employed to support a group of networks. We must, however, avoid oversimplifying power relations to only particular categories, as power relations intersect, and often less explicit power hierarchies emerge within community networks.

1 https://www.apc.org/en/project/local-access-networks-can-unconnected-connect-themselves

Inclusivity was central to the ethos of all the community networks I studied, yet, in all networks, differences between people affected their respective agency in shaping the meanings embedded in their network. Consider, for instance, one aspect of power relations in a community network in a country with a high equality ranking and traditions of solidarity. One man in this network described three groups of local inhabitants: people like him, who moved permanently from the nearest large city within the past five or six years and established their main activity locally; people with weekend homes; and people whose families had lived in the area for generations. In this network, I interviewed many more people who were newcomers, like this man, than people with historical local ties, primarily because most participants were recruited through the community network's initiators' closest social group. One network initiator had family connections in the area but moved to a city where they met other initiators through the free software movement.

Like other newcomers, the network initiators had greater physical and virtual mobility by virtue of their education, class, and varied income sources, as writers, teachers and software developers. Interview participants with historical local ties, on the other hand, were working-class and had manual jobs. A woman user of the network, with historical local ties, explained that there had always been people from far away staying in weekend homes but recently a dramatic increase in newcomers had elevated property prices and filled the area with "strangers" who displayed an unwarranted ownership of the little town.

The community network acts as a bridge between inhabitants, and people with historical local ties said that it had facilitated connections that contributed to new opportunities for business and socialising. People with historical local ties chose to associate with this community network and, as for some other community networks I studied, there were alternative providers of the same services. Interviews also illustrated that the network's initiators actively encouraged members with historical local ties to host meetings to decide about the community network and lead the technical workshops that are more or less mandatory for membership. Nonetheless, people with historical local ties were more reluctant to be interviewed and, unlike the network's initiators, were less forthcoming about certain views, which suggests that the network's initiators have greater influence over the meanings associated with the network.

Some meanings associated with community networks repeated across the networks I studied, and this report illustrates how these meanings inherit from other stories, through elements such as narratives and tropes. Such story elements are recognisable concepts and patterns of ordering that help us understand and communicate about new situations. Narrative selects and puts events and thoughts together into some coherent sequence to convey a particular perspective on a story. Tropes are archetypal narratives that use other familiar stories to make a perspective clear; for instance, the trope of David and Goliath is about competition in which the little guy is the hero.

Whose story counts?

A David and Goliath trope permeates narratives about resisting concentrations of power. Across my research, champions, and some network members, referred to the role of community networks in opposing domination by technology and telecommunications giants, which often linked to other critical attitudes about multinational corporations and monopolistic control. Some of my data, however, suggests that airing views that might be tagged politically liberal and progressive was more comfortable for community network members with greater cultural capital, such as people with university educations or professional jobs. For instance, in the community network that illustrated power relations between newcomers and people with historical local ties, it was the initiators who emphasised resisting, or perhaps evading, aggressive or unaccountable control by technology companies. People with historical local ties, on the other hand, more often associated the network with affordability, and its not-for-profit or communal ethos. In fact, interviews with people in this network revealed different perspectives with respect to corporations and control; for example, newcomers opposed the possible location of a new mine in the area, linked to contesting the extractive nature of transnational companies in general, but people with historical local ties were more likely to mention that a mine brings employment. That is, despite this network's considerable efforts towards inclusivity, a foundational narrative that relates community networks to resisting concentrations of power did not have similar relevance to all network members.

The community network philosophy, as summarised in the Declaration on Community Connectivity,[2] addresses the use of technology to

2 The Declaration on Community Connectivity was developed at the Internet Governance Forum, Guadalajara, Mexico, December 2016, and the GAIA Workshop, Cambridge, England, January 2017. https://www.comconnectivity.org/article/dc3-working-definitions-and-principles

concentrate power, and this overlaps with a larger political project about autonomy, self-determination, emancipation and decoloniality in relation to telecommunications. However, the meaning of concepts such as emancipation varies amongst people at the grassroots in community networks. In one set of networks, in a highly resource-constrained setting, a major transnational technology consulting company funded solar and other infrastructure. In another set, in a country that favours both capitalist development and government involvement in digital participation, the ability of low-income rural women to shop with Amazon.com marked their internet inclusion. Meanwhile, impoverished members of a cooperative that founded yet another community network hoped the network's growth would directly profit their families. The agency of diverse people in effecting narratives about community networks and meanings about, say, autonomy, emancipation and decoloniality in relation to telecommunications, differs. Unequal agency in shaping narratives about community networks can compromise some of the freedoms and rights pursued by the overarching community network movement.

The worth of human connectedness

The next narrative that repeated in the community networks I studied values human connectedness in a certain way. Participants' stories, in interviews and focus groups, often referred to the role of social ties, sociality and sociability in obtaining or achieving something else, such as economic improvements or safety. Accessible communications had enhanced some participants' job prospects through studying for formal qualifications or improving their English language skills, and real-time business-to-customer or business-to-business relationships, such as sharing information about agricultural market prices amongst sustenance farmers, and about components amongst electronic repair businesses. Members of different community networks also mentioned the impact of solidarity on their safety; in one network different people explained that they were able to coordinate to apprehend a burglar; in another, that they had been able to call a taxi to take an old man who had fainted back to the village from a remote field, and coordinate together to save a donkey cart, full of provisions, when it fell down a mountain. That is, participants tended to frame human connectedness instrumentally.

Instrumental narratives about human connectedness also featured in a set of networks in a region that endured war for many years and hosts millions of refugees and displaced people. Severe conflict, and some post-conflict actions, have undermined people's trust in institutions, neighbours and even family members, and the network's initiator prioritised peaceful coexistence in all activities, emphasising traditional practices of people coming together in dialogues to manage disputes, such as about land or water, and organising host-refugee events, such as football matches. The initiator rationalised cohesion and inclusion by explaining that "You won't go anywhere with excluding because tomorrow you might need the people that you exclude." Such a narrative resonates with an, albeit controversial, argument in international development discourse which proposes that social capital, resulting from social ties, enables people to satisfy everyday socioeconomic needs, such as access to advice or money.

Instrumentalist interpretations can be applied not only to social ties but also to people's felt experience of human connectedness, or the emotions, intuitions and morals a person senses in social relationships. At this level, a person's felt experience of collectivity, such as in setting up a network with others, might function in building trust; and a person's felt experience of social expectations about digital participation, say through social media, might function in accruing cultural capital and mobilising social assets. Facilitators in one set of networks, for instance, noticed that feelings of connectedness with children powerfully motivated women to learn to use technology.

Narratives that emphasise the worth of felt experiences of human connectedness according to their utility in solving certain problems or efficacy in predicting certain states of development are useful for justifying in wider arenas, such as evaluating community networks against the Sustainable Development Goals.[3] However, the nuances of participants' more ordinary stories also tell that human connectedness has a different type of worth. Mundane human decision making is not mostly rational, and the intrinsic worth of felt experiences of human connectedness in everyday life is that such feelings exist. Participants told how community networks contributed to averting loneliness, sharing joy with remote family, feeling the presence of intimates through phatic contact, and feeling pride in caring for their community. Users in two networks said that they supported the network not because it enhanced their own access to telecommunications but, rather, because it enabled access for more disadvantaged local inhabitants. Meanwhile, the majority of cooperative members that founded another network said that their achievements

3 https://sustainabledevelopment.un.org/?menu=1300

benefited their children and other youth of their impoverished area, since they do not own phones able to access the Wi-Fi themselves, and they asked with pride that a book be written about this legacy.

In driving the community network agenda, we do not make explicit the intrinsic value of felt experiences when humans connect to other humans. Perhaps a trope of sentimentality discourages us from expressing that the worth of sociality and sociability is quite simply that they exist, and encourages instead the use of rhetoric common in discussions of so-called "last mile" connectivity, such as impacts on poverty, health, education, employment and economic growth. Ironically, this lack of emphasis contrasts with commercial telecommunications providers who readily market products and services, including to low-income populations, by depicting emotional qualities of human connectedness as much as affordability, convenience or mobility. The intrinsic worth of human connectedness is vital to developing and sustaining community networks, with or without supportive policy and legislation and, across the networks I studied, members implicitly or explicitly indicated the felt experience of their contributions, and acknowledgement of their contributions, be they doing technical tasks or reliably attending meetings.

Hidden skills

Clearly the meanings embedded in community networks are influenced by narratives that do not come via one ideology or any coherent set of stories. To the contrary; for instance, international development discourse as often conflicts as overlaps with opposition to neoliberalism and globalisation. The next narrative I find that influences meanings in community networks is, in fact, directly inherited from the telecommunications industry. It concerns the visibility of, and values ascribed to, different types of creative and coordination work in setting up, maintaining and expanding a network; and it actively performs in excluding women. Indeed, this narrative might qualify as a "master narrative", in Susan Leigh Star's terms,[4] or "a single voice that does not problematize diversity, and speaks unconsciously from the presumed center of things." In the education and industry sectors of telecommunications, stories about creative and coordination achievements are dominated by large-scale engineering feats, like satellite technology, as well as software development, and they are acutely gendered; consider how long it takes to find the 10 women amongst 130 men in images returned by Google to the query "telecommunications engineer"! The dominance of men in activities associated with these achievements contrasts with the dominance of women on factory production lines manufacturing phones, or in customer service divisions of telecom companies; and the disparity in pay and labour conditions of these employment areas tells of the value of these women's work.

Conversations with members of different community networks about their achievements in establishing physical infrastructure were dominated by references to certain activities. For instance, erecting structures like towers and poles, and negotiating roofs, mountains and trees has a prominence, partly because the outdoor work involved is publicly visible and the conspicuousness of the equipment makes them accessible referents in conversation. The material visibility of tasks conflates with gender, and the worth attributed to creative work. Women in networks in three countries described perceptions about physical work that excluded women. In a network in the country ranked highest in gender equality of those I studied, women members explained how one was scolded for climbing a ladder while pregnant, and another refrained from ascending a tower because she was concerned that this might be perceived as hindering an important team activity. In a network in the country ranked lower in gender equality of those I studied, one young man technology intern explained that women's strength made them less able to climb towers – I did not resist asking him how, then, did women in this setting routinely manage to walk for miles carrying huge loads of firewood on their heads.

The relevance here is not about particular physical capabilities in building networks, rather it is about visibility, value and gendered roles. For instance, when we asked a group in one network whether it was only men who cleared the land for erecting an antenna, one man said: "Some women helped carrying sand, water and alcohol. It was both community [voluntary] and paid work; the carpenter [a man] had paid work." Meanwhile, a woman in another network observed that dexterity and care in finer physical assembly meant women were better than men in soldering, crimping wires and assembling components; and a woman researcher in yet another network enthusiastically displayed the circuitry of a router prototype. Indeed, in four networks studied, women referred to their own creativity, from weaving baskets and rugs, to sewing and upcycling fabrics, from crocheting to pom-pom making and many other crafts. However, women's finer physical work in networks often disappears

4 Star, S. L. (1999). The Ethnography of Infrastructure. *American Behavioral Scientist, 43*(3), 377-391.

inside homes, fitted between many other tasks, whereas when men undertake finer physical work it is more obvious and focal. For instance, men who use the solar electricity their network provides in electronic repair businesses display their skills in fixing in the open fronts of little shops and promote their business locally, in jingles on the radio. The men also promote their businesses nationally, by inserting adverts into movies they distribute; one of the men overlays African movies with audio in local language, speaking through a female voice synthesiser for characters that are women.

While people in the community networks I visited mostly acknowledged the importance of social coordination, they rarely spoke of it with the reverence or heroism they attached to software and network engineering tasks. In several networks, and even amongst our project research team, men refer to the monetary market value of their technical skills. Women employed by one group of networks explained how the members of a local, traditional, male-dominated governance authority overseeing one network were determined to speak only with technicians, who were men, including about non-technical aspects of network. Such valuing is, perhaps, the reason why young women engineers in a university that instigated another group of networks are reluctant to work at the network's rural sites and prefer to develop software in the lab.

Paradoxically, the worth ascribed to work may offer new opportunities for women's agency in technology. Nearly 40% of the group of community networks in which traditional, male-dominated governance authorities make decisions allocated local administration to young women because the most visible everyday work is secretarial, such as registering subscribers and record keeping. Yet these roles also provide opportunities to develop technical skills, meet others in different villages and, according to one woman employee, had fostered self-confidence in the women who participate. This situation is analogous to the prevalence of women in the field of computing in the 1960s because programming tasks were perceived as comparable to typing or filing; which provokes asking what we need to do to ensure that all types of creative and coordination work are valued, and that diverse people can do all types of creative and coordination work.

Meshworks not networks

The final trope is about our emphasis when we use the word "network". Featuring as much in common speech as in business, and fields of science and the humanities, the "network" is one of the most prominent metaphors of the past 25 years. In telecommunications we apply it to connecting discrete technical and organisational components, like nodes and links, content and services, administrators and users. When we talk and teach about community networks we also tend to describe them as structures in summative and static ways. In one community network, for instance, people learned about lines of sight by holding hands with people whose homes they can see from their own, and I watched members of another initiative visualise the values that their networks are based on by drawing lines to join together points on a large paper graphic. The focus in these activities is on the connection itself, rather than the many ongoing movements that make and sustain those connections, such as the movements of bodies that join hands or the pen that links points on paper. We focus more on the net, and less on the work; we think of network as noun rather than verb. In reality, of course, the technical and social fabric of a community network emerges from, and is embedded in, the details of people's ongoing lives. Connections are made as people move along in life, never stationary in the passage of time. People's paths, as they move along, thread and loop through and between each other, diverging and converging, and twisting and knotting together. Even when community networks are connected to each other across vast distances, and travel between them requires a car, the processes of interconnecting them always involves humans' lives.

Many aspects of fastening and maintaining the social connections underlying a community network cannot be described using the rational and explicit terms of telecommunications vocabularies. These fastenings have emotional and dynamic qualities that are the stuff of human lives. Indeed, the way people lay paths through the world when they produce and experience the material and social infrastructure of community-based telecommunications, suggests that a "meshwork" is a more appropriate word. A community meshwork comprises paths lived by people, and a flow of engagements with the circumstances that they produce and experience. It does not comprise fixed connections between human and non-human components. A "community meshwork" helps to anchor descriptions, plans and actions about connectivity to human temporal and spatial scales, and may help to avoid the ways that monolithic telecommunication systems underserve populations and erase the existence and agency of people, regardless of the sentiments they draw on to market their products and services. The contrast of "community network" and "community meshwork" resonates with different depictions of development; one that links more or less distinct inputs, outputs, outcomes and impacts together in straight lines of causation, and another that notices emergent qualities and interactions of complex systems. The former depiction represents neither the complicated, relational character of a community meshwork, nor the potential agility enabled in the movements people make.

Making new meanings

A narrative about community meshworks, rather than networks, is an alternative that recognises the capacity for new meanings to emerge. Similarly, with regards to the other three stories that I used to illustrate the way narratives work together with power relations, I propose that we may create new meanings by tuning our awareness to their manifestation.

A narrow focus in resisting concentrations of power by telecommunications giants may limit the ability of community networks to respond to the way autonomy, self-determination, emancipation or decoloniality mean different things to different people in different places. Undoubtedly, an enabling policy and regulatory environment is essential for community networks to flourish and achieve more for people in the global South. Yet, strategies to oppose the status quo according to particular political interpretations about the use of technology to concentrate power also function to maintain particular hegemonies of doing, knowing and being. Or in Audre Lorde's words, "The master's tools will never dismantle the master's house."

An example of hegemonic thinking and doing is a tendency to deploy instrumental narratives about human connectedness in claims about community networks' impacts, at the expense of narratives about the intrinsic value of human connectedness. Attending to the nuances of members' and users' felt experiences of human connectedness, and articulating these feelings, intuitions and morals in operational and strategic decisions, may be vital in creating new narratives about community networks.

Another example of hegemonic thinking and doing concerns the worth ascribed to different types of work involved in setting up and maintaining community networks, where a dominant narrative illuminates and values technical tasks. The efforts of some community networks to develop members' and users' technical skills, and the proliferation of devices that are easier to set up and maintain, can help in revising the relative value ascribed to social and technical work.

However, to realise a new narrative about technical and social work in these new reconfigurations, we need to be vigilant to the ways that power relations will constantly act to manifest worth according to existing hierarchies, such as gender or education, and how we are never passive in this process.

During the eight months while I conducted my studies, a great many people in a great many places created new prospects for community networks in general, such as by raising awareness and advocating for more supportive policy, and created new prospects for diverse people to develop them. Perhaps this will yield the time and space for us to enrich narratives about power and telecommunications by better encompassing the ways that people at the grassroots understand and relate to government, transnational corporations, or local entrepreneurialism.

Feminist infrastructure and community networks: An opportunity to rethink our connections from the bottom up, seeking diversity and autonomy

Bruna Zanolli (Vedetas, radiolivre.org), Carla Jancz (Vedetas, Actantes), Cristiana Gonzalez (PPGS-IFCH/ Unicamp), Daiane Araujo dos Santos (Rede Base Comum/Casa dos Meninos) and Débora Prado (Lavits, Labjor/Unicamp)
redeautonomafeminista.org; www.casadosmeninos.org.br

Introduction

In this report we intend to examine three experiences with community and autonomous networks – by focusing on relationships that are established between diverse women and non-hegemonic groups and considering the overlapping of discriminatory systems that can be experienced based on sex, gender, race, ethnicity, class and other axes in an intersectional perspective.[1] These groups have been engaged in developing and crafting new communication tools and infrastructures in their local communities. We want to share our view as researchers and activists that networks constitute an array of relations that go beyond the mere act of sharing and distributing access to a particular kind of technology, including when communities reclaim the use of electromagnetic spectrum, building radio and mesh networks.

From a feminist perspective on technology and infrastructure, we briefly discuss our assumptions in this field considering that technology is neither neutral nor deterministic. The operation of a community network implies relationships between a multiplicity of individuals and social groups with different perspectives, interests and needs, and who are not affected in the same way by socio-technical systems given existing inequalities such as race, class, nationality and gender. This means that technology initiatives and the way that they are framed can produce inequalities and differences that emphasise the political structuring of the social world emerging under the impact of the material internet infrastructure. Community networks also show that technology is a terrain of struggle on which hegemonic forces express themselves through specific design strategies in opposition to non-hegemonic groups that are nevertheless more or less successful in influencing the future form of the network infrastructure with which they are engaged.

Whenever voices, experiences and realities of people with little or no access to the internet are homogenised under the category of "unconnected", their visions, aspirations and creative potential are invisible. Since a feminist perspective plays with different modes of being, it may help to question the basis of knowledge and subject positions that were/are seen in one-dimensional binary opposition in communities, and help to note the contradictions, tensions and ambivalence that characterise different communities, groups and subjects in technology initiatives.

On this basis, we present ongoing experiences in Brazil – *Rede Base Comum*, *Fuxico* and *Radia Pankaru* – all autonomous networks that are challenging androcentrism and new forms of colonialism or exploitation at the very local level. We also advocate that alliances between multiple groups can be expanded and technologies can be re-appropriated respecting local specificities and the different means by which communities articulate their experiences, without placing technical expertise above all other kinds of knowledge.

A claim for heterogeneity

There is a common tendency among "experts" to use the term "last billion" to homogenise those who lack internet and mobile phone access. Such narratives imagine technology as moving not only "from the West to the rest", but from the urban to the rural, from the cosmopolitan to the local, and from the globally networked to the remotely disconnected. Under such a framing the notion that culturally and geographically peripheral sites are in definitive need of "magical" solutions is implied.

1 Kimberlé Williams Crenshaw conceptualised intersectionality to denote the various ways in which race and gender interact to shape the multiple dimensions of black women's employment experiences. Although her theory has aroused controversy, Crenshaw helped to make visible some of the dynamics of structural intersectionality and pointed out that people and groups experience the overlapping of discriminatory systems. She also pointed out the limits in identity politics, affirming that its problem "is not that it fails to transcend difference, as some critics charge, but rather the opposite – that it frequently conflates or ignores intra group differences." For more information, see: https://www.racialequitytools.org/resourcefiles/mapping-margins.pdf

This particular view can emerge even in activism or free software and open technology environments, where hegemonic narratives on networks are often naturalised and presented as universal to the detriment of already existing experiences and local ways of bonding. In this broader context, it is important to think over the risk of using the term "community" as a simple label that not only overshadows the recognition of local technologies and network alternatives, but also reinforces the lack of respect for the voices of those silenced within a collective experience. In order to avoid this risk, we understand it is necessary to consider multiple voices and internal inequalities in community-based projects, which should impact on methods, time planning and especially on the dynamics of project implementation.

The lack of consideration of these aspects often makes the evaluation of the achievements of a community network difficult, or even leads us to measure its success mainly by the number of nodes it connects. In this sense, the pressure for scaling up community networks may, for example, run over latent conflicts on gender-related aspects and undermine efforts to break down hierarchical authority structures where difference between classes, ethnic groups, races and experts and non-experts becomes inequality. It may also reduce the time required to mature social ties and make a community network more bold and diverse. In addition, there is a risk of internet access provision and network stabilisation being considered the primary concern, without any consideration regarding the power disputes over its standards, protocols, software and infrastructure design.[2] In many cases, connection is seen as the most important goal to be achieved in a whole project, instead of one among many others, neglecting discussions and proposals that could be carried out from more intersectional perspectives.

At the end of the day, we often face the absence of actions designed to actively make these spaces (physical or digital) more welcoming and safe for women – for example, the creation of daycare facilities in spaces of infrastructure design and deployment; the implementation of affirmative action to build more representative and less power-concentrated network and community management structures; a collective agreement on an anti-harassment and non-discriminatory policy; or even the awareness of the potential need to remove intimate content disclosed without consent or misogynist content from community servers. All this is revealing of how even collective initiatives sometimes are designed in such a generalised way that they dismiss the potential offered by an intersectional framework with all its complexity, mobility, and untiring political possibilities, and with no consideration for structural violence and discrimination.

It is not surprising that some initiatives thought up by white, cis,[3] non-poor men, who are disproportionately represented in many decision-making spaces, are loaded with assumptions and ideological baggage, and do not contemplate issues such as those mentioned above, even in the field of open and free/libre technologies, when those needs are not part of their universe and daily life. Furthermore, the choices that were made considering this certain group's needs and concerns – which will not be the same for all the participants in any network – are often not presented as one of the possible alternatives, but are hidden under an appearance of "ready-made solutions". This kind of experience can reinforce not only the naturalisation of inequalities, but also, as pointed out in various feminist literatures,[4] reinforce a colonialist and problematic heritage through the universalisation of a particular and privileged condition (and the choice of a privileged narrative, which like its formulator is seen as universal and, therefore, capable of dealing with all inequalities at once).

Considering this assumption, the discussions aimed at making collective spaces and infrastructure designs more "on the ground", welcoming and supportive to different people and their values and practices, are many times perceived as a waste of time or a secondary item on the implementation agenda. All this may foster the feeling that certain spaces are not supposed to be occupied by women, or that certain kinds of knowledge cannot be considered technologies, or even the notion that external experts

2 Vicentin, D. J. (2017). *Internet Governance, Infrastructure and Resistance*. lavits.org/wp-content/uploads/2017/08/P8_Vicentin.pdf

3 According to the LGBTI+ Communication Manual, "cis" (or cisgender) is the term used to describe people who are not transgender. It refers to individuals who identify themselves, in all aspects, with the gender attributed to them at birth. See: https://agenciapatriciagalvao.org.br/wp-content/uploads/2018/07/manual-comunicacao-LGBTI.pdf

4 Some references regarding this discussion are: Harding, S. (1998). ¿Existe un método feminista? In E. Bartra (Ed.), *Debates en torno a una metodología feminist.* Mexico City: Universidad Autónoma Metropolitana-Unidad Xochimilco; Haraway, D. (1995). Saberes localizados: a questão da ciência para o feminismo e o privilégio da perspectiva parcial. *Cadernos Pagu, 5*, 7-41; Rago, M. (1998). Epistemologia Feminista, Gênero e História. In J. Pedro & M. Grossi (Eds.), *Masculino, Feminino, Plural*. Florianópolis: Ed. Mulheres; Ribeiro, D. (2017). *O que é lugar de fala?* Belo Horizonte: Letramento; Sardenberg, C.(2002). Da Crítica Feminista à Ciência a uma Ciência Feminista? In A. A. Costa & C. Sardenberg (Eds.), *Feminismo, Ciência e Tecnologia*. Salvador: Rede Feminista Norte e Nordeste de Estudos e Pesquisa sobre Mulher e Relações de Gênero (REDOR), Núcleo de Estudos Interdisciplinares sobre a Mulher (NEIM), Universidade Federal da Bahia.

will always have better answers to community problems than the community members themselves.

The Brazilian experiences we will address exemplify how the presence and active participation of diverse women, including indigenous women, LGBTIQ people, community leaders living in the periphery of urban centres and in rural villages, feminists, tech activists and NGO participants, impact on the organisation of practices, activities and spaces of power where they are mostly thought of as hegemonic subjects.

In our examples:

- The experiences involve diverse women in the formation of infrastructures and networks, therefore encompassing both gender and technology issues.
- Participants believe in the importance of building both online and offline local networks.
- Participants see as problematic ready-made and easily presented solutions (with English often the embedded language of instruction and implementation), as well as hierarchical knowledge arrangements, pointing out that experiences with any kind of technology should be designed to value local expertise, support learning processes, seek diversity and dialogue with the local context and, through doing this, entrench participants' autonomy.
- The experiences help us rethink our methods and practices, disrupting naturalised assumptions that are built upon the invisibility of different groups and traditional knowledges,[5] which have been supporting community organisation and sustainability concurrently with or even before digital technological knowledge.

The purpose here is not to present these experiences as a solution to the diversity problem, but to reflect on how they break down the invisibility of both local infrastructure relations and social inequalities,[6] highlighting local knowledges and practices and proposing new social alliances. We believe these experiences can help us to move towards the decolonisation of our technological imaginary. In this regard, we understand technology in a broader perspective. As one co-author of this report once defined it: "[B]eing, living, loving, suffering, resisting, organising, cooking... all are ancestral forms of technology," and the infrastructures are "the elements that make technologies operate so powerfully."[7]

We also share Sophie Toupin and Alexandra Hache's perception that:

> One of the main constitutive elements of feminist autonomous infrastructures lies in the concept of self-organisation already practised by many social movements that understand the question of autonomy as a desire for freedom, self-valorisation and mutual aid. In addition, we understand the term technological infrastructure in an expansive way, encompassing hardware, software and applications, but also participatory design, safe spaces and social solidarities.[8]

Rede Base Comum, a local network to act on our living areas

Rede Base Comum (Common Base Network in a free translation from Portuguese) is an urban community network in the district of Jardim São Luiz, located in the south of São Paulo. It is a local mesh network currently managed by the NGO Casa dos Meninos.[9] Although this is not a feminist network by definition nor a collective that organises itself from a gender perspective, it was the women from the community that managed the entire process of its implementation and are now in charge of its maintenance.

The network was born in 2010 with the objective of creating a common space for residents to gather and learn from knowledge exchanges and to share local resources. It was also designed to improve the territory's[10] appropriation of technology while creat-

5 Although incorrect from a strictly grammatical viewpoint, the plural form "knowledges" is used to emphasise the fact that there are multiple kinds of knowledge.

6 Oliveira, D. (2017). Community networks and invisibility regimes of infrastructures and bodies. Paper presented at LAVITS International Symposium "Vigilancia, Democracia y Privacidad en América Latina: Vulnerabilidades y resistencias", Santiago, Chile, 29 November-1 December. lavits.org/wp-content/uploads/2018/04/40-Débora-Prado-de-Oliveira.pdf

7 Zanolli, B. (2017). Podcast: Feminist spectrum and infrastructure. *GenderIT.org*. https://www.genderit.org/node/5029

8 Toupin S., & Hache, A. (2015). Feminist autonomous infrastructures. In A. Finlay (Ed.), *Global Information Society Watch 2015: Sexual rights and the internet*. APC and Hivos. https://www.giswatch.org/en/internet-rights/feminist-autonomous-infrastructures

9 Casa dos Meninos is a social organisation founded in 1962. Since 1999, it has carried out social, educational and cultural activities and programmes, using geo-referencing technology and web development. The NGO's target audience is young people aged 13-29 years old and adults aged 40-75 years old, all residents of the neighbourhood of Jardim São Luiz, in the district of M'Boi Mirim, located in the southern outskirts of the city of São Paulo.

10 We understand territory as in Milton Santos' work: "The territory is the ground and the population, that is, an identity, the fact and the feeling of belonging to what belongs to us. Territory is the basis of labour activities, of residence, of material and spiritual exchanges, and of life, upon which it also has influence. When talking about territory one must, therefore, understand that we are speaking about territory used by a population." See Santos, M. (2003). *Por outra globalização: do pensamento único à consciência universal*. Rio de Janeiro: Record.

ing a sense of community in a densely populated city. Inspired by the ideas of Brazilian geographer Milton Santos, the physical local network is seen as an experience that can help locals amplify the impact of political action in their residential areas, redefining the composition of their immediate territory into smaller, delimited areas where residents circulate, interact and modify collective spaces, such as schools, health units, local markets and others. The network was also built with the prospect of creating a new culture, based on collaboration, solidarity and local exchange practices,[11] challenging the pre-established culture of individualism and competition in our society.

It is noteworthy that the mesh network set up by Rede Base Comum was preceded in an "analogical way" by a community network in the territory created when residents mobilised for the "Daycare Centres for All" protests in 2008. Back then, we organised field work to gather information on the lack of daycare facilities in São Paulo, combined with geo-referencing tools, which showed the community, especially mothers, that a large number of children without daycare access in São Paulo were from their area.

The lack of daycare facilities in São Paulo could give one a feeling of impotence: official numbers indicated more than 100,000 children without daycare in the city at that time. Nonetheless, the geo-referenced information unified the mothers and helped them organise and intervene with a more concrete demand. As a result, the data collected in the field work enabled a local movement that resulted in two new daycare units in the neighbourhood, inaugurated in 2011 and 2013.

Social mobilisation coordinated by the community preceded the organisation of a local digital network, helping us to think of a mesh network as a way of supporting important interactions already existing between residents and relating to their reality and local needs. In this process, a series of meetings and discussions were carried out to define the operation and maintenance of the network structure. In a demonstration on how technology can support a non-hegemonic vision, it was decided that in order to maximise autonomy, the network should be managed by the community itself, including when it came to defining internal policy and content issues. Here when we talk about the "community", we are also talking about ourselves, and women who live in the neighbourhood and were already bonded in "offline" social networks.

Nevertheless, when our group of women looked for network infrastructure experiences and technical expertise, we faced an unwelcoming scenario: the complete absence of training spaces on information and communications technologies (ICTs) in our own neighbourhood, including for-profit technical schools – which led us to conclude that not every area in the city is seen as a place of interest regarding technological expertise. Clearly, this lack of opportunities is a reflection of social and geographical tensions that underline internet infrastructures – the place where people live and their social class resonate in the level of their access to information and knowledge.

As a consequence, the team that today manages the network sought partnerships with other groups involved in digital culture. However, even in the free/libre and open technologies activist field, our first impression was that this space did not belong to us, as we faced a predominance of white males and only a few women working with infrastructure and free networks. Furthermore, these partnerships were sometimes focused on teaching particular content or techniques in a short period of time – a dynamic in which there was little space to connect all that "external" expertise to local aspects.

One of these partnerships, however, occurred differently. During a course on community networks designed for women organised in 2017 by Vedetas – who had set up a feminist server intended to support feminist groups in their online activities, to increase women's security and autonomy and to build feminist technologies[12] – we discovered that feminist practices could influence learning and knowledge production. Once we managed to find out about spaces where diverse women were involved with the construction and modification of network infrastructure, we could observe the differences in the teaching methods and in the environment, and realised we should take into consideration what we were producing and thinking for Rede Base Comum in order to meet our needs when building our own infrastructure.

One aspect that drew our attention was that the partners who work with feminist infrastructures consider the reality of "on the ground" practices of the local population when introducing digital tools. In other groups that we had contact with, it was always stated as an imperative and general truth that it was important to use free software in community networks, for example. We know the political

11 With the network we want to use and develop applications and webpages that allow the community to share local resources, from the exchange of local knowledge, texts and audiovisual content to open channels for the promotion and organisation of shared physical items among the local population, such as books and building materials.

12 https://vedetas.org

importance of its use; however, we cannot always act in ideal conditions and in this case specifically most of the residents in our territory were still more familiar with the use of proprietary software than with free software interfaces. In their very first contact, a digital environment that was too different could divert people from the use of the local network.

In the feminist infrastructure workshops we participated in, partners presented the use of both proprietary and free tools in parallel, which helped us to understand in practice the differences that make the use of free software so important. This simultaneous use served as a period of adaptation to a free software environment. This has helped us to adapt the workshops we have conducted in Rede Base Comum, in which we apply a similar methodology, introducing free tools in a more practical, gradual way and looking at their differences in relation to the proprietary software. As a result, digital tools were presented as a popular resource and not a tool designed by and for experts.

Another difficulty we had was in relation to the application of ready-made solutions. In some partnerships with digital culture groups there was an insistence on using already developed tools and "plug-and-play" solutions in our network. We know that this is the result of a positive intention and effort to "facilitate" the life of those who are starting their projects with community networks, but in addition to using platforms and tools already developed elsewhere, we have always affirmed that it would be important for us to learn how to develop some of our own tools. We had long conversations with some partners in this regard, but they did not understand our need to appropriate this knowledge. Feminist groups had a better understanding of this need, as the importance of autonomy is a prominent part of their discussions. Today we are designing workshops and training projects which aim to have women programmers and developers in our own group and territory building some applications for our network locally. We consider this to be one of the main priorities in the organisation of our time, resources and projects, even if this means slowing down the expansion of our network to neighbourhood areas.

Fuxico: Weaving connections beyond the internet

Fuxico[13] is an autonomous and portable device made by Brazilian women to connect people who are present in the same physical space in order to promote content, exchange experiences, and foster collaboration. It creates a wireless network – disconnected from the internet – with the objective of sharing digital content in real time and in a completely anonymous way.

Motivated by the desire to explore the theme of autonomous feminist networks, in 2017 a group of four female hackers from Brazil and Mexico coming from different feminist social infrastructure and technology initiatives – Kéfir,[14] Periféricas[15] and Vedetas – met at the AWID international forum.[16] This participation resulted in a one-year project with the purpose of exploring what was mapped as four points of tension in feminist infrastructures: "consent and intimacy", "situated knowledge and memory", "seeded connectedness" and "autonomous decision making".[17]

Through this project we worked with 230 women[18] from events and workshops we facilitated in São Paulo and Salvador, and also agriculturists from Vale do Ribeira and rural artists at EncontrADA,[19] a self-organised event carried out in the São Paulo countryside. The work itself had very diverse approaches, such as setting up temporary prototypes of autonomous mesh networks in events, to presentations, talks and longer courses on the subject of autonomous infrastructures.

During the project we realised that merely a theoretical presentation about the theme or practice

13 redeautonomafeminista.org/fuxico

14 https://kefir.red

15 www.perifericas.com.br

16 AWID is an international, feminist, membership organisation committed to achieving gender equality, sustainable development and women's human rights. https://www.awid.org

17 These women hackers also explored the points of tension in feminist autonomous infrastructures at the 2017 Internet Freedom Festival. See the event's wiki at: https://internetfreedomfestival.org/wiki/index.php/Getting_gender_inclusive_from_the_ground_to_the_cloud and the podcast "Feminist autonomous infrastructure: Technomagical fires to warm your hearts" at *GenderIT.org*: https://www.genderit.org/node/4921

18 At the first meeting, held in 2017, it was possible to see that there is diversity among the participants: there are women of different ages, most of them between 20 and 40 years old (there are no elderly women); there is racial diversity, especially among whites and blacks, and there are Brazilians with Asian ancestry who work with Asian feminism. People also come from different areas of activity: graffiti artists, designers, lawyers, digital marketing communicators, journalists, physicists, high school students, academic researchers, and feminist activists from different groups. They also come from different places in Brazil: there are women from the city of São Paulo, from the interior of the state of São Paulo, from Pará, Bahia, Pernambuco – although most are living in São Paulo at the moment, even if temporarily. Some are mothers, some not; there are cis and trans women, queer and non-binary people, heterosexuals, bisexuals and lesbians. The concern with diversity is expressed not only in the composition of the participants, but in the organisation's efforts to try to anticipate demands to ensure that the space is really welcoming for different women. Before the course starts, for example, the organisation surveys the participants via email or telephone to map out demands for daycare, food aid and transportation, and accessibility for people with disabilities.

19 https://encontrada.hotglue.me/?historico

with routers and antennas was not enough to deepen the discussion about the possibilities of women using and sharing local services. In some of the field experiences, the very idea of digital communication was thought of as only being about the internet, something that this initiative sought to deconstruct.[20]

We then decided to take a concrete example, choose some technology that the participants could touch and experience, but that was not an unoriginal ready-made solution or an imported tool. Being an external solution – as mentioned, in English by default – and with little regionalisation in the "user experience" design and objectives, this imposition without establishing any dialogue with participants on the local context and on their understandings about technology usually acts against community and individual autonomy in our experience. Even free software with its solid goals of supporting civil society and the free democratic access to knowledge, like MediaWiki[21] or PirateBox,[22] if applied in an unchanged and unreflective way to any context, would appear as a foreign technology and only intensify the distance women feel in anything related to technology.

Therefore, we chose to work with a heavily customised version of PirateBox, a 2011 free software project defined as a "DIY [do-it-yourself] anonymous offline file-sharing and communications system built with free software and inexpensive off-the-shelf hardware".[23] PirateBox is an operating system for Raspberry Pi,[24] which creates a wireless network – not connected to the internet – to exchange digital content such as images, videos, audios, documents and conversations, prioritising anonymity.

Initially the interface was translated into Portuguese and the pirate's visual identity that is part of the product was replaced by an image that served as a locally meaningful analogy. The intention was to give meaning to the experience by bringing elements that were part of the women's daily life: for farmers, we used the idea of "seeding" – a space for exchanging indigenous seeds, where the content exchanged could, by analogy, multiply as "Creole seed".[25] For women on the outskirts of Bahia state, we used the name of a little flower called *fuxico*,[26] which is an image used in a very common craft technique from the interior of the Brazilian Northeast that is more than 150 years old. By using pieces of fabric left over from sewing, the women create individual pieces that look like fuxico. They then sew numerous fuxicos together, producing clothes, decorations, tablecloths and bed covers. We called this localised solution "Fuxico", alluding to other techniques developed by Brazilian women.

These analogies seek not only to create a sense of belonging in the participant experience, but to give a new significance to what we understand by technology. Women often do not see technology as a "female thing" and all things considered, "feminine", in our culture, is often classified as anything but technology. One of our goals as feminist hackers is to change this correlation and question the politics of such classifications and the scale of power they represent. As the neuroscientist and artist Christine Liu claims: "Knitting is programming. Sewing is engineering. Baking is chemistry. Women have been STEM [science, technology, engineering and mathematics] pioneers longer than they've received credit for."[27]

The use of metaphors and analogies expressed here reflect the view of Diana Maffia[28] who, in formulating her critique of the hegemonic sciences, reveals that the production of "truths" is based on false notions of objectivity and neutrality, which require the use of literal language and the exclusion of emotion. In doing so, metaphors, far from having value for knowledge, create obstacles to meaning. What was experienced with these women is exactly the opposite: the metaphors are valued and build bridges – often they are perfected by the participants, who also formulate their own comparisons to test the knowledge learned. The truth here appears, therefore, in the sense proposed by Maffia, that what is legitimated by different perspectives will be

20 There are, for example, other forms of digital communication that we call attention to in our collective, such as digital television and radio and intranets.

21 https://www.mediawiki.org/wiki/MediaWiki

22 https://piratebox.cc

23 Ibid.

24 According to the Raspberry Pi Foundation, the Raspberry Pi is an affordable and capable little computer, which can be used in electronics projects and for many of the things that a desktop PC does. https://www.raspberrypi.org

25 The manual of feminist practices produced in 2018 by SOF – a Brazilian feminist organisation – says: "Creole seeds are those grown and maintained by traditional peoples and communities throughout generations, perpetuating the natural wealth of our lands. Through agro-ecological crops and seed exchanges, they remain alive." This is an activity of resistance in a country where the agribusiness industry poses a threat to traditional ways of tending the soil and Brazilian people's health as a whole. For more information, see: https://www.hrw.org/pt/news/2018/07/20/320493

26 For some examples of fuxico, see: https://www.artesanatopassoapassoja.com.br/artesanatos-com-fuxico-passo-passo

27 https://twitter.com/christineliuart/status/856729454013366272

28 Maffia, D. (2005). Epistemología Feminista: por una inclusión de lo femenino en la ciencia. In N. B. Graf & J. Flores (Eds.), *Ciencia, tecnología y género en Iberoamérica*. México DF: Universidad Autónoma de México/Plaza y Valdés.

true (in this case, from the perspective of the women who attended the activities) – it creates meaning that is not finished, but which can be renegotiated. The Fuxico project is a collective reflection on the expressions and wishes of Brazilian women and was designed to connect people present in the same physical space, such as feminist events and collective venues – but not limited to them. In addition to the original PirateBox features adapted to this context, Fuxico includes by default educational content about feminism, autonomy, technology and stories about women. The device also includes a manual on feminist digital security techniques and practices – since any digital experience that seeks to include women and other social minorities must consider that discriminatory violence is structural and could always occur in both online and offline spaces.

The original PirateBox project, as many free software projects do, focuses a lot on trust, anonymity and the absence of control or censorship mechanisms. As feminists we worry that the lack of built-in functionalities to delete content that proves to be violent or harassing is a flaw that will distance women and other social minorities in fear of online violence. In the context of digital territories as facets of real-life territories, cryptography and the possibility of adding security layers to specific spaces and files, and the ability to choose who you want to share content with, is a form of resistance. The guide[29] we share by default includes information on holistic tactics and alternative digital security software that the users should familiarise themselves with, but in the future we wish to add media governance and cryptography features built into the Fuxico interface.

In retrospect, this project encouraged us to work on the issue of autonomous networks from a different perspective. The aspects that unite and create networks, as well as common narratives and experiences, were more important than the range a certain kind of technology had. Rather than thinking about the extension and quantity of nodes, we reverse engineered that logic by working in the smallest as possible sphere: a single router – a single box with the potential to bring together narratives, ideas, knowledge and desires. Once this common territory is established, the intention of the project is to encourage women to gradually expand these connections, adding routers and antennas as far as it will make sense for them.

Rádia Mulheres Pankararu

The indigenous territory of Pankararu is located in the backwoods of the state of Pernambuco, in the Northeast region of Brazil. The population that lives in the territory is approximately 7,200 and although their land was demarcated in 1942 they still fight against outsiders to make them leave the territory and to have their land rights ensured.[30]

Rádia Mulheres Pankararu (which means Pankararu Women's Radia)[31] was started in January 2018 and was the result of the effort of many women, but mostly the women associated with AMIGP (*Associação Mulheres Indígenas Guerreiras Pankararu*, which means Pankararu Indigenous Women Warriors' Association), a couple of radio lovers,[32] and Thydewá,[33] a partner NGO. The initial idea was to organise a workshop on basic electronics with women, supported by Fundo ELAS, a local fund for women's rights.[34] However, after talking with them we discovered that they had an old dream of having a local low-power FM (LPFM) radio station. So we arranged for experimental low-power equipment and provided workshops on basic maintenance and audio content production using free/libre and open source software (FLOSS).

The signal reaches almost the whole indigenous territory and the radia is open for women's participation only (they are still thinking about how men can participate without taking the action out of women's hands). Before the radia, the only two FM radio stations that could be heard in the territory were commercial stations from cities located in

29 The publication, called the *Practical Guide to Strategies and Tactics for Feminist Digital Security*, aims to provide women with greater autonomy and security on the internet by presenting strategies and tactics of digital defence for feminists. The content is directed to women in Latin America and was developed considering different women: blacks, trans, lesbians, organised activist movements of women or those who act individually in the network, whether from urban areas or rural peripheries, and with different levels of access to technology. Each subject is connected to real cases of online violence and has practical information on how to deal with adversity in similar scenarios and understand where to focus on security efforts. The second part of the guide is devoted to the use of mobile phones and how to have a safer device. There is also information on what to do on social networks to counter hate speech and the unwanted spreading of intimate content. The guide is available in Portuguese at: https://feminismo.org.br/guia-pratica-de-estrategias-e-taticas-para-a-seguranca-digital-feminista

30 More information on the land conflicts faced by the Pankararu can be found in the following articles in Portuguese: https://jornalistaslivres.org/povo-pankararu-luta-para-concretizar-demarcacao, https://jornalistaslivres.org/pankararu-seguem-na-luta-para-ocupar-seu-territorio .

31 We use the term radia with an "a" instead of radio with an "o", reflecting the feminine vs. masculine word ending typical of Portuguese, because the sexism in our language is an important issue and has an impact on how infrastructure and machines are perceived as a male field.

32 Luiza Cilente as the coordinator of the project and Bruna Zanolli as radio and studio technician.

33 www.thydewa.org; see also the project Pelas Mulheres Indígenas (For Indigenous Women): www.mulheresindigenas.org

34 www.fundosocialelas.org

Source: Associação de Mulheres Indígenas Guerreiras Pankararu - Pernambuco, Brazil

the surroundings. The day after the radia did its first broadcast, there were people knocking on the door of one of the women's homes at 7 a.m. asking what time it was going back on air!

AMIGP is an association of women that fights against domestic and sexual violence and advocates for sexual and reproductive rights for indigenous women. The association is mostly comprised of indigenous teachers and local articulators,[35] as well as homemakers and students. They renovated a small room at the back of the association's headquarters, where the radia is currently located. In May, another project came to the space, providing a FLOSS laboratory for young women.

Although the radia project is still new and it is hard to measure its achievements regarding social change, it is undeniable that the community was strengthened by the LPFM radia and its voice was amplified. We believe that the radia, where content production is shared, strengthens community bonds and helps with collective thinking. Because of this, the dynamics of an LPFM radio are more collective than the use of the internet on individual devices, and can be an ally when we think of the sustainability of collective relations, representing a broader view on community networks and autonomous infrastructure. We believe that an LPFM radio is, in this context, a form of community network, meaning that it provides a way of communication and exchanging thoughts and ideas. The internet connection in the indigenous territory is restricted (there is satellite only, and it is expensive and not reliable), so having a local radio station was a specific choice that was made considering the low budget of the project. At the same time, the communication problems to be addressed in the Pankararu community were better attended to by broadcasting, being able to amplify the voices of the community leaders who are fighting for indigenous women's rights

Additionally, an LPFM radia entirely led by indigenous women is unprecedented in Brazil, a country that struggles with concentrated management and patriarchal laws when it comes to radio spectrum. The concentration of radio spectrum use in Brazil is a historical issue and, although there is a law regulating licences for community radio networks, it is very difficult to get a concession due to bureaucracy. Sadly, the technical restrictions imposed by the concessions also end up lowering the potential of setting up a community radio initiative. The present law establishes that only one frequency is available for community radios to use nationally – and this remains the case regardless of the size of the city and its population. This is also limited to 25W of power or a maximum 1 km of range, which is really low, especially considering big cities with tall buildings and rural areas that are vast.[36] There is also a need for a legal entity to apply for a concession, and due to the fact that there is only one frequency available, communities within the same range of 4 km become competitors for the same concession.

There are many other examples of how the hindrances imposed on community radio in Brazil

35 A local articulator is a person who has a communication role in the community and is politically active at a local level. It is a common term in Portuguese.

36 Changes in the law are currently being discussed, but the changes, although significant, fall far short of addressing the real demands in the field. For example, the power allotment would go up to 150W and it would be possible to have two frequencies per municipality. More information can be found (in Portuguese) at: https://www12.senado.leg.br/noticias/materias/2018/07/10/aprovado-projeto-que-aumenta-potencia-das-radios-comunitarias

Source: Associação de Mulheres Indígenas Guerreiras Pankararu - Pernambuco, Brazil

impact on communities. The net result is that more radio stations operating without concessions are closed down every year than new concessions granted, showing that the level of access to a concession falls far short of attending to people's demands.

The difficulties to access the spectrum are amplified when looked at through an intersectional lens, meaning that even among the groups that have been awarded concessions, women and minorities are still marginalised and occupy secondary posts. It is evident that in most community radio stations in Brazil, women still occupy positions related to secretarial duties and cleaning, and are very rarely in positions of leadership and technical management. The fact is that the whole decision-making process of the allocation of the radio spectrum is still practically all male driven, from legislators to regulators, leading to the lack of diverse and feminist perspectives regarding the possibilities of use of spectrum as a resource. This patriarchal domain makes spectrum access difficult and hinders the increase of diversity in community radio networks. Sadly, because of the facts mentioned above, nowadays LPFM radio stations in Brazil are mostly used by evangelical groups with religious and economic purposes rather than cultural and political objectives. New kinds of technologies that use radio spectrum and could decrease social gaps are not being explored or are even undermined in Brazil. This is the case of digital radio, which can enlarge the access to frequencies and even send other kinds of data like images and videos, in addition to audio. Digital radio could be a communicational solution for communities in remote areas that lack access to basic services and the internet.

To destabilise, but not conclude

More than producing definitive answers, we expect with this report to create destabilisations that enable us to rethink community networks, and to challenge the current internet world order by following a different method, considering the local context and demands from a feminist perspective on infrastructure.

Shared experiences in processes of building and maintaining community networks, or other collective forms of ICTs, have the potential to challenge the way in which broadband, wireless or radio connections have been implemented in rural areas and the periphery of urban centres. Usually these experiences either attend exclusively to commercial interests or aim at simply, quickly solving the problem of the digital divide through internet access.

It is unquestionable that the experience of community networks has the potential to address the invisibility of infrastructures.[37] At the moment these networks usually occupy an abstract place in our imaginary but their impact on our day-to-day lives and the power relations they establish gain a bolder materiality. From our perspective, community and autonomous networks are not limited to a shortcut for development or a "breadcrumb" offered to "disadvantaged" groups to access the internet. They represent a possibility of establishing connections based on grassroots rules and in which technologies can be re-appropriated, or creatively used or repurposed.

However, it must be noted that any process of interaction with technology carries constraints and conventions that may reproduce hierarchies and inequalities even in collective processes. In other words, even community networks can reproduce norms that alienate women and non-hegemonic groups from spaces of power and autonomy. In addition, some collective experiences based on

37 Bowker, G., & Star, S. (1999). *Sorting Things Out: Classification and Its Consequences*. Cambridge: MIT Press.

"one-weekend workshops" and flash projects of network implementation run the risk of being detached from local demands, failing to build long-term autonomy and strengthen communities.

In this broader context, the presence of diverse women in the design and management of infrastructure and networks – in addition to making these spaces more democratic – seems essential to challenge the androcentrism and colonialism that contaminate our knowledge and practices in the global South. In the experiences we quickly presented here, the presence of diverse groups and the absence of pressure for a quick incorporation of a certain technique or for network expansion are in line with the time needed to mix knowledges and to mature human connections towards making our networks more communitarian, free and autonomous.

The presence of diverse women is fundamental on several levels, from the implementation and management of a local network to institutional spaces of decision making. The lack of popular and community access to the radio spectrum, for example, is another important community network issue that it is clearly a matter of human rights and should be addressed carefully by politicians and legislators, and hopefully in the most intersectional way possible. The digitalisation of the FM signal could be a way of broadening access to this common good, considering that the digital signal occupies less bandwidth than an analogue signal, and there is technology to make better and more dynamic use of spectrum using software-defined radio (SDR) and cognitive radio technologies. Brazil, however, is far from digitalising the radio signal and promoting a more democratic, less bureaucratic and profit-based model of access to the radio spectrum.

Finally, as we have argued, we see evidence that mere access to new technology could reinforce rather than reduce inequalities. This observation seems important to break the invisibility not only of technological infrastructure, but also of the asymmetries of power that are clear from an intersectional perspective. This means proposing alliances that do not erase differences, but instead value the power of diversity. It also means a commitment to an active effort to link digital expertise to women's different grassroots technologies and skills already in use, and to local daily life, aiming to create a welcoming and safe environment for those who are outside the hegemonic norm.

Decentralising culture: The challenge of local content in community networks

Nicolás Echániz and Florencia López Pezé
AlterMundi
www.altermundi.net

Introduction: A bit of history

Our first steps with community networks go back to 2003, with the beginnings of BuenosAiresLibre. At that time, free networks were phenomena of big cities (Buenos Aires, Montevideo, Seattle, Portland, Berlin, Rome, etc.). They were set up mainly by "geeks", the community that had ties to the free software movement.

The networks were experimental spaces and were usually set up to offer access to self-contained web servers, FTP repositories, games, etc.; that is, they functioned as metropolitan intranets, and reflected the interests of those who set them up. At that time, these networks did not have their own public IP resources, autonomous system numbering, or peering agreements with neighbouring networks, and they were not linked to internet exchange points. In general, their members resolved their individual internet connectivity needs through commercial providers.

Some networks of this first era evolved, such as guifi.net in Catalonia,[1] but many went down in history. The freenetworks.org website[2] maintained information about free networks in the world for years, but today it is no longer online.

AlterMundi, like other similar organisations,[3] focused its efforts outside the big cities and took the model of community networks to disadvantaged and digitally excluded areas. These communities have huge socioeconomic and educational differences compared to big cities and the most concrete communication need is to achieve internet connectivity.

Our perspective is that community networks should be, mainly, a vehicle to allow the "unconnected" to connect themselves. Over time and thanks to some successful examples, the perspective on community networks in the areas of internet governance – national and international – began to change. Longstanding organisations such as the Association for Progressive Communications (APC) and Internet Society (ISOC) developed plans and strategies focused along these lines: a Dynamic Coalition was created at the Internet Governance Forum,[4] a Special Interest Group was started in ISOC,[5] and a project on local access networks was started in APC.[6] Community networks started to be seen as an effective solution to reduce the digital divide and became an important issue on the agenda of various relevant actors.

This report focuses on the role that community networks play in creating an inclusive and culturally diverse internet. We discuss the concept of right to access, suggesting its limitations. We then introduce an alternative notion of the "right to co-create the internet". We also share some experiences that give us perspective on the history and future of community connectivity as a fundamental enabler to the right to co-create the internet.

The right to co-create the internet

The right to access communications is one of the fundamental notions defended by sectors of civil society devoted to issues concerning the digital divide and connectivity in excluded areas. At AlterMundi we believe that this notion has to be challenged, and ask ourselves if defending it is not, in some way, supporting the agenda of the concentrated powers of the internet ecosystem and reproducing its operating models. Is it possible to think about the construction of the internet from another paradigm?

Other social movements help us reflect on this problem. Let's think of the terms "food security" and "food sovereignty". The first is the term that governments and food corporations use to refer to the condition that exists when "all people, at all times, have physical, social and economic access to sufficient, safe and nutritious food which meets their dietary needs and food preferences for

1 See the Catalonia country report in this edition of GISWatch.

2 https://web.archive.org/web/20171204140813/http://freenetworks.org:80/member-affiliates

3 Atalaya Sur, Coolab, Zenzeleni, Gram Marg, etc.

4 https://www.intgovforum.org/cms/175-igf-2015/3014-dynamic-coalition-on-community-connectivity-dc3

5 cnsig.info

6 https://www.apc.org/en/project/local-access-networks-can-unconnected-connect-themselves

an active and healthy life."[7] Governments are primarily responsible for guaranteeing this right and must not interfere with individuals' efforts to "earn a living". They must also "create opportunities" for those who do not yet enjoy the "right to food".

However, other actors use the term "food sovereignty". Food sovereignty is the right of peoples, communities and countries to define their own food policies that are ecologically, socially, economically and culturally appropriate to their circumstances. It also claims food as a right.[8] That is to say, the dispute is not the right to eat but the right of the communities to food self-determination and therefore to define their own food policies. This means building a new social relationship that questions oppression and inequalities.

The concept of "food security" is similar to the "right to access communications", but there is no clear concept analogous to "food sovereignty" in the field of communications. In this area, there is a broad grouping that seeks to build another paradigm and does not identify with the "right to access". The argument is that the widespread use of the term "access" is not casual. It implies looking at the person as an individual user, in a consumption relationship which is unidirectional; people access services and content that they do not control and that do not belong to them. This is a model of "feed them" access.

Organised and self-managed communities that build their own infrastructures and technologies to meet their communicating needs work in a different direction. What they do is understood more clearly as a right to co-create the internet. From this perspective, they do not constitute themselves as consumers, but as empowered citizens. They are also cultural actors who manifest themselves online using the ability to produce, control and host their content and services, efficiently solve local communication challenges and share their culture, while still accessing, at the same time, the global network under equal conditions as peers.

In this view, mobile networks, public Wi-Fi hotspots, internet balloons or planes, and other state and private initiatives that look okay from the perspective of "access", fall short as a vehicle to fulfil the right to co-create the internet.

To reach the internet or build the internet?

We understand that the internet is, like culture, our bodies or land, a territory in dispute. And we understand it in a multidimensional way: with a physical dimension (infrastructure, standards and network protocols), a logical dimension (services and applications), and a cultural dimension (contents, messages). The important thing, for those who intend to intervene in this dispute, is to understand that in the three dimensions there are strategies, practices and technologies that strengthen the co-creation of the internet, or in contrast, reinforce the idea of mere "access".

For AlterMundi, we understand that each new community network is a new part of the internet and it is necessary to work so that each of them strengthens local culture and popular organisation.

So, those of us who work for the development of community networks: How will we guarantee that their growth does not result in just adding consumers for the large, concentrated content and service providers? How will we help preserve and increase cultural diversity? How will we strengthen the people's processes of local organisation instead of invading them with global idiosyncrasies?

Experiences with local content

The struggle to counteract the expansion of the global monoculture is unequal. Powerful actors operate in all dimensions – physical, logical and cultural – generating an inclined field where the birth of local alternatives is difficult.

Our experience with the networks of the Paravachasca Valley in Córdoba, Argentina, has been and continues to be a complex challenge. In this region, five community networks interconnect villages through their own infrastructure, with more than 120 km of backbone links and about 100 nodes that make up the mesh networks of the villages. The set of networks has its own autonomous system number (ASN)[9] and global IPv4 and IPv6 resources. The bandwidth in the networks is symmetric and only limited by the capacity of the links. All connected devices receive a global IPv6 address and a dynamic name resolution system allows each connected device to be reached by its hostname.

We can say that the physical dimension enables an environment in which locally hosted services and content have no impediments to flourish and be accessible both to the community and to others outside of the community who have access to the internet. However, having fertile land is not a guarantee of being able to produce locally.

In the first years, while community networks were being deployed in the Paravachasca Valley, experiments with content generation and sharing of culture and communication were carried out locally.

7 www.fao.org/economic/ess/ess-fs/en

8 https://viacampesina.org/es/seguridad-soberania-alimentaria

9 https://en.wikipedia.org/wiki/Autonomous_system_(Internet)

For a long time, QuintanaLibre – one of the five community networks – ran a captive web portal with relevant information for the community: public transportation schedules, a map of the area, a list and description of local organisations and institutions, cultural activities, a section of advertisements for local products and sales, etc. Chat and voice over IP (VoIP) services and streaming of the local community radio were also provided. The services, implemented on free software platforms and low-cost hardware (embedded computers and display-less notebooks), were successful during the first years. Much of the cultural dimension of the network was expressed through these local, alternative services and information.

However, different circumstances determined that these services stopped working or being used over time. It is striking that much of the "failure" to sustain these alternatives in the upper layers (logical dimension) was largely due to success in the lower layers (physical dimension). The increase in capacity in the links that connect these community networks with the rest of the internet, which in principle would represent an advantage, was promoting practices that tend to favour global alternatives over local. A clear example: people use global music and video on demand services instead of downloading content to local devices and sharing them; even when that content, such as music and children's programmes, was clearly displayed and accessible. The growth of content delivery networks (CDNs) with increasingly closer caching nodes also tilts the court in the same direction. Even local cultural products tend to be shared through global systems such as YouTube, Facebook, and WhatsApp groups.

On the other hand, the emergence of smartphones as a dominant device (over 80% of the connected clients) and the ubiquitous use of WhatsApp as a tool for individual and group text, voice and video communication – made more feasible by the improvement in connectivity with the outside world – resulted in a difficult competition to overcome for the local community network services, especially for chat and voice calls. The implementation of local services using free but centralised platforms was also a weak point. A single damaged server represented the loss of one or more services for the entire network. The low demand for these services meant that local technicians lacked motivation to replace hardware and repair services, which remained off-line.

It is interesting to note that although the tools used by neighbours to communicate are controlled by global corporations, the groups that have been set up in the main continue to play a role of local organisation: network maintenance, political action, cultural activities, disaster response, car-pooling, local commerce, etc. That is, the physical dimension is controlled and deployed by the community, the logical dimension is controlled mainly by global corporations, but the cultural dimension still maintains a strong component of local empowerment.

Nevertheless, it is true to say that in regions where the quality of the collective connectivity to the internet achieves levels comparable to the community network performance, the successful implementation of local services imposes more challenges than certainties. In contrast, where connectivity with the global network is scarce or non-existent, creative initiatives have been born and have grown to represent, in all dimensions, a significant part of communication in cultural life. An example is Rhizomatica and REDES in Mexico, who founded *Telecomunicaciones Indígenas Comunitarias* together with the community operators. The absence of fixed or mobile telephony services served as a stimulus for the creation of a mobile telephony community network that has expanded to cover more than 63 locations through 15 networks that are self-managed by indigenous communities. Today they have more than 3,500 user-members.[10]

Cuba, where internet connectivity is very limited and expensive, also offers practical examples. As an alternative to internet connectivity, creative initiatives were born, such as "*El Paquete*" (the package), a sneakernet[11] that every week distributes varied content such as videos, music, website downloads, software, etc., across the island. The "package" – or content – is stored on high-capacity hard drives that are copied and then distributed by land transport. The people are responsible for transportation, replication and then the partial distribution of content to each user. There is another similar project administered by the state called "*La Mochila*" (the backpack) that distributes educational content. Yet another case worth mentioning is the Street Net or SNet, which consists of a metropolitan network built by the neighbours. This network has hundreds of nodes and covers tens of kilometres in the city of Havana and surrounding areas and provides access to local services such as forums, games and content.

In northern Argentina, in Jujuy, the organisation Atalaya Sur together with the local community is building a community network in a region that lacked access to the internet. The Chasqui network provides local IP telephony services, video and

10 See the Mexico country report in this edition of GISWatch.
11 https://en.wikipedia.org/wiki/Sneakernet

book repositories, text messaging and a social network, all based on free software. Since the arrival of a 3 Mbps connection to the internet, the use of WhatsApp has begun to spread, although the community is still choosing the local social network to share content.

Why all dimensions matter

There are numerous other examples of networks and systems that facilitate communication and provide local services in regions with little or no connectivity to the global network. But, when they are fully integrated with the internet, how will these systems adapt? How can they take advantage of the experience and continue to empower their communities? When local alternatives compete directly with the offer of powerful corporations and their systems, will the creativity and ingenuity that made them possible adapt and survive? Is it worth asking these questions?

Returning to our initial comparison with the field of food, we understand that true food sovereignty tends to be expressed in all its dimensions, which we could simplify as:

- Control over/access to the land for those who farm it.
- Appropriation of tools, techniques, seeds, supplies, etc.
- Fair and sustainable production and distribution of healthy food.

We would not hesitate to fight for food sovereignty if Bayer/Monsanto controls the intermediate layer (seeds, techniques, supplies). So why would it be acceptable or reasonable that in our co-creation of the internet, the logical layer, the systems and applications that host our culture and transport our messages, are controlled by the likes of Facebook or Google?

We understand that this is where we must clearly demarcate the limit between the defenders of "access" and the promoters of internet co-creation. At times we find ourselves in scenarios where actors who structure their business in the logical dimension of the internet appear as partners, defending positions that resemble ours. Their business, their platforms, live in this layer which is the most difficult to appropriate with local alternatives. While we work to deploy more networks and expand the physical layer, more people will participate in the cultural layer and will do so through the platforms that these actors control. So, if community networks do not face the problem in all its complexity, we will be reproducing logics that we wanted to modify. The tools we use to transmit culture and to communicate will determine the type and reach of our messages as much as the seeds we grow will determine the food we can harvest.

The decentralised repository of culture

We are convinced that a powerful response to the questions we have been asking is to develop and generate the conditions (technical, social, educational, budgetary, etc.) for the appropriation of distributed and decentralised tools that take advantage of existing intercultural scenarios, enhancing cultural diversity through peer-to-peer communication within and among communities.

Because of this, we began developing the decentralised repository of culture.[12] It is a free tool to deepen the experiences of appropriation of technology in community networks beyond the physical dimension, in a real, efficient and valuable way that allows sharing and distributing culture with a counter-hegemonic logic. The fundamental principle of the repository's design is simple: decentralised but organised.

Historically,[13] the systems used to share culture directly between users have either taken the form of peer-to-peer (P2P) filesharing protocols and clients or forums and specialised systems, mostly websites. The former, although they are more resistant to attempts at takedown, do not allow an elaborate and efficient organisation of the corpus of contents they host. Centralised systems, such as forums and specialised sites, allow the organisation and categorisation of content, but they are fragile in the face of attacks, both cyber and legal, and their accessibility necessarily depends on the quality of the internet access to the centralised system.

The idea of "decentralised but organised" represents the best of these two approaches. In the decentralised culture repository, the metadata that makes content organisation possible is replicated along with the content itself. Pieces of the repository that become fragmented still maintain their classification locally. The repository is a natural partner of community networks with little connectivity that use sneakernet techniques to transport information from and to the outside communities that do not have access.

Each fragment of the culture repository, which consists simply of a number of instances connected in a network, makes sense in itself. This allows for a transparent evolution between the different stages of connectivity of a community network.

12 https://github.com/Altermundi/openrepo-desktop

13 https://en.wikipedia.org/wiki/Timeline_of_file_sharing

Our hope in AlterMundi is that this newly born project evolves over time to become a vehicle to re-appropriate portions of the logical dimension in the same way that previous developments such as LibreMesh[14] and LibreRouter[15] are currently enabling the re-appropriation of the physical dimension for numerous communities.

If we think about the global scenario, where the unconnected have the opportunity and the role of connecting themselves, it is vital and necessary to deploy community network infrastructures but also to complement those processes with an ecosystem that is coherent and does not reproduce the systematic exclusion and oppression that the unconnected have suffered so far. What we have to contribute as social actors has to go in the direction of developing tools that enable communities in their role as free, sovereign and empowered subjects to produce and share culture. In this way, conditions are created so that the right to co-create the internet can be engaged, appropriated and inhabited by all the people who coexist in the great global network, creating, in all dimensions, their own internet.

14 https://github.com/libremesh

15 https://librerouter.org

Country reports

COUNTRY REPORT INTRODUCTION

Cutting a line of sight for community connectivity

Alan Finlay

The 43 country reports gathered here were selected based on our working definition of community networks, as "communication networks that are built, owned, operated and used by citizens in a participatory and open manner". While this definition was a useful starting point, the country reports also illustrate that what we think of as community networks can be a lot more messy, and that a number of local-level networks that self-identify as community networks are closer to hybrid or blended models of community access.

The networks differ in their purpose, their governance and sustainability models, their politics, their stakeholders, their relationship to the state and the economy, their size, and even their technological set-up, or what they "do" (some networks are intranets, and do not offer access to the world wide web). Even the notion of "community" is quite loosely applied. Compare, for example, the story of what is considered the first community network in Ecuador – in a rural community of some 50 people – to guifi.net (Catalonia), also a "community network", but "with tens of thousands of working nodes, and hundreds of volunteers, professionals and public administrations involved."

The country reports, as a result, offer a rich entry point for comparing local access initiatives across the globe that self-identify as community networks to better understand points of comparison, agreement and departure. Although by no means comprehensive, the result can be read as a raw survey of community networks in different contexts. To complement and enliven the points of comparison, country reports such as those from Peru and Venezuela offer arguments *contra* community networks, due to factors such as the cost of equipment, prohibitive laws, and the proliferation of mobile connectivity. Similarly, for different reasons, community networks are not operating in China and Seychelles, whose country reports can be read here.

The political agency of community networks

A number of reports speak to the social and political agency of community networks, as well as actors in those communities. Sulá Batsú (Costa Rica) argues that "community networks should not be conceived as small or weak organisations; they can be constituted as large, sustainable and influential organisations that are in the hands of the people they provide services to." Similarly, Sarantaporo.gr (Greece) shows how the historical social and political agency of communities needs to be recognised, and drawn on – in this case, the rural and cooperative movements in Greece in the early 20th century. The authors write that community networks should be seen as constituted by "participants [as] rational social actors rather than docile consumers" – communities, that is, have the "potential to muster collective power that can bring about social change." Several reports refer to an economics of "solidarity", where those who cannot pay for connectivity are subsidised by those who can. Volunteering is encouraged as a form of active citizenship.

A sophisticated theoretical politics drives a number of initiatives – particularly those in Europe. In Italy, ninux.org "started as a 'geek experiment', and maintained this approach throughout its evolution. This gave it a specific ethical and ideological purpose, and allowed it to actively contribute to the spirit and development of the European community network movement."

A sense of agency is critical to this ideological purpose – a "do-it-yourself" attitude is referred to in many reports, with Sarantaporo.gr referring to its governance system of 10 people as a "do-ocracy".

Freifunk in Germany is energised by a hacker ethic, and enacts new forms of citizenship in its "free wireless network activism" connecting over 300 refugee shelters and centres:

> In parallel to the traditions of established hacker organisations like the "Chaos Computer Club", the Freifunk initiative provided the socio-material practices to problematise the infrastructural politics of refugee shelters and reception centres, but also sought to actively reconfigure them.

This intervention is part of Freifunk's ongoing work to establish it as a "legitimate form of 'digital volunteering', which includes a sustained engagement with public institutions and a struggle in legally backing up its own emerging practices."

Community networks can also be "political" in a normative sense, whether to counteract internet shutdowns in the Democratic Republic of the Congo (DRC), or to circumvent surveillance. While China "forbids significant organisation outside the purview of the state," one author could not complete a report because of a country context that was described as "extremely sensitive" – it was not safe, in this environment, to publicly disclose information about community networks.

"Step-by-step" tech

Most community networks discussed here follow the mesh network methodology – literally creating a mesh through "organically" connected routers or "nodes". The network can then have as few as one connection to the internet that is shared by the community.

One usefulness of a mesh network is that it can expand in a step-by-step way as more nodes are added as they are needed – and authors advise that communities should not rush the process. Zenzeleni in South Africa calls this a "model of slow co-creation". As WirelessPT in Portugal puts it, it is also a system that can "self heal" when combined with software that detects breaks in the mesh when a node goes down, automatically looking for the nearest working node to keep the network functional.

Antennas are sometimes used to relay the signal over longer distances, including the backhaul internet connection to/from the nearest town or city, or extending it to nearby communities. These are mounted on towers and other prominent points. The number of antennas needed typically depends on the geography of the region.

In one interesting description of hands-on practical methods (Ecuador), line-of-sight antennas are set up at dusk so that the neighbouring city can be clearly identified when the evening lights get turned on:

> They advised us: Climb up to high places at dusk to identify potential links, and then just try the most obvious link in the fastest, least expensive way possible. Fastest in terms of just buying an antenna instead of making antennas yourselves, and least expensive in terms of using a friend's internet connection instead of contracting your own. In order to take a first step, let go of the idea of building a network for five communities all at once – maybe that will happen, but it's not the first step. Start with a single link, and that small, practical step will teach you things that enable you to grow the network later.

Open source firmware for routers such as LibreMesh or OpenWrt is commonly used, with the Wi-Fi signals typically transmitted on unlicensed 2.4 GHz and 5.8 GHz bands. Backhaul connectivity is normally through licence-exempt or licensed wireless links, or fibre. (Fantsuam Foundation in Nigeria tried satellite but found it too costly.)

Sometimes the backhaul connectivity is donated through partnerships (e.g. with universities, or through service providers who partner with a project looking for an expansion of their customer base at the local level). Networks are also considering the potential of TV white space (TVWS) for connectivity – in an interview, a pilot project underway in Tanzania is discussed here.

Networks in both Mexico and the Philippines experiment with and promote mobile connectivity for their community access solution.

In the absence of an electricity grid, or unstable power supply, community networks rely on generators, solar energy or, in some cases, hydropower. Although these solutions give communities more control over their power supply, Alternative Solutions for Rural Communities (ASORCOM) in Myanmar found that alternative power solutions can also be vulnerable:

> There was no national electrical grid in the project area, so the project had to depend on hydropower and solar. During the rainy seasons, hydropower stations were washed away by flash floods, and solar power was made ineffective by heavy cloud cover. The network did have a backup generator, but the diesel needed to run it led to extra costs for the communities. This meant that the use of the network was limited in the rainy seasons.

Not all of the community networks discussed here have access to the internet – while a number have developed an internet-intranet ecosystem, in the case of networks such as Mesh Bukavu (DRC), content such as Wikipedia, ebooks, and computer science and English course material is downloaded onto the intranet. The community can also chat to others through an instant messaging system. Similarly, one of the important functions of Network Bogotá in Colombia is crime prevention, and the security cameras are an integral part of the network set-up. These examples of network use highlight that setting up a community network is not just about connecting to the outside world via the internet, but about a community using technology to attend to its local, sometimes more practical needs.

Governance from the ground up

Community networks are a matter of perspective – suggested by the substitution of the phrase "first mile" for "last mile" to signify the technical challenge of reaching citizens and homes. They work from the community outwards, rather than from the vantage of the state or the service provider inwards. "Access" is not just about access. As the authors writing on ninux.org put it: "If the whole community network movement turns into a 'connectivity factory', its original and innovative push will be strongly reduced."

Reflecting this, most of the reports emphasise a form of community ownership – these are, in the main, all bottom-up, grassroots initiatives, and the empowerment of local communities and members of the communities is a shared concern. While government involvement is sought in several networks, and the private sector is sometimes seen as a partner, community ownership is a cornerstone of most of the projects discussed here. Participatory governance models are typically promoted, with ownership by the community being fundamental to the long-term sustainability of the network. While there are numerous variants of the community network model, collective approaches to governance can be considered a defining feature of community networks.

However, participatory models are not always easy to manage or sustain. ASORCOM in Myanmar shows how they can test our assumptions of community and collaboration and shared notions of the "common good". Keeping a sense of "community" in community networks can be hard work:

> Sometimes users would fight amongst themselves. Some users downloaded videos and games that affected connectivity for everyone. Sometimes people wanted to charge their neighbours for connecting to their router. We have had to offer counselling to resolve these disputes. We have also had to install software to monitor and control the system.

Colnodo (Colombia) describes how confidence in new forms of community participation can wear thin without the proper commitment from its proponents:

> This delay has begun to frustrate the participant communities. Some leaders have withdrawn their support and, as a consequence, the managers of the initiative have lost legitimacy, given that the communities perceive this delay as a breach of their commitment to the project.

Sensitivity to local processes is important, as Macha Works (Zambia) argues:

> In the process of engaging the community, the organisation exercises sensitivity to local contextual frameworks and understandings, for instance, regarding time and space, affecting both the practice of human interaction and the assessment of realities.

This is, the authors argue, "important to ensure the long-term sustainability of the intervention."

Particular attention should be given in community networks to the empowerment of marginalised groups or individuals, whether through the formation of governance structures, training interventions, or other community empowerment programmes. In India, women weavers are taught how to upload their designs onto the internet, and "barefoot engineers" are trained to set up antennas and perform other tasks typically seen as "men's roles".

Catalonia offers an example of an advanced governance model, with clear roles and procedures (that can be used elsewhere). Two key questions need to be asked: What is the objective of the network? And, is this a shared objective? This "helps to focus [...] efforts" and "increases certainty" by reducing the "likelihood of misunderstandings and conflicts." In the way that "different communities [have] different goals," and "determining who that community was and their goals created the profile of the network" in the Caribbean, governance models can vary, and depend on the objective of the network, the size, and the stakeholders involved.

Getting the right support

Legislation governing community networks is uneven, and frequent calls are made by authors to have community networks recognised in law, and to cut away at the regulatory red tape that inhibits their operations. This includes licensing exemptions for the 2.4 GHz and 5.8 GHz band and TVWS, and supporting community networks through universal service funds. As the Internet Society Kyrgyzstan Chapter found, registration requirements can break the spirit of a start-up initiative:

> The main obstacle that made us lose all our hope was the requirement to register the use of frequencies. We thought that we could use

certain frequencies, as long as nobody else was using them. When we found out that we needed to register them and that it takes half a year just to go through the application, we were devastated...

Part of the advocacy challenge is for governments to recognise the practical and real contribution that community networks make towards achieving their own development targets. Instead, as WirelessPT found, successful projects sometimes fall prey to the whims of political opportunism:

> Policies in favour of community networks had never existed. The idea of sharing resources in a community was always looked down on with prejudice or at least seen as something that could not make money and was therefore unimportant. Any potential political champions one could find would always want public credit and visibility for their personal brand in exchange for their support, sometimes demanding control and trying to dictate how the project would work.

However, others find policy makers more ready to support them. In Nepal, for example, the government has been responsive to the needs of community networks, following a period of heavy restrictions due to the country's civil war:

> A second regulatory obstacle was that to become an ISP in Nepal, it was necessary to pay a huge licence fee. NWNP [Nepal Wireless Networking Project] lobbied the regulatory body, the Nepal Telecommunication Authority (NTA), to reduce the licence fee. As a result the NTA issued a new law that made the licensing procedure simple. It also reduced the fee to just 100 Nepalese rupees (around USD 1) a year for those who want to work as rural ISPs.

Stakeholders can include the state and private sector actors. While POPDEV Bénin argues that government community centres should be strengthened through participatory governance and mesh network infrastructure, in South Africa the Department of Telecommunications and Postal Services announced its intention to support and work with Zenzeleni during a parliamentary budget speech. Similarly, Gram Panchayats (village administrations) and the government's Common Service Centre (CSC) programme are essential collaborators in Gram Marg community-led networks in India.

Private sector partnerships are typically secured for connectivity. In the Philippines, the VBTS-CoCoMoNets project describes its partnership model as an "innovation":

> Our first major innovation is our public-private partnership for sharing cellular spectrum with a large mobile operator. Given the absence of regulatory support and spectrum access for community cellular networks in the Philippines, we found it necessary to find a partner that shares the project's vision and that would allow the community network to operate under their frequency licence. We found that partner in Globe Telecom, a major telecommunications company in the Philippines. Since our sites have a smaller subscriber base than what they would consider viable, our community network deployments are placed under their corporate social responsibility programme.

Adaptability and resilience

Community networks appear to be highly adaptable. They connect municipalities in Catalonia, and the urban slums of Kenya. They "work" in the high mountain passes of a sparsely populated natural reserve in Georgia, and in the Amazon rainforests. They are adapted to geography, socio-demographics, and scale – they "work" whether the network has 35,000 nodes, or only a few. They help rebuild broken communities after civil war, and connect refugees in temporary shelters to their families back home. They are used by urban professionals and grassroots weavers, activists, farmers, refugees, the poor and tourists.

As Nigeria's sectarian violence shows, they can be vulnerable, torn down. But they can be resilient too. In the United States, the Red Hook Initiative (RHI) community network was the only communication channel left standing following the devastation of Hurricane Sandy:

> When Hurricane Sandy struck New York in October 2012, flood-prone Red Hook was devastated. Cell phone service was down and internet service went out in places. The neighbourhood was dark, with chest-deep water in the streets – but with its small mesh network, RHI was still able to connect to its staff and communities in parts of the neighbourhood that had no communications or power at all for weeks after the storm. RHI organised volunteers using the mesh to help distribute supplies to elders and others unable to leave the public housing towers in the neighbourhood, and gave the community a voice online to broadcast what was happening. People all over the world following RHI's Twitter feed put together online shopping lists and shipped supplies to Red Hook.

Much of the resilience of community networks is due to the attitude and experimental energy of the actors involved in setting them up. "Be revolutionary and dare to take a chance," writes Miguel Vieira from WirelessPT, who had to figure out his network solution from scratch:

> My first trip to Moitas Venda [in Portugal] to start the initial deployment was the hardest. I had only three weeks to fix and deploy old broken hardware that was left abandoned by the previous community wireless project, and I had no skills or knowledge on how to do manage it.

"One of the key characteristics of ninux is its hacker nature," write Leonardo Maccari and Claudio Pisa:

> Ninux.org [...] was the initiative of a computer science engineering student, Nino Ciurleo. Nino had grown technically in the ham radio community as well as the Italian hacker scene and was influenced by the punk do-it-yourself attitude.

Similarly, an early music streaming network in Australia, TS Wireless, existed because of the energy of enthusiasts who simply wanted to try something new. "TS Wireless may not have sustained an online community for more than six months, yet there was community around us, tweaked by a crazy idea all along," writes Andrew Garton. "It was there, and still exists, through the network of software developers, web coders and designers, passionate *wardrivers* and NetStumbler aficionados."

He adds: "We didn't bridge any digital divide, we didn't fill a development void nor provide critical information where it could not otherwise be reached. We experimented with a new idea..."

"Our network exists because we want it to exist," state the authors from the small community network in Ecuador. "We build it, we maintain it, and we use it – and sometimes we break it, we argue about it, we insult it when it goes slower than we like or cuts off entirely, and we get frustrated about it... but mostly it works and we are thankful."

ARGENTINA

CONVERGENCES: ALTERMUNDI'S EXPERIENCES AND CHALLENGES

Asociación Civil AlterMundi and Red Comunitaria QuintanaLibre
Jésica Giudice
https://altermundi.net, https://librerouter.org

The real and the ideal

Access to information and communication is a right for all human beings, and it is the state that must guarantee our rights. In some cases, the state fulfils the role of providing access to information and to means of communication directly; sometimes it is met through the work of different social actors; and sometimes it is not fulfilled at all, and it is the people who end up resolving their basic needs and, therefore, enabling their rights.

José de la Quintana is a small town of some 2,000 people in the mountains of the province of Córdoba, Argentina, which does not have a local government, and the regional government does not participate actively in the community life. The neighbours and the organisations of the town have had to meet more than one need on their own – for example, organising festivals, repairing the streets, cleaning up the riversides, creating and maintaining the cemetery and a cultural centre, and much more.

The town has families that have lived there for generations, as well as "newcomers" to the community, those who return periodically to their rural holiday houses, and also some seasonal tourists. There is a school for each level of education: kindergarten, elementary and secondary school. There is no dominant type of work in the town, no factories, no companies. Many of the inhabitants work in the nearby cities (15 to 30 km away) or in the provincial capital (some 60 km away), which is also the second largest city in the country. Currently, two wireless internet service providers (ISPs) offer their services there, and their offices are based in towns more than 20 km away. To provide internet access in a way that everyone would like, with the stability that everyone would like, and at a price that everyone can afford, seems a utopia.

QuintanaLibre: A network seedbed

In 2011 a group of neighbours decided to start a small network to share the internet link one of them had. QuintanaLibre was conceived.

That is how it started, but the idea quickly proved too small and more neighbours wanted to join. For this there were two strong drawbacks: the narrowness of the bandwidth and the maintenance of the local network.

In order to have more bandwidth than the initial 512 Kbps, we spoke with the two ISPs in the area. We hoped to arrange a discounted collective purchasing agreement with one of them in exchange for reducing requests for technical assistance, but this proposal did not succeed. We ended up upgrading to a 2 Mbps connection, which was the best one available.

However, a barrier became apparent when we wanted to combine our community project perspective with that of a commercial enterprise. Although it seemed (and still seems to us) that both approaches are absolutely compatible, and that they even empower each other, the ISPs did not want to take the risk, or begin the journey of discussing ways of collaborating with our network. We also believe that they thought our project could not prosper and survive over time: it was, at best, a nuisance.

In parallel, we thought it was necessary to design a network model in which local people could take care of its maintenance. The logic was: all the nodes should be the same, so that their maintenance would be similar; we all get together and learn how the node is maintained, and those who have difficulties can ask a neighbour who has already learned by fixing their own node. This is how we distribute the maintenance load.

Following this path, we called friends with technical experience in networks, who helped us define the technical aspects of mesh networks for small, digitally excluded populations like ours. Then we dealt with the obstacle of deploying a point-to-point link to the nearest city in the absence of a cooperation agreement with the ISPs.

The first network of community networks in the country

There were a number of things we still had to do before we were in the position to set up a mesh network that we could rely on. We defined a hardware reference (routers, antennas, casing, etc.) and developed our first mesh firmware, which greatly simplified mesh deployment.

These first steps were decisive for the expansion of our network and for connecting to the other community networks that were emerging in the region: AnisacateLibre, LaSerranitaLibre, LaBolsaLibre and NonoLibre. The first network of community networks in Argentina!

These networks all took QuintanaLibre as a reference point to get started. However, each community organises and manages its network in different ways. Only AnisacateLibre was initiated by a person with a technical background; the rest of the communities gathered their courage, organised and informed themselves and consulted with us when they had problems. They learned more and more from their own experiences, and two or three members in each village ended up going deep into the technical side of the network.

Today we are in contact, we are friends, and sometimes they ask for help; but we also help each other and work together in improving the backbone network that interconnects us. This organic growth allowed us to form a community, despite the distance and the fact that we are from different localities.

This union also strengthens us internally and externally. It is easier to advance when there are more of us; the impact in the region is greater and we can share the achievements of each network.

However, this growth brought with it the need for more bandwidth. We managed to solve it, first with a residential connection managed by AnisacateLibre and then through an agreement with Silica Networks,[1] which donated a symmetrical 20 Mbps connection for a two-year period. Then we signed another agreement with the National University of Córdoba[2] to take advantage of its idle bandwidth.

Currently, QuintanaLibre has more than 60 interconnected nodes that cover the territory of the village and San Isidro, the neighbouring community. This network also includes the secondary school and the cultural centre and offers coverage in public spaces, some streets and two important bus stops. The rest of the networks have between 15 and 25 nodes, but all of them are also planning a major expansion at this time.

Building our own router

For AlterMundi, the spread of this idea and the collective enthusiasm gave rise to a project to create our own hardware. Despite all the difficulties and the complexity of the process, we managed to design and produce the LibreRouter.

1 https://www.silicanetworks.com

2 https://www.unc.edu.ar

By developing our own hardware and software (the first of many developments!), we no longer have to deal with the endless caprices of the market, nor reverse-engineering to enable the disabled functions of commercial hardware, nor the constant need to adapt free software.

Now we decide how our most important hardware is constituted and how the systems and applications that control and assist it are developed. In short, we have increased the technological sovereignty that community networks can offer.

During this process, what stands out above all is the will and tenacity of the network members who sustain and give meaning to all of this effort.

In all this, what about the state?

At the time of writing, the National Communications Agency (ENACOM)[3] published its first regulation on community networks.[4] This is the first time that community networks have been defined by the state. For now, this resolution enables us to request a licence for non-profit operators, exempt from payment of fees. It also affirms the importance of supporting and promoting community networks.

Although this is a great step forward, it also poses new challenges and makes obstacles more visible in pursuit of the realisation of a more complete regulatory framework that understands, recognises and favours the emergence and development of community networks.

On the part of ENACOM, it is important that it recognises an error in its drafted definition of community networks. The initial idea of the state was to limit the scope of regulation to networks located in localities of no more than 5,000 inhabitants. However, this limitation was included as part of the definition of what a community network is, leaving several pre-existing community networks out of the concept. One of the challenges that this error reveals is the need to agree on a shared definition of what is (and is not) a community network, which serves as the basis for any other field. Developing such a shared definition is already work that is being done at the Latin American Summit of Community Networks[5] and it is expected that the definition will be periodically reviewed.

Soon we will test the licence application process for non-profit operators. We will also request resources from the Universal Service Fund when a

3 https://www.enacom.gob.ar/

4 Resolución 4958/2018. https://www.boletinoficial.gob.ar/#!DetalleNorma/190061/20180817

5 cnsig.info/cumbre/lac/2018/08/30/Cumbre-Latinoamericana.html

new call is opened. This fund is nourished by a small proportion of the profits of the large telecommunications companies, and is reserved for solutions in neglected areas such as ours. The method of allocating these funds, through the Non-Refundable Contributions (ANR), must be adapted so that community networks can participate.

This first regulation is a small step forward, but is not comprehensive of the universe of situations that need to be regulated to protect community networks. It is necessary to complement the regulation, for example, covering issues such as the use of spectrum, access to infrastructure, free peering agreements, hardware homologation for our devices – starting with the LibreRouter – and other aspects mentioned in the International Telecommunication Union's Recommendation ITU-D 19,[6] cited in the considerations of the regulation.

It will be a long process until the state and other actors recognise and capitalise on the efforts of community networks and consider us allies in the task of connecting the disadvantaged regions of Argentina.

In general, there is an incompatibility between the organic structures that have developed in these disconnected communities, and the mechanisms that the state uses to try to reach them. One of the most frequent reasons for the failure of these initiatives is the lack of anchoring in the community, which also results in solutions that its inhabitants cannot understand, adopt and sustain over time.

Community networks evolve in these different aspects at the same time. The coordination, the design, the technology, the hardware, the software, the policies, the management, the maintenance... everything progresses in the way and at the rhythm of each community. This characteristic makes the technological and human network more resilient.

In summary, we want to work for community networks to cease to exist in a legal gray area that makes it difficult for new initiatives, projects and businesses to rise locally.

Sustainability of the right to information and communication, a collective responsibility

The most important aspect of sustainability is not whether or not a community network can generate revenue, or how it can survive through contributions from the community. The most important aspect is that the complex structure that guarantees the right to information and communication must be sustainable. By directly contributing to the materialisation of a non-negotiable right, we deserve to be integrated as part of this structure.

From a socioeconomic perspective, community networks should be considered a sustainable option because they save the state expenses needed to guarantee a right that the state is clearly unable to fulfil.

The collective work of these networks resolves moral debts that the state has with rural communities and other vulnerable and excluded areas.

Action steps: Collaborating, assisting, consulting and helping each other

AlterMundi and community networks in Argentina have some specific proposals to make:

- Allow us access to resources from the Universal Service Fund.
- Consider the use of LibreRouter in state connectivity projects.
- Enable community networks to have free transit through the Federal Fibre Optic Network (REFEFO)[7] and other state infrastructures.
- Prioritise localities with community networks when the national or provincial government designs projects and developments.
- Encourage direct communication between the state and the managers of the community networks in a way that enables mutual recognition.
- Develop laws and regulations for community networks in consultation with representatives of community networks.
- Reduce administrative expenses through, for example, doing away with stamp duties and other red tape that burdens community networks.
- Reduce the financial reporting responsibilities for community networks.
- Facilitate access to credit and financing for community networks.
- Promote collaborative projects between the state and community networks.

6 Recommendation ITU-D 19: Telecommunication for rural and remote areas. https://www.itu.int/rec/D-REC-D.19-201003-I

7 https://www.argentina.gob.ar/modernizacion/comunicaciones/planfederaldeinternet

AUSTRALIA

TS WIRELESS: THE RISE AND STUMBLE OF A COMMUNITY NETWORK FOR ELECTRONIC MUSIC MAKERS IN MELBOURNE

Media and Communications, Swinburne University of Technology
Andrew Garton
www.swinburne.edu.au

Introduction

It was a Thursday winter evening in Melbourne. Around 30 or so people had either walked from nearby tram stops or pedalled across the city to Toy Satellite's[1] two-story warehouse in North Fitzroy. Some, their faces lit by open laptops, were using Netstumbler[2] to find us. We had emailed GPS coordinates and the name, or SSID, of our now freely open wireless hub. The more intrepid would use our network's signal strength to get them to the front door. It was 11 July 2002 and everyone was gathering for the launch of TS Wireless, a joint project with London-based Free2Air, a free community wireless network. The plan was to stream royalty-free music produced by local artists 24 hours a day to anyone within a three-kilometre radius. It took around 48 hours to have our server hacked and the whole operation halted! But it started off great!

In the beginning we shared

In the beginning BBSes, or Bulletin Board Systems, were the earliest publicly accessible computer, the first one known to have gone online in Chicago in the United States, on 16 February 1978. The Computerized Bulletin Board System (CBBS)[3] was based on software written by Ward Christensen and Randy Suess, considered the fathers of public access networks. The internet's precursor, ARPANET,[4] was still in its infancy. People would dial in to BBS computers, exchanging software, documents and graphics. Data rates were slow at the time, commonly 300 characters per second, and modems were devices known as "acoustic couplers" which were mounted onto telephone handsets; the ear piece would receive data while the mouthpiece would send it.

Individual BBSes were the earliest equivalent of a website, each supporting communities of common interest. Perhaps the first BBS style of network established for artists was Robert Adrian X's ARTEX,[5] an electronic mailbox for sharing ideas and organising intercontinental telematic artworks. A Canadian who spent his entire adult life in Vienna, Adrian X had pioneered politically charged artistic practices within electronic and broadcast networks. ARTEX foreshadowed "store and forward", the share and copy potential of the future internet. It was a kind of precursor to the ubiquitous Dropbox, a micro-cloud storage utility when clouds still hung in the sky. It was cheap too, costing artists a few cents a day for data storage only.

By the 1990s the internet was well underway, and artists were not only wanting to share: they were being found. I held the view that an increasing number of people began looking for music beyond radio and music stores. There was plenty available if you knew where to look. The global reach of the internet meant it was easier to share far more music than anyone could possibly hear, and so much more than anyone had ever heard on a radio or found in music shops. The problem, in my opinion, was not piracy that would afflict the music industry, it was diversity. There was so much more diverse music and so many artists appearing online than there would ever be on any of the industry charts listing the 100 most popular recording artists of the day – the measure by which royalties were not only dispersed, but guaranteed.

1 Toy Satellite was a multidisciplinary producer of works for screen, installation and performance spaces. Founded in 1995, Toy Satellite was incorporated into APC.au, a non-profit organisation providing hosting and consultancy for communities in Australia and Southeast Asia. This added an international dimension to Toy Satellite's work. It was shuttered in 2005, leaving a vast portfolio of interactives, museum installations, publications, soundworks, video and online initiatives. https://toysatellite.org

2 NetStumbler is a tool for Windows that facilitates detection of wireless LANs using the 802.11b, 802.11a and 802.11g WLAN standards. It first appeared in 2001, with development of the app continuing up to 2004. It was the de facto tool for "wardrivers", "warbikers" and "warwalkers" – people who searched for Wi-Fi networks with a wireless-equipped device together with a GPS device to record and share the location of unsecured wireless routers. www.stumbler.net

3 Christensen, W., & Suess, R. (1989). *The Birth of the BBS*. www.chinet.com/html/cbbs.html

4 Advanced Research Projects Agency Network.

5 ARTEX (Artists' Electronic Exchange System) was conceived in 1979 as a simple "intercontinental" email service for artists. It was used by some 35 artists worldwide between 1980 and 1991. alien.mur.at/rax/ARTEX

We saw an opportunity. How could we share all this independently produced music to people who didn't know where to look for it? How could we make this happen legitimately and accessible to anyone for free? No sooner had we published a white paper describing TS Wireless than we found a partner in the London-based wireless network host, Free2Air.[6] Enthusiastic about our project, they arranged for an aerial antenna and several metres of specialised cable to be shipped out to us. Things were moving quickly. Now we had to find a server, the technical nous to pull it all together, and permission from the body corporate to mount an antenna onto the roof of the building. There was no shortage of music to share, no shortage of skills to make it happen and a whole lot more to learn. TS Wireless was underway!

Streaming music locally

By 2002 streaming audio had become easy, video less so. However, streaming anything over wireless was still in its technical infancy. With the help of an international network of open source software developers and visiting and local Wi-Fi expertise, we gave it a good, decent, thoroughly robust crack. But before we had the server up and running we encountered our first and most challenging problem.

6 free2air.org described itself as "open distributed public network infrastructure." It was set up in London by Adam Burns in 2001, growing to support a freely accessible and widely distributed cluster of wireless networks and projects. This included ambienttv.net, NewsFilter and RayLab, East End Net, Air Shadow, Wireless Ant Farm and TS Wireless.

The royalty collection agency

The original plan was not to stream ambient music – which we ended up doing – it was to share locally produced music in all its myriad forms. Along with Free2Air, we wanted to make bandwidth a community concern, not a commercial one; we wanted to make access to local music a community concern, not a commercial one. We wanted to share our infrastructure to offer local households and businesses access to music that had yet to find an audience. We wanted to create an online community that could evolve around its own interests with music as the fire around which we would gather. But to do so we had to pay a licensing fee to the Australian Performing Rights Association (APRA). We immediately saw another opportunity. APRA did not.

Around 2000/2001, on behalf of their member composers, authors and songwriters, APRA expanded its performance licensing usage to encompass any venue in which music was publicly heard or performed. This included hair salons, cafés and many work spaces. Even Toy Satellite received a notice from APRA urging us to pay an annual fee for any music we played in our studio. Internet service providers (ISPs) were not immune from such fees either. Any music stored or streamed online was considered a public performance of said works.

We contacted APRA and asked how we could list any of the music heard over TS Wireless so that artists would be paid royalties. If we were to pay an annual performance licence fee, surely we could submit a playlist ensuring all local artists, their own members, would have royalties distributed to them. They replied stating they had not the means to do so. APRA, a national royalty collection agency with an annual turn-over in the millions, did not have the means to allocate royalties to any of the artists being played in any of the venues nor streamed from any ISP being charged an annual performance licence fee. Wow! So, who does? The answer was simple. Royalties are distributed to the popular artists of the day on the assumption that most, if not all music being heard would be popular music; otherwise it wouldn't be popular. That meant that a small number of artists were the beneficiaries of most of the performance licence fees whether they were heard in any local café, which they were not, or streamed over TS Wireless, which they absolutely were not.

We offered APRA an opportunity.

We were prepared to share with APRA all TS Wireless playlists. We would even write software to email them the playlist in a form that could be imported into their database or spreadsheets. These playlists would be filled with all sorts of fascinating metadata, enabling APRA to apportion royalties to all local artists on our server. That's *all* artists.

APRA replied stating they did not have the means to interpret that data into meaningful outcomes for artists legitimately being heard on our servers nor in any of the venues that would sign to TS Wireless. We had considered adding a feature where custom playlists created by TS Wireless subscribers could be sent to APRA. But given APRA could not make use of this information, no matter what form it could be sent to them in, we had to consider our options. How do we proceed with TS Wireless knowing the music we would have to pay APRA the rights to stream would not result in any additional royalties to the Melbourne music makers we would host?

It was at this point that we decided to only stream generative, ambient music for which APRA had neither the means to register it as being "composed" by anyone, nor an argument to warrant a performance licence. In short, music that was constantly changing was not in APRA's interests to have on their books. Our generative pieces were works that were never heard the same twice; some of them were hours, days and even weeks in length. APRA did not have the means to support such original works no matter how they were composed nor by whom. Even before we got underway, we knew an audience for generative ambient music would be limited. We forged ahead nonetheless.

The team

Visiting Australia at the time was Spanish-born Alberto Escudero-Pascual, who brought with him extensive wireless networking expertise. He had the technical skills and real-world experience and he grew fond of Melbourne, often saying it was the first place he had visited where no one asked where he had come from. At the time Alberto was an associate professor at the Royal Institute of Technology, Sweden, where he was completing a PhD in mass electronic surveillance systems. Alberto knew a thing or two. He also knew how to modify Orinoco network cards so that they would converse with Linux operating systems.

Remote expertise came in the shape of the quietly spoken Adam Burns, who once told me it was through the study of high mathematics that he had come closest to God. Adam was the founder of Free2Air and had worked with me at Pegasus Networks, Australia's first publicly accessible ISP.

Our local team included web coder par excellence Justina Curtis. Justina wrote up most of our documentation, coded our front-end and hosted

with unyielding generosity the many late nights it had taken to construct the entire project. Justina brought expertise in training in information and communication technology (ICT) media literacy, and interaction and accessibility design. She was also a co-founder of Toy Satellite. Every project needs a social binder and Justina was ours.

Supporting us all with no end of bad jokes, rigorous system administrative skills and an ability to work insanely long hours and remain not only focused, but thorough, was Grant McHerron. Both he and our in-house technical director, Bruce Morrison, were determined to not let any problem impede them. Bruce had also worked with me at Pegasus Networks and, like Grant, had the gift of tolerance and fortitude. It is impressive to see people work at problems as if they were puzzles, applying game theory within their deliberations. Another member of our team was Linux aficionado Dennis McGregor. I can't recall much about Dennis other than the long hours we spent together stitching the entire project together with good humour and collaborative ease.

Network design

TS Wireless was accessible from within a three kilometre radius of our 2.4 GHz omnidirectional aerial mounted on the roof of our studio in North Fitzroy, Melbourne. We made use of the full capacity of the aerial, which provided the most extensive coverage for the technology available at the time.[7] Yes, we did get permission from the body corporate to climb over the top of several premises. We made sure we could provide maximum line of sight to all participants. This included local members of Melbourne

7 Had we wanted to extend the reach of our network we would have required repeaters, or wireless network amplifiers, which would necessitate far more infrastructure than the notion of "local" warranted.

Wireless,[8] arts practitioners (the Melbourne Fringe Festival had their office space nearby), our neighbours and random visitors.

The antenna was mounted to an existing television aerial with U clamps. Waterproof electrical tape was used to insulate the connection between the cable and antenna mounting. Thirty metres of heavily insulated cable was wound around the perimeter of the roof and into the building through a rear window where it was plugged directly into Hermes, our wireless server. Hermes also administered an ADSL connection to our broadband provider as well as a firewalled gateway to the internet.

Hermes was installed with a basic setup of Mandrake Linux.[9] We used Firewall Builder,[10] an open source application, to create and manage the firewall, and X-Windows[11] from which non-tech team members could run diagnostics tools such as KOrinoco[12] to generate signal strength charts. A second server, Yuri, was set up to run the generative music software, SSEYO Koan Pro.[13]

For streaming we used ffserver[14] and FFmpeg,[15] the latter for its ability to do on-the-fly encoding to multiple formats concurrently, thus avoiding being bound to platform-specific, or rather, biased codecs. FFmpeg does capture and encoding and then outputs to ffserver and/or to file(s); ffserver then provides streams to clients connecting over a network. Due to limited resources and the heavy processing demands of multiple format encoding, we chose to use the MPEG codec for both audio and video.

We encoded two MPEG streams, one of a high quality, and another of a lower quality for those with less bandwidth. The encoder component of FFmpeg had the ability to output to multiple destinations, so we sent one stream to a local ffserver, to an ffserver in Sweden, and to a hard drive for archival purposes.

Free2Air.org had also mirrored our two MPEG streams – or MP2 stream – by running a client from the server in Sweden and then re-serving it over an Icecast[16] server in London. A web server threw up a web page for users who would find a link to the live audio stream, with FAQs and a contact form. It was pretty much a BBS with internet server software and hardware. It was a private internet. It had all the features of the internet without the reach to it. This was a useful backup to our wireless streaming experiment, because, at the time of deployment, it was illegal to provide access to the internet from a wireless node, that is, piggy-backing off a commercial ADSL provider. The Australian Communication Authority regulated all spread spectrum radio communication, including the 2.4-2.4835 GHz radio-frequency band our wireless network would operate in.

TS Wireless on air

Back at the launch, curious and eager, everyone was invited to have their hands scanned as they entered the building. These were hurriedly compiled into a video and projected as large as we could possibly make it onto the interior walls of the Toy Satellite studio at the moment TS Wireless was officially launched. Adam Burns was video streamed live from London gifting us with his wisdom, describing access to free and reliable bandwidth as a fundamental right. Alberto Escudero-Pascual concluded the evening with an impassioned appeal for community as the locus of new ideas, what artist Brian Eno describes as "cooperative intelligence."[17]

Once up and running, at our peak, TS Wireless supported 50 simultaneous users. We have no idea who these users were. We had no need to know; but we knew audio streams were frequently served.

This was after we had secured our servers ensuring they had no internet capacity whatsoever. Initially we had an internet connection via a local ADSL provider. This meant we could have our audio streams mirrored by servers anywhere in the world, including Free2Air. But if you recall, dear reader, within 48 hours of our launch we had been hacked.

8 A non-profit club set up to help the community in setting up publicly owned and operated residential area networks. www.wireless.org.au

9 Mandrake Linux was first released in July 1998, gradually evolving into an extremely user-friendly version of the Linux operating system. Its creator, Gaël Duval, was particularly focused on both first-time Linux and non-tech users.

10 fwbuilder.sourceforge.net

11 X-Windows is a graphical user interface (GUI) for operating systems such as Linux and Unix. It was developed at the Massachusetts Institute of Technology (MIT) in 1984.

12 KOrinoco is an open source application used to configure and monitor wireless LAN PC cards. korinoco.sourceforge.net

13 Koan Pro was developed by UK-based brothers Tim and Peter Cole in their studio on the River Thames. It was released in 1994. In 1997 I used the Coles' software in the first streamed broadcast conducted by the Australian Broadcasting Corporation (ABC). Titled "Sensorium Connect", the sound work was produced in collaboration with the performance artist Stelarc and consisted of a composition six weeks in duration.

14 https://www.ffmpeg.org/ffserver.html

15 FFmpeg's own website claims it is "able to decode, encode, transcode, mux, demux, stream, filter and play pretty much anything that humans and machines have created." It's true! It can. It's as widely used today as it was in 2000 when it first appeared. https://www.ffmpeg.org

16 Icecast is a streaming media server that is commonly used for internet radio and jukebox services. It was created by Jack Moffitt and Barath Raghavan in 1998 and released in 1999. The authors described it as an "open source audio streaming server that anyone could modify, use, and tinker with." www.icecast.org

17 Steven Johnson interviews Brian Eno for the public education blog Be You: https://redefineschool.com/brian-eno

In short, someone had literally sat in a car opposite the studio and pulled gigabytes of data through our internet connection. It nearly sent us broke. Australia has some of the most expensive broadband charges on the planet. If we breached our cap, which we had after our launch, we were charged for every single byte of data that moved across it. We learned that if someone wants bandwidth badly enough they will come with every means at their disposal to nab it.

A second location was proposed by a small business in a neighbouring suburb where we could trial a semi-commercial operation amid shopkeepers. Every shopkeeper had been approached by APRA to pay that annual licence fee. As such, many were keen to see local music makers benefit from this expense. TS Wireless provided a model many were keen to subscribe to.

Smith Street Wireless was to launch in 2004, but yet again APRA was unable to work with our playlists. Subsequently no one would subscribe to Smith Street Wireless if licence fees paid to APRA did not result in royalties to local artists. Smith Street Wireless was doomed.

Reflections

We didn't bridge any digital divide, we didn't fill a development void nor provide critical information where it could not otherwise be reached. We experimented with a new idea to find that no one was particularly interested in a perpetual music streaming service. Free too! Portable MP3 players were as commonplace then as smartphones are now. Most people we knew were content with curating their players.

There was also no interest in the metadata we offered to share with royalty collection agencies, in spite of all the public spaces, businesses and venues paying annual performance licence fees. This left independent music makers with fewer means to accrue royalties in a marketplace adapting to new technologies. But we did, as we had done so many times prior, find that the community we had sought to nurture had been with us all along.

The community that preceded the network

TS Wireless may not have sustained an online community for more than six months, yet there was community around us, tweaked by a crazy idea all along. It was there, and still exists, through the network of software developers, web coders and designers, passionate wardrivers and NetStumbler aficionados. These are the people we rarely see, who had created some of the more experimental and wildly innovative networks of their time. We had worked in remote villages in Africa, Southeast Asia and Indochina. We provided training and advice to regional and rural telecentres in Australia, including creating some of the earliest websites for community groups, non-government organisations and small businesses in the country. We had, at personal expense, established the means for anyone interested to learn about wireless networks and how they may foster community, belonging, nurturing curiosity and innovation where it may otherwise languish. Our experiment may have failed on paper, but it succeeded in bringing together a group of people who shared an experience that we had all grown from, and the ripple effect of that gathering expands still.

Epilogue

Sixteen years since TS Wireless rose and stumbled, APRA draws annual performing rights fees from hotels, pubs and taverns, restaurants and cafés, fitness centres, numerous work spaces such as doctors' and dentists' surgeries, corporate reception areas, service stations, salons and nightclubs, motels and other forms of accommodation. They have special licences for skating and ice rinks, community bands and choirs, recreation and leisure centres, schools, universities and colleges, sporting events, local government authorities, transport systems, funeral directors and funeral providers.

APRA now applies music recognition technology described as a "digital fingerprint" that each piece of music carries which is matched against a database containing the work's metadata "enabling the owners of each matched work to be identified and paid accordingly."[18] They still have no clear mechanism for allocating royalties from all the performance licence fees they collect from public spaces, but they do have a term for where such fees reside: "distribution by analogy". This means that "licence fees are added to an existing distribution pool that is most similar in terms of its music content." I'm not sure I understand what this means, but what I do know is that if any of my music is played at Frankie God of Hair where I get my hair cut, I'll get a decent trim, but not a single cent in royalties.

What remains of TS Wireless are its antenna and cables sitting in a Melbourne storage unit. The servers Hermes and Yuri and all our wireless workstations were taken to an e-recycling centre, the CRT monitors thrown into an awesome open bin along with routers, modems, metres of ethernet

18 APRA Distribution Practices, APRA AMCOS, November 2017. apraamcos.com.au/media/Distribution-Rules-and-Practices/APRADistributionPractices.pdf

cable, keyboards, spare motherboards, and memory cards.

There was nothing graceful about this kind of closure, but there was comfort knowing that much of this gear would find reuse elsewhere.

Action steps

This is a story about how internet service providers, or intermediaries, are being coerced into licensing agreements for the content accessible through them. If the water we drink is foul we do not prosecute the plumber, an intermediary: we head to the source of the contamination. Moreover, we do not require the plumber, an intermediary, to pay a licence fee for the water that will flow through the pipes they install, nor do we license the pot, also an intermediary, in which our rice is cooked.

Author and activist Cory Doctorow's third law for the Internet Age is "information doesn't want to be free, people do."[19] This is about a free and open internet. This is about what we stand to gain from an internet that continues to give us the means to learn from each other through sharing, how the works of artists inhabit and become part of our private lives and our personal stories.

APRA is not at fault here. They are merely protecting the interests of a system that protects its apex inhabitants, from a small number of publishers to the artists that sustain their businesses. What can change are the laws that protect these interests. To change these laws, from how copyright functions to international trade agreements, we need to know how they function and what alternatives we can advocate for.

As such, what follows is a reading list for how best to inform yourself so that you may come to your own conclusions, and so that you may contribute to the critical debates around these issues. I have then suggested ways you can participate.

Reading list

Free Speech: Ten Principles for a Connected World, Timothy Garton Ash, 2016, Atlantic Books, ISBN 9781848870925

Information Doesn't Want to Be Free: Laws for the Internet Age, Cory Doctorow, 2014, ISBN 9781940450469

Digital Rights Resources for Children and Young People, Angela Daly, Digital Rights Watch, https://angeladaly.com/resources/digital-rights-resources-for-children-and-young-people

Ways to participate

Modernising Australia's copyright laws, Department of Communications and the Arts, Australian Government
https://www.communications.gov.au/departmental-news/modernising-australias-copyright-laws

Electronic Frontiers Australia
https://www.efa.org.au/get-involved

Digital Rights Watch
https://digitalrightswatch.org.au/join

19 Doctorow, C. (2014). *Information Doesn't Want to Be Free: Laws for the Internet Age*. San Francisco: McSweeney's.

BENIN

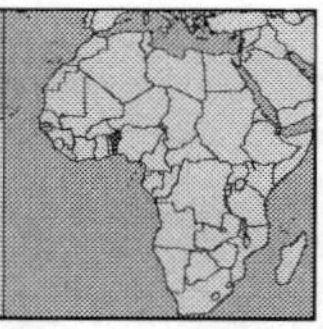

REACHING OUT: EXTENDING THE PARTICIPATORY FUNCTION OF COMMUNITY CENTRES IN BENIN

POPDEV Bénin
Sênoudé Pacôme Tomètissi

Introduction

Access to information and communications technologies (ICTs) is still an issue in Benin, especially in rural areas where both fixed and mobile technology is limited. To improve accessibility and to address inequities, the government has developed several public policies and projects implemented by a dozen of its public agencies. Through the Universal Access to Information and Communication Technologies project, two state-owned companies, the Benin Agency for Universal Electronic Communication and Postal Services (ABSU-CEP)[1] and Benin Postal Company[2] joined forces to set up community centres in four localities with no or limited ICT access.

This report highlights the need for both a human rights and participatory governance approach to be taken in these centres – both prevalent approaches in the global community networks movement – in order to improve the benefit of the centres to the communities. It concludes by suggesting that government community centres offer a useful starting point for a community network that is run and maintained by the community. Given the proliferation of these centres across the developing world, it is useful to consider how they can be a bridge to low-cost, autonomous community access to the internet, with countries like India offering useful models for consideration.

Policy, economic and political background

In 2003, Benin's government developed a strategy for the revitalisation of ICTs in the country.[3] In 2016, a Declaration of the Digital Economy Sector Policy was developed and adopted during the weekly cabinet meeting by the newly elected president Patrice Talon.[4] The declaration highlights six core projects in the digital sector as part of Talon's four-year action plan for the country.[5] Following the development of the plan and the declaration, the Digital Law of the Republic of Benin[6] was promulgated in April 2018. It builds on domestic laws and on the African Union Convention on Cyber Security and Personal Data Protection.[7] Three categories are applicable to broadcasters and service provider activities: the licence regime (where a licence is needed to operate), the authorisation regime (where authorisation is needed from the regulator to act as a broadcaster or service provider), and a third category where no authorisation is necessary.

In principle, community networks do not need authorisation to operate, but there are some limitations. The law states that the establishment of any electronic communications network or the provision of any electronic communications service that does not fall under the licence or authorisation regimes is permitted if the service declares itself with the regulatory authority.[8] This service will also be subject to compliance with any legal and regulatory provisions that are in force. However, independent private networks and devices that have low power or offer short-range connections do not require any declaration.[9]

Benin's human rights status

In 2017 Benin committed to implement some 191 recommendations during the United Nations Universal Periodic Review (UPR).[10] Some of these recommendations relate to arbitrary detentions, extrajudicial executions, excessive use of force by the security forces, freedom of expression and of

1 www.absucep.bj
2 laposte.bj
3 https://researchictafrica.net/countries/benin/Plan_%20TIC_%20Benin.pdf
4 www.fraternitebj.info/conseil-des-ministres/article/conseil-des-ministres-du-26-5484
5 www.cir-benin.com/images/Pdf/PAG_Portefeuille_des_Projets_par_Secteur.pdf
6 sgg.gouv.bj/doc/loi-2017-20/download
7 https://www.au.int/web/sites/default/files/treaties/29560-treaty-0048_-_african_union_convention_on_cyber_security_and_personal_data_protection_e.pdf
8 ARCEP, Autorité de regulation des communications électroniques et de la poste. https://arcep.bj
9 Where necessary, the regulatory authority sets emission and frequency bands used by low-power and short-range devices (Article 55 of the Digital Code). https://sgg.gouv.bj/documentheque/lois
10 https://www.upr-info.org/sites/default/files/document/benin/session_28_-_november_2017/responses_to_recommendations_upr28_benin.pdf

the media, and the arbitrary suspension of media outlets.[11] In 2018, *La Nouvelle Tribune,* a privately owned daily newspaper, was suspended by HAAC, the institution in charge of media regulation in Benin.[12] However, delegations participating in the dialogue under the Benin UPR indicated that the adoption of the Government Action Plan 2016-2021, the laws and policies to protect children, as well as measures taken to improve access to social services were likely to strengthen human rights in the country.[13]

State-owned centres for communities

The digital economy in Benin is subject to numerous challenges, including a lack of infrastructure and investment by electronic communications and business actors. In 2015, the ICT sector in the country represented about USD 300 million, and contributed 6% to Benin's gross domestic product.

The 2013 government strategy on ICTs[14] stated that there were 1.2 million inhabitants who still did not have access to ICT services. Marginalised groups included rural populations, women, youth and people with disabilities who did not have access because of their social, cultural and economic status, suggesting that the extension of universal service must take these factors into account. In May 2014, ABSU-CEP adopted the national strategy and shortly after that, together with the Benin Postal Company, launched the community telecentres in four municipalities: Glazoué and Malanville in 2015, and Azovè and Ouaké in 2016.

There is no internet or mobile service in Glazoué, Malanville, Azovè and Ouaké. The municipalities also have poor connection to the electricity grid, with many areas suffering frequent power outages.

The centres were expected to impact on about 411,000 people, offering them access to the internet and multimedia services, technology-based financial and postal services, and e-governance services, while also introducing them to the use and potential of ICTs in general.[15]

While the postal company provided space for the centres, ABSU-CEP recruited a local manager for each centre and equipped each of the centres with a solar power system to run 15 computers, one multifunction printer (used to print, scan and photocopy), and satellite internet connectivity. The centres are now under the responsibility of the respective municipalities with technical support provided by ABSU-CEP and the postal company.

The centres also offer services such as training for women, students and community groups, and training on document scanning, computer usage, photocopying and web-based research. They provide basic financial services to the population through a quick and cheaper money transfer service in collaboration with the postal service and e-banking, as well as access to basic energy services through their solar power system, such as recharging mobile phones or flashlights.

Most training is offered free of charge but users pay for services: printing is USD 0.045 and access to the internet about USD 1 for two hours. These fees contribute to covering the cost of running the centres.

In March 2016, ABSU-CEP adopted the National Programme for Universal Electronic Communication and Postal Services,[16] which continues to support the idea of the community centres for local-level access.

Limitations of the state-run model

While the state-run model for community access clearly has benefits for the community, there are two key areas where it is limited: the lack of participation in the initiatives by the community, especially the most marginalised; and a lack of a clear human rights-based approach to the access initiative, which means it is not necessarily used to its fullest potential to enable the rights of the community.

Need for an inclusive governance and management approach

According to the 2016 National Programme for Universal Electronic Communication and Postal Services, the purpose of community centres is to empower marginalised rural communities that

11 www.lanationbenin.info/index.php/societe-2/146-societe/16630-3e-rapport-sur-l-examen-periodique-universel-du-benin-la-societe-civile-s-impregne-des-recommandations

12 www.agenceecofin.com/regulation/2907-58938-benin-le-site-de-la-nouvelle-tribune-egalement-suspendu; www.rfi.fr/afrique/20180525-benin-le-quotidien-nouvelle-tribune-suspendu-autorites

13 www.bj.undp.org/content/benin/fr/home/presscenter/articles/2017/03/20/de-l-eau-potable-pour-les-m-nages-ruraux-au-b-nin.html; www.africanchildforum.org/clr/Legislation%20Per%20Country/benin/benin_children_2007_fr.pdf

14 MCTIC. (2013). *Etude pour la définition de la stratégie d'accès universel télécom et tic et du plan d'actions pour la mise en œuvre de projets pilotes en république du Bénin.* Cotonou: MCTIC. www.absucep.bj/fichiers/telechargeables/rapportFinal_SU_Volume1.pdf

15 For population data see www.insae-bj.org/12_departements.html?file=files/enquetes-recensements/rgph/RGPH4; ABSU-CEP (2016). *Rapport annuel d'activités.* Cotonou: ABSUCEP. www.absucep.bj/fichiers/publications/94wlua_absu-ceprapport2015publicversion.pdf

16 The programme covers 2016-2019, available at www.absucep.bj/images/projets/PNSU.pdf

do not have access to ICTs: young people, rural women, many of whom are unfamiliar with new technologies, and people living with disabilities, who encounter practical difficulties to access them.

The strategy aimed to address their needs specifically – needs that have not been taken into account by commercial service providers. Yet one has to ask how this is possible if these groups are not involved in the governance and day-to-day management of the centres?

With regards to women, the 2016 programme draws on the National Gender Promotion Policy (PNPG),[17] which is in line with the United Nations Convention on the Elimination of all Forms of Discrimination Against Women (CEDAW).[18] The PNPG is mentioned in the programme as a way to enable women's access to ICTs. Although the PNPG does not explicitly mention women's access to ICTs, some of its five strategic axes include women's effective access to education, literacy and decision-making structures in all spheres. It also ensures the empowerment of women and mainstreaming of women's issues in municipal plans. Adapting the strategy of the PNPG to ICTs and putting women at the centre of community initiatives is therefore in line with state plans. Learning from successful projects that have trained groups of women with low literacy levels in their native languages using ICTs can be helpful and make centres more open and interactive.[19]

It is the perspective of this report that including vulnerable groups in the governance of the centres will further improve the centres' value to the communities. The problem with the current structure is that they are managed by local staff under the supervision of municipalities and the postal company. While access to services is equal for everybody, involving vulnerable groups such as women's organisations, young people and people living with disabilities can increase their interest and participation.

On the occasion of International Girls in ICT Day, Aurélie Adam Soulé, the minister of the digital economy and communication in Benin, argued that the low representation of women in ICTs comes from the fact that the sector is wrongly considered as a domain for men. For her, initiatives to promote women's involvement in ICTs therefore need to be encouraged and women's groups need to be empowered so as to enable equality of access.[20] This could be done, among other strategies, by using affirmative action schemes in telecentres.

Participatory governance of community centres, although often complicated, can improve the use and usefulness of community services offered, make the centres more relevant to the needs of the local community, and even contribute towards their financial sustainability in the long term.

Involving the community as rights holders

At the same time, the community centres should be framed within human rights discourse. A human rights-based approach to their governance will help promote the development and dissemination of local knowledge, and increase community dialogue and action.[21] A human rights approach can be understood to foreground participatory approaches to decision making, while focusing on empowering communities in a non-discriminatory way. A human rights approach also places an emphasis on the economic, social and cultural rights of the community, and emphasises the well-being and dignity of rights holders as individuals. Through a participatory approach, the centres would empower women, young people and people with disabilities to advocate for policy reform, and to lobby duty bearers to meet their obligations.

Conclusion: Democratising access infrastructure at the local level

The aim of setting up the community ICT telecentres was to increase access to the internet and its services for marginalised groups such as women, young people and people living with disabilities. Since then, they have been helpful to many people. Among other things, the centres help people with low literacy levels edit and print their documents, and communicate with each other via Skype – introducing valuable new ways of engaging with the world to the communities. Local schools also use the community spaces for field work and other school exercises. As public infrastructures, they are managed as common goods and are community oriented. As all municipal infrastructures, their implementation and management details are accessible to everyone. They are under the supervision of the local government representatives,

17 http://www.inpf.bj/IMG/pdf/politique_nationale_promo_genrebenin.pdf

18 https://www.ohchr.org/Documents/ProfessionalInterest/cedaw.pdf

19 One example is the ICT training for women's groups through a partnership between the local NGO APHEDD and Boite A Innovations. www.bai.alphaomedia.org/gal_news18.php

20 https://www.lanationbenin.info/index.php/actus/159-actualites/16075-journee-internationale-des-jeunes-filles-dans-les-tic-encourager-la-cible-a-evoluer-dans-le-secteur

21 Figueroa, M. E., et al. (2002). *Communication for Social Change: An Integrated Model for Measuring the Process and Its Outcome.* omec.uab.cat/documentos/com_desenvolupament/0154.pdf

all elected by the community for a period of four years.

But to what extent can they be considered communal while their governance and management do not involve the most vulnerable groups in the community? How accessible are these centres if the infrastructure is not used for functional literacy, for example? All people regardless of their cultural, political or social status are part of the community and should be empowered to play a role as part of the social ecology of that system.

Democratising community ICT centres can contribute to their development. This also extends beyond governance to the democratisation of the infrastructure. The people's contribution to extending their internet access through mesh networks, and consequently to their geographical, economic and physical accessibility, can make them truly *community* networks. There are examples in other parts of the world – such as Community Service Centres in India – where government centres are used to bridge the digital divide along with community network roll-out. Given the proliferation of government e-centres across the world, they present an opportunity to leap-frog the digital divide; they can be a useful starting point for community networks, which do not always have to break the soil first.

In law, community networks can be implemented without any authorisation. However, their implementation in rural areas, especially in remote regions, means that solar power will need to be used, making the overall set-up costs, including the ICT hardware, costly. The setting up of such a network can also face many other challenges, such as the sustainability of its management structure, the openness of its governance structure, and local dynamics that emerge when collective projects are set up. These challenges, however, do not detract from the potential of community centres that may be the unique way to bridge the digital divide for vulnerable people, and may already have started to deal with some of the challenges that community networks will face on the ground.

Action steps

To be more open, community access initiatives need to learn from the global community network movement and:

- Adopt a human rights-based approach to their roll-out plans and strategies. The community should be involved in all aspects of access initiatives, which should be accountable and transparent.
- Create open and inclusive steering committees that include right holders so that they participate in decision-making processes. In particular, women, young people and people with disabilities should be included in governance processes.
- Use local access networks as an opportunity to foster interaction on local development at the local level, and to boost community dialogue and action.

In addition, local internet access initiatives should:

- Train facilitators at the ICT centres in local languages to encourage participation of people with low literacy levels.
- Involve community-based organisations in the design and implementation of any access networks, including government projects, so that the development potential of the access initiative is maximised.
- Consider extending internet access in the community through implementing a Wi-Fi mesh network or using any other suitable networking technology.

BRAZIL

HOW COMMUNITY NETWORKS IN MARANHÃO ARE HELPING *QUILOMBOLAS* PROMOTE THEIR RIGHTS

Instituto Nupef
Oona Castro, with the collaboration of Carlos Alberto Afonso and Paulo Duarte
https://nupef.org.br

Introduction

Although access to broadband internet has been growing in Brazil, one still finds a huge access gap in some regions and among certain population groups. This is the case of many *quilombola* communities in the state of Maranhão, situated in the Northeast region of Brazil. A *quilombo* is a settlement inhabited by descendants of Afro-Brazilian slaves who escaped from plantations before the abolition of slavery in Brazil in 1888. These communities are mostly located in rural areas, where quilombolas preserve their culture and the environment. They still need to fight for lots of basic rights – such as access to lands, to public services such as health, education and communications. In these areas, telecommunications operators barely provide access to the internet and, when they do, the service is usually of very low quality and not affordable.

This situation motivated Instituto Nupef – Núcleo de Pesquisas, Estudos e Formação, in partnership with local communities, to develop a pilot programme creating two community mesh networks in these communities, connecting quilombolas from different municipalities.

This report discusses the process of setting up these networks, and points to the benefits for the local communities.

The context

In 2017, two networks were implemented in Maranhão: the first one, which has been operational since June last year, in the municipality of Penalva, and the second in the municipality of Cajari, in operation since December 2017. Both cities received an internet link in August.

Both Penalva and Caraji contain part of the Baixada Maranhense Environmental Protection Area,[1] a sustainable use conservation unit created in 1991. The municipalities are officially considered "extractive reserves", according to Brazilian law. A river divides the two municipalities, and, depending on the season, this can be dry or in flood, with the water at different levels due to rains or river tides.

The estimated population of Cajari, according to the 2010 census, was about 18,751 inhabitants.[2] Nevertheless, the population in the community where the network was implemented was about 300 people. The population of Penalva, according to the same census, was estimated at 34,246 inhabitants,[3] but the community in the rural area has about 26,000 people, and there are about 1,000 people living in the area served by the network.

Brief history: Choosing the communities and the model

The two communities in Penalva and Caraji were chosen based on an analysis of their socioeconomic conditions and after talking to different groups. Among these was a project called New Social Map of the Amazon,[4] located in the social anthropology programme at the federal University of Amazonas. The project builds knowledge about the Amazon region and its inhabitants by mapping information that regular cartography usually ignores, such as conflict areas, and relations between peoples and territories. As a result, their work serves as a tool for strengthening social movements. The project identified communities for Instituto Nupef to start talking to.

A few conditions were necessary to have a community network implemented: the first and most important was the local need and motivation to get one in place and ensure continuity. The second and very crucial condition was that a community governance association had been set up in the communities that we could engage with. Affordable connectivity provided by the market also needed to be absent, while the geography needed to be amenable to a mesh network being set up. After some deliberation, we decided that the best

1 https://en.wikipedia.org/wiki/Baixada_Maranhense_Environmental_Protection_Area

2 https://censo2010.ibge.gov.br/sinopse/webservice/frm_urb_rur.php?codigo=210250

3 https://censo2010.ibge.gov.br/sinopse/index.php?dados=21&uf=21

4 novacartografiasocial.com.br

communities to run the pilot were the quilombos of Bairro Novo in Penalva and Camaputiua in Cajari.

After initial discussions, Instituto Nupef planned how to set up the networks. The first idea was to install a tower that would connect the two mesh networks set up in the communities. However, the costs related to setting up the tower were too high. The city hall in Penalva had a telephone tower that was in use and we considered the possibility of using that to connect the radios; but in the end it proved to be cheaper and easier to contract a satellite internet link that would connect the network access points to the internet.

Setting up the networks

Building the networks involved about 20 people from inside and outside the community. Nupef's team managed the process. With the help of community volunteers it also mapped the areas to determine the spots where radios should be installed, offered technical training to community members so that they could fix common problems, and developed software and hardware solutions for the networks. The community volunteers helped engage other people in the community, and also assisted with some technical issues such as installing the routers.

The first step was to do a site survey (or geo-referencing) in each community, establishing the coordinates needed to create the mesh. In the quilombo of Bairro Novo, seven radios were used, covering a perimeter of about 520 metres and an area of about 17,770 square metres. In the quilombo of Camaputiua, which is a much smaller community, five radios covered a perimeter of 1,020 metres and an area of about 20,600 square metres.

In Penalva, 75 users received a password to connect to the internet. They then shared their passwords with other community members, reaching a total of over 465 connected devices.

In Cajari, where 118 families live, 20 people received passwords and use the network regularly. Besides that, 200 people have been benefiting from intermittent internet access, and have been issued with daily passwords when necessary. Many of these are people from other communities that commute to Cajari in order to use their network.

Using the network, Camaputiua's quilombola association has helped over 400 community members get their national identity numbers (CPFs) and other documents which are required for participation in social and government programmes, economic transactions, and securing individual rights (such as support for family agriculture, funding opportunities, opening bank accounts, and the recognition of land access rights).

Applications and systems

As the networks are connected to the internet through satellite links, they both have access to standard internet services such as VoIP telephony, email, social networks and video streaming (initially this was provided using Elastix).[5]

Nupef's team used LibreMesh on the routers, an OpenWrt/LEDE-based firmware for wireless mesh nodes. Community members were taught how to change passwords and other system configurations, and even how to perform some advanced tasks like saving customised configuration on the devices using the LibreMesh "cook" function.

Sustainability

It is important to note that a key decision with regard to the sustainability of the project was to make use of common and easily accessible equipment. The initial purchase of equipment was made by the Nupef team. A computer, routers, cables, boxes to protect the radios, and other hardware and accessories were bought, as well as extra routers and other items so the community had spare equipment available in case of the need for replacement (they have already replaced some). The equipment purchased is easily available in the capital of Maranhão, São Luis, in case the community needs to buy new hardware.

Through the two community associations, local managers (who are leaders in those communities) have been collecting very small amounts of money each month from community members to pay for the satellite link and make improvements to the project. Bairro Novo's community association has already been able to buy a printer – as many users said they needed one. In Cajari, the community is smaller and so is the group of regular network users, so they have not been able to make improvements to the network so far. However, they are managing to cover the costs of the satellite link.

Women in the network

Women are active in both networks, especially in Penalva, where the leadership of two women from the community was key for achieving success: Maria Nice Machado Aires and Geovania Machado Aires have played and continue to play very important

5 Elastix is a unified communications server software that offers tools such as IP PBX, email, instant messaging, fax and other funcionalitites. Available at: https://www.elastix.org/pbx/small-business-phone-system

roles in setting up the network and in its sustainability. From the very beginning they demonstrated strong interest in getting the community to embrace the new project.

Maria Nice was born in 1954 in Penalva and represents communities such as quilombolas, babassu oil harvesters, smallholders and other traditional communities. Geovania, 30 years old, a quilombola and a teacher with a master's degree, became the number one assistant in the network in Bairro Novo and coordinates the project locally. She handles both technical and management issues. While Maria Nice and Geovania are mainstays for the community network in Penalva, both contributed enormously to the set-up of the network in Cajari too, by connecting us with local leaders (they introduced us to a local leader there, Ednaldo Padilha, also known as "Cabeça") and by providing organisational support.

Through testimonies from community leaders we learned that women have benefited significantly from the networks. Mothers have been able to apply for maternity leave government support, while many have embraced the project as they were used to paying about 150 Brazilian reais (over USD 35) a month so their children could go downtown to do homework and research, apply for public exams and programmes and take care of other school-related needs in LAN houses[6] – that is, they paid for transportation and for using computers at the centres by the hour. Now they pay 20 Brazilian reais (less than USD 5) per month through a cooperative model to help the local association keep the networks running so their children have access to the internet within their communities whenever they need it.

Legislation and enforcement

The regulatory framework related to telecommunications and spectrum control and allocation is made up of a wide variety of laws, resolutions and norms, defined by the Congress and the federal government as well as Anatel, the regulatory agency. There are many bills under debate in the Congress, among which is a bill to change the General Law of Telecommunications[7] (*Lei Geral de Telecomunicações*). Spectrum allocation is the responsibility of the Ministry of Communications (for radio and TV) and Anatel. A complex mosaic of different laws apply to different services related to the use of the spectrum. The National Frequency Plan in Brazil[8] dates from 2016[9] and dozens of resolutions by Anatel regulate the use of frequencies and radio communications, as well the provision of broadband services. Describing and analysing all this legislation would result in a paper itself – and a very complex one, since there are lots of gray areas and rules to be defined.

In our context, what is most important is that according to Resolution No. 506/2008,[10] broadband wireless access systems for local networks were exempted from authorisation when providing access to no more than 5,000 people. However, this resolution was revoked by Resolution No. 680/2017.[11] While the exemption remains, Anatel seems to require, as of 2018, registration on Mosaico,[12] a platform meant to manage spectrum allocation.

There have been controversial decisions by different courts on the issue of spectrum. In April 2018, a decision by the Superior Court of Justice stated that "the principle of insignificance did not apply to cases of clandestine transmission of internet signals through radiofrequency that is characterised as a criminal act in article 183 of Law No. 9472/1997."[13] The principle of insignificance is one that judges apply whenever they analyse the infringement as one that did not cause a significant harm. This decision conflicts with a decision by the Supreme Federal Court from 2017, according to which *habeas corpus* had been granted to acquit a defendant accused of practising clandestine telecommunications activities, based on the principle of insignificance.[14] This may affect the principle of legal certainty and impact on innovation and social changes due to fear of being punished for creating networks that are not protected by the law.

The networks in Penalva and Cajari are not yet registered on the Mosaico platform or with any other government body. However, these networks operate within the confines of the law, under the exemption granted to wireless networks for fewer than 5,000 people in the 2.4 GHz and 5 Ghz frequencies.

6 LAN houses are commercial initiatives that provide access to the internet, similar to cybercafés.

7 www.anatel.gov.br/legislacao/leis/2-lei-9472

8 www.anatel.gov.br/Portal/verificaDocumentos/documento.asp?numeroPublicacao=346331&assuntoPublicacao=null&caminhoRel=null&filtro=1&documentoPath=346331.pdf

9 A new one was approved in June 2017, but was later revoked in January 2018, due to debates over criteria, frequency bands, and other issues.

10 www.anatel.gov.br/legislacao/resolucoes/23-2008/104-resolucao-506; also available in English: http://www.americascompliance.com/wp-content/uploads/2011/08/506-English_1.pdf

11 http://www.anatel.gov.br/legislacao/resolucoes/2017/936-resolucao-680

12 https://sistemas.anatel.gov.br/mosaico/login/login?service=http%3A%2F%2Fsistemas.anatel.gov.br%2Fmosaico%2Fportal%2FportalInternet

13 https://www.estrategiaconcursos.com.br/blog/sumula-606-stj

14 *HC 138134/BA, rel. Min. Ricardo Lewandowski, julgamento em 7.2.2017. Informativo STF 853.*

Action steps: The future and final considerations

The networks in Penalva and Cajari are meant to serve as a vital tool for the quilombola movements – those who struggle for the rights of these communities. Feedback has been extremely positive. With the success of the community networks in Penalva and Cajari, people in surrounding communities now want the networks to be expanded to give them access, and have requested Instituto Nupef's support to find the resources and to implement the new networks with them.

For Instituto Nupef, despite its significant experience in implementing information and communication technology (ICT) projects for populations with no access to the internet, this pilot improved the organisation's know-how on social and technical aspects related to building networks in rural areas. Nupef is now in dialogue with other organisations and social movements interested in implementing similar projects in their communities, including babassu oil harvesters.

On the advocacy level, our experience has reinforced the importance of keeping our eyes wide open in the field of community connectivity, though it is a considerably vast one that requires a dedicated group of people watching and trying to influence decisions with a social and public-interest perspective. It is not an easy task, as there are resource constraints related to people, money and time; but it is definitely a key one if we want to see the enforcement of communication rights in Brazil.

Finally, we propose setting up a programme in Brazil on community networks that will serve as a dynamic reference point for the application of networking at the community level. The purpose of this programme will be to produce and share knowledge and information on community connectivity, and to develop the capacities of individuals and organisations. In particular we are interested in the innovative use of ICTs which may support sustainable development, human rights, social justice, good governance and democratic values.

This involves not only creating and maintaining a periodically updated repository of rules and regulations that affect spectrum policies and practices which impact on networking initiatives in local communities, but also building knowledge and capacities by directly implementing local community networks where they are socially needed and where the market does not provide the means to make access happen.

BRAZIL

OFF-THE-GRID, AUTONOMOUS COMMUNITY NETWORKS IN BRAZIL

Fonias Jurua Project
Anna Orlova and Rafael Diniz
https://fonias.submidia.org/en

Introduction

To understand how high-frequency radio community networks were developed in the Brazilian Amazon, it is important to reflect on the geographic characteristics and historical and socioeconomic background of Brazilian Extractive Reserves. By drawing on these factors we explain the novelty of our technical solution – building local autonomous connectivity in the Amazon rainforest using digital radio in the high-frequency band – and how it is the most appropriate solution in this context. We put into perspective aspects of available infrastructure and the local context as the main factors defining the solution that can best serve the needs and wishes of the local population to provide information and communication solutions.

Socioeconomic and historical background

The Amazon region, which extends through many South American countries, is one of the least developed in terms of information and communication technology (ICT) infrastructure, as in these areas there is insignificant investment in infrastructure. For many communities here the only available communication technology is the high-frequency radio transceivers that are practical and affordable and have already been in use for decades.

Although the first Amazon digital radio network[1] using the high-frequency radio band in the Brazilian Amazon forest was created in Acre in 2014-2015, high-frequency radio transceivers – bi-directional radio transceivers that allow communication directly between the two transceivers without any intermediate points – existed and were used by the local population for a long time before that. Locals remember the use of high-frequency radio transceivers by the *patrãos* (bosses) who owned and managed the rubber plantations in Brazil. Until the early 1980s, the *seringueiros* (rubber tappers) were the many generations of migrants from the Northeast region of Brazil, along with some indigenous people populating the Amazon rainforest who were often enslaved and forced to work in the rubber tapping industry. Looking to end this oppression and hardship, rubber tappers mobilised with the labour movement and with environmentalists. This led to the liberation of rubber tappers and the establishment of the first (legally recognised) extractive reserve in the Brazilian Amazon in 1990 – the Alto Juruá Extractive Reserve.[2] As a result of this struggle, rubber tappers claimed their right to live, work and organise their lives themselves on the land where previously they were forced to work in harsh conditions.

Geography of the Alto Juruá Extractive Reserve

The Alto Juruá Extractive Reserve is located in the state of Acre, on the border with Peru, and comprises an area of 506,000 hectares of forests and rivers. It is a federal administrative territory set up as a natural conservancy that allows traditional communities to live and earn their livelihoods, including through natural resource extraction, inside the conservation zone, based on the assumption that their traditional way of life favours and enhances the protection of natural areas.[3] Therefore, inhabitants can live from the land and rivers of the extractive reserve; however, they also face the need to find sustainable ways of farming and maintaining their lives inside the reserve without depleting the natural environment. This form of coexistence with the environment would guarantee protection of the forest and of the people living inside the reserve.

For years the main challenge for traditional and indigenous communities of the Amazon forest was being isolated and scattered throughout the forest on their own without any access to communication. The only available means of transportation is by

1 https://fonias.submidia.org/en

2 www.icmbio.gov.br/portal/unidadesdeconservacao/biomas-brasileiros/amazonia/unidades-de-conservacao-amazonia/2776-resex-alto-jurua

3 Antunes Caminati, F., et al. (2016). Beyond the Last Mile: Fonias Juruá Project – an HF Digital Radio Network Experiment in Amazon (Acre/Brazil). In L. Belli (Ed.), *Community Connectivity: Building the Internet from Scratch. Annual Report of the UN IGF Dynamic Coalition on Community Connectivity*. https://internet-governance.fgv.br/sites/internet-governance.fgv.br/files/publicacoes/community_connectivity_-_building_the_internet_from_scratch.pdf

rivers and there is no electricity power grid inside the reserve, which leaves people with the option of either gasoline-powered generators or solar energy. Therefore transportation and energy are expensive[4] and not accessible for the majority of the extractive reserve population.

Precarious public infrastructure in the reserves also makes it almost impossible to access public services. For example, many families receive financial support from the government programme *Bolsa Família*,[5] through which a monetary allowance is provided to the female head of the household for every child, on the condition that the children attend school. Schools in the extractive reserves are supposed to provide transportation for the children inside the reserve to bring them to school. In practice, however, this often does not work, and instead children are staying at home. This has a direct effect on girls in particular: without any primary education they have no option of continuing their education or being employed outside of their immediate surroundings, and are left with the only choices of early marriage (at the age of 13 or 14) or staying at home to look after the children and the household.

Today Amazon rainforest reserves are endangered. With massive territories without any oversight or protection from the army or state, extractive reserves represent an easy target for illegal extractive activities, such as logging, hunting, and the mining of precious metals and other natural resources like oil. After the decline of the rubber tapping industry in Brazil, rubber tappers also needed to find ways of sustaining their lives inside the reserve, and recently many have opted to leave for urban areas in hope of better lives and jobs.

All of these aspects were taken into consideration when our community work began. There was a clear need for a means of communication and information technology that could help locals monitor their territory, mobilise and coordinate their actions, exchange information with the municipal centre and receive assistance with basic services like health care and education. Taking into consideration the lack of any infrastructure and the long distances between communities separated by an impenetrable rain forest and their expressed wishes to use radio transceivers, the solution for community connectivity in this context was to build an autonomous, affordable solution for connectivity using old, existing infrastructure.

The Amazon high-frequency digital radio network in Acre

A high-frequency digital radio network as a solution to provide community connectivity has been implemented and is currently operating in the Brazilian Amazon forest in two states: in the Alto Juruá Extractive Reserve in Acre and in the Terra do Meio region in Pará state.

The Alto Juruá Extractive Reserve has a network of seven radios inside the reserve and one main hub-station in the city of Marechal Thaumaturgo. This network is a result of a long-term collaboration that started in 2013 between traditional and indigenous communities in Acre, and researchers and professors from the University of Brasilia,[6] Sao Paulo State University[7] and University of Campinas.[8] The network was developed as part of the academic research project "Fonias Juruá" to provide information and communication infrastructure to rural Amazon communities that are under-served by regular and commercial information and communication networks. It was based on the requests for two-way voice radios from 24 traditional communities in the Alto Juruá Reserve and aimed at engaging locals in the process of political participation and empowerment through joint sustainable experience. At the moment the estimated number of users is around 500 people.

It took four years from the beginning of talks with the community to purchasing equipment and taking the first trip to the extractive reserve where the first six radio stations were installed with the collaboration of community leaders and locals living inside the community. Five of the radios were installed inside the extractive reserve and one station was installed in the city of Marechal Thaumaturgo, which served as a hub to connect and exchange information between all the stations. The idea was that Marechal Thaumaturgo would provide socially and politically important information and news to the extractive reserve that would serve as an incentive to foster communication and the exchange of information inside and outside of the reserve. It was also necessary for people living inside the reserve to talk to their relatives living in the city of Marechal Thaumaturgo, exchange information about local production and prices of goods and services, ask for medical assistance and advice about social services, and access other locally important information. Another important aspect of communication was to report illegal activities taking place in the reserve, such as illegal logging, hunting and mining.

4 Petrol prices are one and a half to two times higher compared to other states of Brazil.

5 www.mds.gov.br/assuntos/bolsa-familia/o-que-e; https://en.wikipedia.org/wiki/Bolsa_Fam%C3%ADlia

6 www.unb.br

7 www.unesp.br

8 www.unicamp.br/unicamp

Today the Acre high-frequency radio network is composed of eight two-way radios – a point-to-multipoint broadcasting set-up allowing every station in the network to receive the transmission and to communicate among each other. This system is autonomous, low maintenance and easy to use by any member of the community after basic training. The solution is composed of standard high-frequency transceivers, common wire horizontal dipole antennas that are positioned to work in near vertical incidence skywave (NVIS) mode, and software-defined radio (SDR)[9] techniques for digital communications. As there is no power or electricity infrastructure available, each radio station is powered by a solar panel and batteries, making it an environmentally friendly set-up.

Since its inception the network has been in regular use by the local communities without any major problems. In 2016 we successfully accomplished trials with a digital transmission system based on the Digital Radio Mondiale (DRM) standard (this was first attempted in 2015 but was only partially successful). The solution for digital transmission is made up using an embedded computer, an interface to the radio and the SDR software. We managed to send text files and images over the radio in the 80-metre HF band (3545 kHz) to locations 100 km apart.

Comparing the two networks: A look at the Terra do Meio Digital Radio Network

The network in the state of Pará has no formal name, but we refer to it here as the "Terra do Meio Digital Radio Network". The network operates with a main station set up at the NGO Instituto Socioambiental[10] (ISA), with other independent stations (around 10 stations) in the urban area of Altamira that provide special services and communication. In the rural area of the Amazon forest there are at least 50 stations, meaning that, like the Fonias Juruá network, this network can be considered predominantly rural. The network has been set up by ISA, SDR Telecom[11] (a company founded by a member of the Fonias Juruá Project) and the local people of the Terra do Meio region. The network has more than 4,000 users.

In comparison to the Acre community network, the Terra do Meio Digital Radio Network is significantly larger, with around 60 radio stations spread across the region. The Terra do Meio region is also bigger and therefore the network connects many more communities, creating a high demand for radio use. The daily use of radios is vital to the local population. For example, in some cases when people are travelling to another place in the region they take their radios with them and upon arrival they assemble the radio to be able to talk. At one meeting we witnessed a queue of people waiting to use the radio to talk to their relatives and acquaintances, for personal or business reasons. This potential of the network would not be possible if the network was small. As a result, the size of the network plays a significant part in its usefulness.

Despite the very similar social and geographic contexts, the way the two networks operate is also different in terms of ownership (the way the radios are owned and shared), as well as the economic incentive to use the radios. Here the economic capacity of community members plays a defining role in the ways the radio network develops and extends inside the forest territories, as well as how community members attribute importance to the use of radios. From our observations we can say that in Pará, the use of the network is more economically motivated. People in the communities run daily business over the radios and therefore use it more regularly and more frequently. This factor appeared to impact on the maintenance of the network.

In Acre we observed only one community where the maintenance of the radio was given importance by the members of the community, as opposed to the maintenance support received from the project team. In this community, the head of the household where the radio station was installed had a personal interest in maintaining the station because he was using it for running his own business – in other words, there was an economic incentive. He purchased a new battery at his own expense to keep the radio working when the initial one broke down. This suggests that networks that enable the economic agency of communities have a stronger prospect for sustainability.

Empowering women

There is a clear division of labour in the Alto Juruá Extractive Reserve that was established throughout the years and was defined by the way of life inside the reserve. Most hard physical work like hunting and farming is done by men, whereas women tend to stay at home to cook and take care of the family and the household. However, when it comes to the decision-making process, women take equal part and participate actively.

When one community had a meeting to decide and vote on the house where the radio would be installed, one woman openly voted against her

9 https://en.wikipedia.org/wiki/Software-defined_radio
10 https://www.socioambiental.org/pt-br
11 https://sdrtelecom.com.br

husband when he proposed to have the radio in their house. In another community women were very proactive at the community meeting, expressing their concerns and asking questions. They became so engaged that they even participated in the process of installing antennas and helping to dig holes and install wooden posts along with men – something that only men did in other communities. The first volunteers to test and use the radio were teenage girls, who appeared less shy than some of the young men who resisted using the radio for the first time in front of everyone else.

Because they tend to take care of household duties, women naturally stay closer to the radio during the day and seem to be more keen to talk and share news with others compared to men.

Action steps

Today one challenge for community networks operating over the high-frequency radio band is the lack of licences for community use. In the Brazilian Amazon many networks or high-frequency radios operate without a licence. Partly this happens because some are not aware that they should have licences as network providers (for example, recently some radios were seized by Anatel – the Brazilian telecom regulator – in the city of Altamira), whereas in other cases the remoteness of the Amazon region and lack of any form of oversight gives a certain freedom for the use of spectrum without licences. There are ongoing discussions and requests from network operators and researchers submitted to the ministry of communication for community-use licences to be formalised; however, so far the ministry has refrained from taking action towards a new type of licence being introduced. Instead it has asked for working examples of communities operating community networks as a basis for further discussion.

Community networks that want to operate mobile phone networks also cannot use any mobile phone radio bands for their own needs. There are no community licences for this purpose. Nevertheless, Anatel allows, for example, Wi-Fi internet providers to operate without licences, with certain restrictions such as a maximum of 5,000 users and with only certified equipment (Resolution No. 680, 2017, Anatel).[12] Even in these cases, however, community networks fall under the same category as small for-profit network providers.

Today there is a lack of telecommunication access and infrastructure for millions of Brazilian citizens: people living in isolated rural areas without commercial telecoms coverage, and the poor and socially disadvantaged that cannot afford expensive commercial telecom services. As a result, there is a clear necessity to declare and convey the needs of non-commercial network operators for a new type of telecommunication licence for community use at the policy-making level.

Since 2015 a group of implementers and supporters of emerging GSM (mobile telephony), high-frequency radio and Wi-Fi community networks have been working together to advocate for community connectivity at the level of policy making and to establish and articulate the concept of "community network" in the Brazilian telecommunication law. However, there have been no significant achievements so far.

12 www.anatel.gov.br/legislacao/resolucoes/2017/936-resolucao-680

CANADA

FIRST NATIONS COMMUNITY NETWORKS AND THE FIRST MILE CONNECTIVITY CONSORTIUM

First Mile Connectivity Consortium
Rob McMahon and the First Mile Connectivity Consortium
www.firstmile.ca

Introduction

First Nations are leading local and regional community networking initiatives in Canada. From trenching fibre backhaul along gravel roads to installing Wi-Fi antennas on the roofs of homes and businesses, community champions and regional technology organisations are connecting rural, remote and Northern regions. These projects respond to persistent divides, demonstrating ways that communities drive infrastructure development while retaining ownership and control of networks and services. Many of these indigenous community networks have been in operation for years, and range from local internet service providers (ISPs) to complex regional organisations that provide services to multiple communities.

Along with the passion of community networking champions, indigenous community networks involve an array of other activities, such as seeking policy support, engaging in technical development, and responding to high demand for digital services. Lower-cost equipment can now be easily transported and set up by local technicians. If adequate backhaul is in place, these networks offer robust internet services to households and organisations. Once service becomes available, people quickly adopt online education, health and e-commerce applications, social media platforms like Facebook and Twitter, and mobile apps. For example, the Atlantic Canada First Nations Help Desk based in Membertou First Nation is developing a series of language apps to protect and revitalise the Mi'kmaw language.

Adoption in these communities reflects a "whole community" understanding of and approach to digital technologies for social and economic development that counters models focused on "individual" and "household" metrics. A broader ecology of community support enables local innovators to build, shape, manage, operate and utilise these tools – activities reflected in networking projects such as the Mamwapowin Technology Society in Maskwacis, Alberta.[1] Although it is located less than an hour from the province's capital of Edmonton, few commercial internet services are available in Maskwacis. Many residents are unable to afford residential services, which can cost up to CAD 132 (roughly USD 100) a month – an expensive service in a region with few employment opportunities.

In 2016 Bruce Buffalo, a member of Samson Cree Nation, set up a free Wi-Fi hotspot that redistributed his personal internet connection to several local neighbourhoods. The self-taught technician invested his own money to build a wireless system with two switches and an online access portal. Since then, Bruce has expanded the network, upgraded his wireless equipment, and founded Mamwapowin as a non-profit. He is now working with Maskwacis Cultural College to secure a 100 Mbps backhaul link, which will provide bandwidth he can use to connect Wi-Fi users to public safety resources, community services and online education. Once connectivity has been enhanced to support video, Bruce would like to create a community tech hub where he can bring all ages together to learn how to use the internet for education opportunities, business development, and to explore technology in an open learning space. Bruce is also engaged in public outreach about his project, and is currently fundraising to expand it. He has given presentations to groups including the Canadian Internet Registration Authority (CIRA) to share his project as a potential template for others.[2]

Mamwapowin Technology Society is just one recent example of many digital innovations taking place in First Nations across Canada – many more indigenous community networks have been in place for years, with some operating since the earlier days of the internet, almost 30 years ago. In the next section, we highlight some of the political, economic and policy conditions that support and constrain their work.

1 Max, C. (2018, 22 June). Maskwacis man making Internet accessible to the community. *Wetaskiwin Times*. www.wetaskiwintimes.com/2018/06/22/maskwacis-man-making-internet-accessible-to-the-community

2 https://cira.ca/newsroom/events/cira-meet-calgary

Enabling digital self-determination: Political, economic and policy contexts

In Canada, indigenous community networking advocates have fought hard to establish and sustain policy and regulatory frameworks that enable digital self-determination. Almost 20 years ago, the National Broadband Task Force released a report (2001) that identified the digital access divides facing rural, remote, Northern and indigenous communities, and proposed a national broadband strategy to connect all communities – with priority placed on affordable access.[3] Five years later, the Canadian Telecommunications Policy Review Panel report (2006) noted that community networks provide both technological and social infrastructures for digital technology access, adoption and use.[4] Starting in the mid-1990s and continuing to the present, government departments have established funds to encourage the deployment and operations of infrastructure in under-served regions.[5]

But despite billions of public dollars invested, many indigenous communities still lack adequate, affordable access. Canada still does not have a national broadband strategy, and First Nations, Inuit and Métis communities continue to be challenged with limited connectivity. Many people living in indigenous communities are struggling to pay the high costs of accessing and using digital technologies – though research also illustrates they are eager adopters if connectivity is affordable, reliable and meets their needs.

In recent years the government of Canada has paid increasing attention to the challenge of rural and remote digital divides, and to the First Nations community networks working to address it. To drive infrastructure deployment, the Canadian Radio-television and Telecommunications Commission (CRTC) established a new basic service objective (BSO) for broadband in December 2016.[6] This new standard recommends speeds of 50 Mbps download/10 Mbps upload, among other conditions. It was established after years of advocacy by indigenous and public interest groups in CRTC proceedings concerning access in Northern territories, including 2012 hearings on Northwestel's Modernization Plan[7] and a 2014 inquiry on satellite services.[8] Unfortunately, the government of Canada and the major telecom providers serving remote and rural communities continue to deliver connectivity services and infrastructure capable of less than the CRTC's recommended speeds.

With regard to affordability, the CRTC has generally encouraged competition, though it has stepped in to regulate retail rates in some remote/Northern regions that lack competition.[9] The importance of affordable services was raised in proceedings leading to the BSO decision – with a debate emerging between parties calling for an affordability subsidy for individual consumers, and groups who responded that regulations should instead focus on encouraging funds for community networks. The "whole community" approach noted earlier encourages a community-wide funding regime that supports the required networking solutions for everyone.

The government has also set up a series of broadband funding programmes. Most recently these include the CND 500-million Connect to Innovate (CTI) programme managed by Innovation, Science and Economic Development Canada (ISED) to establish regional backhaul networks,[10] and a broadband fund currently being set up by the CRTC.[11] The CTI fund contributed to several major projects in Canada's North, including the Mackenzie Valley Fibre Link in the Northwest Territories,[12] the Dempster fibre project in the Yukon,[13] and funding for satellite infrastructure in Nunavut.[14] Despite these gains, critics note these projects are operated by incumbents rather than community networks, and lack a public definition of "open access" requirements.

Importantly, after pressure from indigenous and public interest groups, infrastructure funding programmes were made available to non-profits,

3 Industry Canada. (2001). *The New National Dream: Networking the Nation for Broadband Access: Report of the National Broadband Task Force.* publications.gc.ca/collections/Collection/C2-574-2001E.pdf

4 Industry Canada. (2016). *Telecommunications Policy Review Panel Final Report.* https://www.ic.gc.ca/eic/site/smt-gst.nsf/vwapj/tprp-final-report-2006.pdf/$FILE/tprp-final-report-2006.pdf

5 Blake, S., McMahon, R., & Williams, D. (2016). *A Guide to Federal Funding for Indigenous Broadband in Canada.* First Mile Connectivity Consortium. firstmile.ca/wp-content/uploads/FMCC-Guide-to-Federal-Funding-for-Indigenous-Broadband-in-Canada.pdf

6 CRTC. (2016). Telecom Regulatory Policy CRTC 2016-496: Modern telecommunications services – The path forward for Canada's digital economy. https://crtc.gc.ca/eng/archive/2016/2016-496.htm

7 CRTC. (2013). Telecom Regulatory Policy CRTC 2013-711: Northwestel Inc. – Regulatory Framework, Modernization Plan, and related matters. https://crtc.gc.ca/eng/archive/2013/2013-711.htm

8 CRTC. (2014). Satellite Inquiry Report. https://crtc.gc.ca/eng/publications/reports/rp150409/rp150409.pdf

9 CRTC. (2013). Op. cit.

10 https://www.canada.ca/en/innovation-science-economic-development/programs/computer-internet-access/connect-to-innovate.html

11 CRTC. (2017). Telecom Notice of Consultation CRTC 2017-112: Development of the Commission's broadband funding regime. https://crtc.gc.ca/eng/archive/2017/2017-112.htm

12 https://mvflproject.com

13 https://yukon.ca/en/dempster-fibre-fact-sheet

14 Innovation, Science and Economic Development Canada. (2017, 14 September). Communities across Nunavut will benefit from faster Internet. https://www.canada.ca/en/innovation-science-economic-development/news/2017/09/communities_acrossnunavutwillbenefitfromfasterinternet.html

cooperatives and community-based organisations. Examples of First Nations CTI projects include initiatives led by Matawa First Nations in Ontario,[15] Clear Sky Connections in Manitoba,[16] and Arrow Technology Group in Alberta.[17] The deployment, operations and sustainability of these community networks require a complex balance between local innovation, regional cooperation, and supportive policy and regulatory conditions. As noted in the 2017 GISWatch country report for Canada, residents of affected communities must be involved in the provision of digital services.[18] They should not be restricted to act only as consumers of broadband – they can also contribute as producers, owners and operators. Such opportunities utilise broadband not just as an enabler for economic development in other industries and services, but also as a locally owned and managed resource in and of itself.

This perspective reflects "first-mile" solutions in the design, development and operations of telecommunication infrastructure and services – that is, those which invest in connections and organisations rooted in affected communities and regions. Such an approach is proposed for communities that lack a business case for private sector companies to build and operate required infrastructure and services. Locally oriented first-mile solutions contrast with "last-mile" initiatives that focus on upgrades to urban-based infrastructures in the hope that they will eventually serve remote and rural regions.[19] The drive to secure affordability and access through self-determined first-mile broadband development has supported the emergence of indigenous community networks in Canada governed by, and directly accountable to, the communities they serve. For example, the Kuhkenah Network (K-Net) began their local broadband network development work in 2000.

Without a means to interconnect with regional backhaul networks, local initiatives remain points of isolation. Local providers can interconnect through "open access" infrastructure, but a clear definition of open access is often lacking – leaving backhaul providers indirect means to obstruct adequate interconnection, such as by charging high rates for access. This issue was debated at a 2017 CRTC proceeding, after which the Commission denied an application by Northwestel requesting forbearance from the regulation of its Wholesale Connect service in the communities served by the Mackenzie Valley Fibre Link network, since forbearance would not be in the public interest of users in affected communities.[20]

Community wireless providers face a related challenge: while spectrum allocations are regulated by the federal government and sold in auction, licences typically cover huge areas at high cost – leaving smaller providers unable to purchase and utilise wholesale spectrum.[21] Historically, Canada developed the Rural Remote Broadband Systems (RRBS) policy to support spectrum utilisation by small providers servicing rural and remote regions.[22] However, while RRBS "offers a clear case study of a bold policy initiative to bring service to difficult regions and introduce new smaller players into concentrated wireless markets," over time it failed, due to "speculative license holders, poor economies of scale and, most importantly, faint government support."[23]

Another challenge relates to the sustainability of community networks. Larger-scale services support broadband-enabled applications such as

15 Northern Ontario Business. (2017, 6 October). Matawa First Nations lands nearly $70 million for high-speed internet. https://www.northernontariobusiness.com/industry-news/aboriginal-businesses/matawa-first-nations-lands-nearly-70-million-for-high-speed-internet-734490

16 Innovation, Science and Economic Development Canada. (2018, 9 January). More than 100 rural communities in Manitoba will benefit from faster Internet. *Cision*. https://www.newswire.ca/news-releases/more-than-100-rural-communities-in-manitoba-will-benefit-from-faster-internet-668464203.html

17 Innovation, Science and Economic Development Canada. (2017, 11 July). Western Canada to benefit from faster Internet. https://www.canada.ca/en/innovation-science-economic-development/news/2017/07/western_canada_tobenefitfromfasterinternet.html

18 Riahi, A., Lambert, M., & Couture, S. (2017). Indigenous disconnect: Canada's divide in internet governance. In A. Finlay (Ed.), *Global Information Society Watch 2017: National and Regional Internet Governance Forum Initiatives (NRIs)*. https://www.giswatch.org/en/country-report/internet-governance/canada

19 Strover, S. (2000). The First Mile. *The Information Society, 16*(2), 151-54; Paisley, L., & Richardson, D. (1998). Why the first mile and not the last? In L. Paisley, & D. Richardson (Eds.), *The First Mile of connectivity: Advancing telecommunications for rural development through a participatory communication approach*. Food and Agriculture Organization of the United Nations (FAO). www.fao.org/docrep/x0295e/x0295e03.htm; Whiteduck, T., Beaton, B., Burton, K., & O'Donnell, S. (2012). Democratic Ideals Meet Reality: Developing Locally Owned and Managed Broadband Networks and ICT Services in Rural and Remote First Nations in Quebec and Canada. Keynote paper for the Community Informatics Research Network (CIRN) Conference, Prato, Italy, November.

20 CRTC. (2017). Telecom Decision CRTC 2017-300: Northwestel Inc. – Request for forbearance from the regulation of Wholesale Connect service in the communities served by the Mackenzie Valley Fibre Link network. https://crtc.gc.ca/eng/archive/2017/2017-300.htm

21 First Mile Connectivity Consortium. (2014). Submission to the Consultation on Policy Changes in the 3500 MHz Band (3475-3650 MHz) and a New Licensing Process in Rural Areas. Reference number (DGSO-003-14): Canada Gazette, Part I, August 2014. firstmileconnectivity.myknet.org/wp-content/uploads/sites/1687/2014/10/FMCC-Comments-3500MHz-Consultation-Oct7.pdf

22 Taylor, G. (2017). Remote Rural Broadband Systems in Canada. *Telecommunications Policy* (in press). dx.doi.org/10.1016/j.telpol.2018.02.001

23 Ibid.

FIGURE 1.

First Nations e-Community Framework

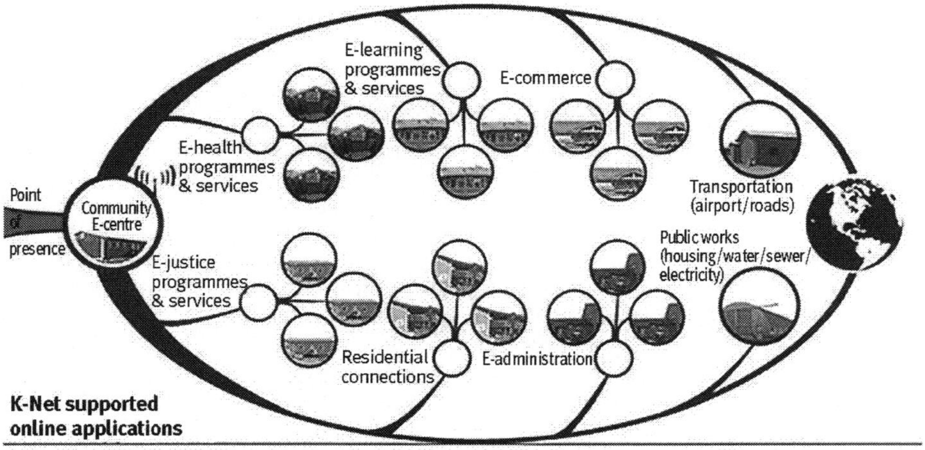

K-Net supported online applications

Online meetings	E-justice	Webcasting
Videoconferencing	Mobile services	Teleradiology
High speed internet	E-learning	Telemedicine
Voice over IP (VoIP)	E-administration	E-commerce

e-health and education that also provide steady revenue streams, even in small populations and dispersed communities. To interconnect communities and regions, local systems can link to regional backhaul networks in ways that allow organisations to balance autonomy with cooperation, such as through sharing bandwidth across regions while supporting local service delivery. For example, Opitciwan First Nation and the regional First Nations Education Council (FNEC) in Quebec set up a fibre optic network to share infrastructure and network support services with public service that sustain the ongoing operation and maintenance of the network. This work reflects the e-Community Framework developed by Judy Whiteduck in 2010 and endorsed by the national Assembly of First Nations.[24] The framework, illustrated in Figure 1, reflects the important role of community organisations, services and activities in sustaining broadband connectivity. These organisations meet many essential needs, provide sustainable local employment, and support community, social, political and economic development. The First Mile Connectivity Consortium (FMCC)[25] introduced the "whole community" approach to telecom development and funding programmes to ensure communities are able to plan and sustain their local networks to address their needs and priorities.[26]

In the next section, we discuss some of the work that community networking advocates are doing to bring these elements of the e-community together, focusing on the work of the FMCC.

Indigenous regional community intermediary organisations in Canada

On-the-ground innovators like Bruce Buffalo build and operate community networks. But without a means to aggregate on a broader scale, even with funding and regulatory support, local networks can remain insular pockets. Through regional community intermediary organisations, local projects gain a means to exchange knowledge and expertise, achieve economies of scale, and gain more political and economic clout.

In Canada, broadband-focused regional community intermediary organisations include First Nation Councils such as Keewaytinook Okimakanak (which established K-Net – see below),[27]

24 Whiteduck, J. (2010). Building the First Nations E-Community. In J.P. White, J. Peters, D. Beavon, & P. Dinsdale (Eds.), *Aboriginal Policy Research: Learning, Technology and Traditions*, 95-103. Toronto: Thompson Educational Publishing. apr.thompsonbooks.com/vols/APR_Vol_6Ch6.pdf

25 firstmile.ca/fmcc-2

26 Beaton, B., McMahon, R., O'Donnell, S., Hudson, H., Whiteduck, T. & Williams, D. (2016). Digital Technology Adoption in Northern and Remote Indigenous Communities. Prepared for Innovation, Science and Economic Development Canada. First Mile Connectivity Consortium.

27 knet.ca

regional governments such as the Kativik Regional Government,[28] and non-profit advocacy groups set up by Northern residents, such as the (now-defunct) Nunavut Broadband Development Corporation. These organisations perform a range of functions, including broadband deployment, operations and maintenance, purchasing, IT support and training, legal and advocacy support, and so on. They play prominent roles in both advocating for and administering ICT infrastructures and services. They represent (and are governed by) groups of communities, acting as mediators between local interests and external entities like government funders or corporations.

Some of these regional organisations have been providing services for decades. For example, K-Net was founded in 1994, during the early years of the internet. Its services, which encompass the whole community framework discussed above, now reach over 80 communities across Ontario and in other provinces. It partners with First Nations, government programmes, telecom transport providers and other groups to facilitate the largest indigenous network across Canada, and probably the world (see Figure 2).

In 2013, several regional community intermediary organisations united to form a national association called the First Mile Connectivity Consortium. The FMCC's members collaborate closely to engage in policy and regulatory advocacy, as well as in research, outreach, and digital literacy initiatives. The FMCC's 11 current members (listed from West to East to North) are:

- First Nations Technology Council (British Columbia)
- First Nations Technical Services Advisory Group Inc. (Alberta)
- Manitoba First Nations Education Resource Centre Inc.
- Clear Sky Connections (Manitoba)
- Broadband Communications North (Manitoba)
- Keewaytinook Okimakanak K-Net Services (Ontario)
- Western James Bay Telecom Network (Northeastern Ontario)
- Matawa First Nations Management (Ontario)
- First Nations Education Council (Quebec)
- Atlantic Canada's First Nations Help Desk (Atlantic region)
- The Native Communications Society of the Northwest Territories.

FIGURE 2.

K-Net network map

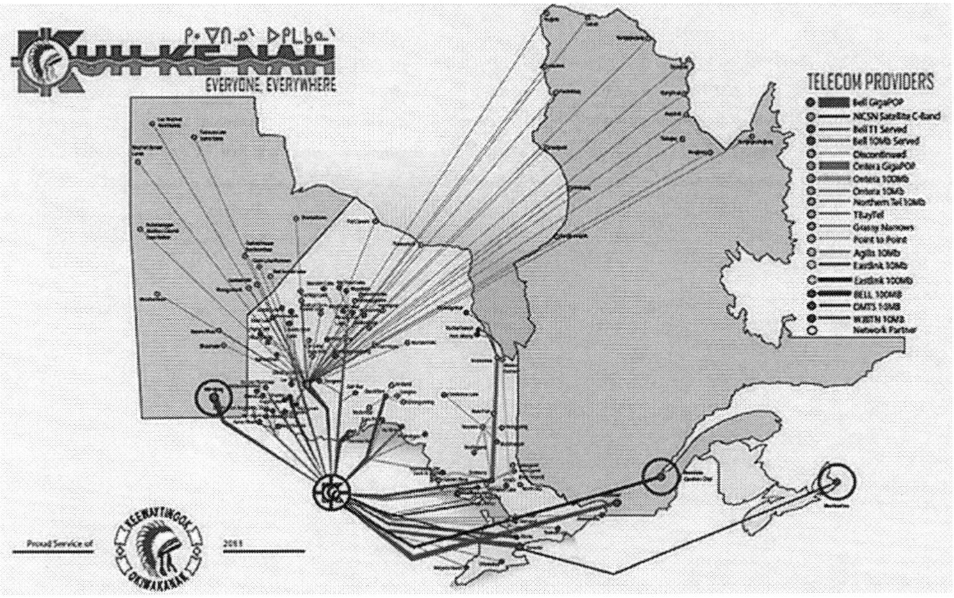

28 Information about the government's internet provider Tamaani is available at: tamaani.ca

Along with participating in the CRTC hearings and ISED funding programmes noted above, and contributing to other government consultations, FMCC members are currently working to support connectivity planning and monitor broadband quality and usage. These initiatives reflect efforts to develop and share appropriate digital literacies that make effective use of telecommunications services and facilities in rural, remote, Northern and indigenous environments. This definition of "digital literacy" extends beyond an individual's ability to use a computer, software like Microsoft Office, or social media – it also supports the planning, management and maintenance of telecommunications infrastructure and services.

This work also aims to train community technicians – an important area of economic development in these regions. For example, the Eeyou Communication Network,[29] a regional non-profit network operating in northern Quebec, has trained technicians, and 33 graduates have found work in ICTs in the Cree communities. In British Columbia, the First Nations Technology Council recently established the Bridging to Technology programme, which focuses on digital skills training and professional development.[30] Other digital literacy initiatives include workshops provided by the Gwich'in Tribal Council in the Northwest Territories and a digital literacy camp developed by Piikani First Nation in Alberta.

Digital literacies also include the ability to conduct internet performance monitoring tests. This is a key issue given the gaps in existing coverage in indigenous regions highlighted by parties like the CRTC, which established a monitoring initiative called SamKnows, and CIRA, which developed an online portal to measure internet performance.[31] The FMCC is working with Northern communities and their regional organisations to pilot a first-mile approach to internet measurement. In northern Ontario, community members are invited to use the KO eCommunity Facebook group to share ideas on how they are using technology in their communities, and to plan together for the future of community networks. In another project, several FMCC member organisations and Cybera (Alberta's not-for-profit technology accelerator) will use CIRA's platform to establish a methodology to conduct ongoing community-based internet performance measurement monitoring.[32] These projects aim to highlight the important role that community members play in monitoring and enforcing CRTC basic service obligations.

Conclusion

2018 is an exciting year for indigenous community networks in Canada. Politically, rural and remote broadband is a growing public issue. For example, in April 2018, the House of Commons Standing Committee on Industry, Science and Technology presented its report *Broadband Connectivity in Rural Canada: Overcoming the Digital Divide*.[33] Among other points, the report provided recommendations to encourage non-traditional network operators to apply for federal funding, including, but not limited to, cooperatives, non-profits, partnerships and local governments. Similar calls for community networking support are seen in CIRA's report *The Gap Between Us: Perspectives on building a better online Canada,* which "captures the experiences, opinions and proposed solutions of 70 grassroots organizations across Canada working to make the internet better for Canadians."[34]

Another key development is the second Indigenous Connectivity Summit, which will be hosted by the Internet Society in Edmonton, Alberta and Inuvik, Northwest Territories in October 2018.[35] The summit provides a space for advocates of indigenous connectivity to discuss and debate community networks, and the policy and technical conditions that might enable their widespread growth and sustainability.

Looking ahead, the government of Canada has announced it will launch a review of the Telecommunications Act and the Broadcasting Act.[36] A seven-member panel has been appointed to the review, and public consultations have been announced for September 2018.[37] This review will update and modernise the legislative framework,

29 www.eeyou.ca/en/home

30 technologycouncil.editmy.website/talent-development/bridging-to-technology

31 https://performance.cira.ca/

32 Cybera. (2018, 11 June). Testing the Gaps in First Nations' Internet Access. https://www.cybera.ca/news-and-events/news/testing-the-gaps-in-first-nations-internet-access

33 https://www.ourcommons.ca/DocumentViewer/en/42-1/INDU/report-11/page-ToC

34 CIRA. (2018, 13 June). CIRA report: Grassroots organizations share gaps and solutions for Canada's internet. https://cira.ca/press-releases/cira-report-grassroots-organizations-share-gaps-and-solutions-canada-internet

35 https://www.internetsociety.org/events/indigenous-connectivity-summit/2018

36 Canadian Heritage. (2018, 5 June). Government of Canada launches review of Telecommunications and Broadcasting Acts. *Cision*. https://www.newswire.ca/news-releases/government-of-canada-launches-review-of-telecommunications-and-broadcasting-acts-684595661.html

37 Innovation, Science and Economic Development Canada. (2018, 28 June). Update on the Broadcasting and Telecommunications Legislative Review Panel. https://www.canada.ca/en/innovation-science-economic-development/news/2018/06/update-on-the-broadcasting-and-telecommunications-legislative-review-panel.html

support the principle of net neutrality, and address how to best promote competition and affordability for internet and mobile wireless.

These consultations provide opportunities for indigenous community networks to contribute to policies and regulations impacting on broadband infrastructure and services. Importantly, this work must recognise the range of telecommunications development approaches documented through the First Nations Innovation (FNI) research project at the University of New Brunswick, which concluded in the spring of 2018 after more than 12 years partnering with indigenous regional intermediary organisations across Canada. As a final task of the FNI research project, the partners collaborated in a book celebrating and highlighting the research and advocacy undertaken over the years. This free online publication is titled *Stories from the First Mile: Digital technologies in remote and rural Indigenous communities*.[38]

Action steps

We suggest the following steps to support the continued development and sustainability of indigenous community networks in Canada:

- Continue to advocate for adequate, affordable, accessible connectivity infrastructure and services (local and backhaul) in all regions of Canada.
- Continue to push for an enabling environment of policies and regulations that support digital self-determination through community-driven initiatives grounded in First Nations concepts such as the e-community, the "whole community" approach to broadband adoption, and "first-mile" connectivity.
- Build on the efforts of groups like the FMCC and its member organisations, CIRA and the Internet Society to ensure adequate consultation and engagement in decision-making processes.
- Ensure that community networks are supported through local and regional initiatives, including affordable access to wholesale, open access bandwidth.
- Continue tracking infrastructure deployment and monitoring internet performance, including through community-based projects.
- Support appropriate digital literacy initiatives that fit the unique contexts and desires of learners based in rural, remote, Northern and indigenous communities.
- Use the whole community model to plan and support local and regional funding and development initiatives serving remote and rural communities as an alternative to the individual/household model.
- Monitor the implementation and outcomes of recent government initiatives, such as the CRTC's basic service objective, the CRTC Broadband Fund, and ISED's Connect to Innovate programme.

38 firstmile.ca/new-book-stories-from-the-first-mile-digital-technologies-in-remote-and-rural-indigenous-communities

THE CARIBBEAN

ISOC'S TASK FORCE FOR LATIN AMERICAN AND CARIBBEAN COMMUNITY CONNECTIVITY (TFLAC3) AND CARIBBEAN COMMUNITY NETWORKS

Internet Society Chapters in St. Vincent and the Grenadines, Guyana, Dominica, Trinidad and Tobago, Puerto Rico, Barbados and Panama and ISOC Blockchain Special Interest Group
Renata Aquino Ribeiro
https://netcollective.wordpress.com

Introduction

This report offers an overview of community networks in the Caribbean and of the northern region of South America. It is based on the work of the Task Force for Latin American and Caribbean Community Connectivity (TFLAC3),[1] a forum for discussion about community network projects in the Caribbean Communications Treaty (CARICOM) region. It holds webinars and meets at internet governance events, including national and regional Internet Governance Forum (IGF) initiatives (NRIs).

The countries that are part of the forum are represented through their Internet Society (ISOC) chapter members: Barbados, Trinidad and Tobago, Dominica, Guyana, St. Vincent and the Grenadines, Panama, Puerto Rico, and Brazil's Amazonian and Northeastern Region, via an ISOC Blockchain Special Interest Group (BSIG) member.[2]

The community networks discussed in this forum are various and in different stages of implementation. Some are urban and connected to community centres, others are rural and connected to universities and NGOs. Many are still in the planning stages. In the Small Island Developing States (SIDS) countries, these networks were described as having gone through rebuilding since most of the countries were affected by hurricanes and other natural catastrophes. Together, all the members of the chapters in the countries that are part of this forum represent an average of 200 members.

The clarity in approach and thinking and the experimentation necessary to build community networks in complex scenarios can be learned from the TFLAC3 participants. The forum also highlights the importance of having allies to support community networks, as well as forums for discussions such as the NRIs,[3] LACNIC[4] and LACNOG,[5] and the Dynamic Coalition on Community Connectivity (DC3),[6] the dynamic coalition at the global IGF which deals with community networks.

Setting up TFLAC3

TFLAC3 was created at the ISOC chapters workshop at the 2017 Latin America and the Caribbean Regional Preparatory Meeting for the Internet Governance Forum (LACIGF). This was a one-day pre-event meeting where participants of several chapters could choose topics for debate, form groups, and build a collaborative project on a theme. The Caribbean representatives came together as a group because of the diverse languages they shared – with Spanish, English and French, among other languages, all spoken in the region. People from the Amazonian region, which comprises nine countries, with its multitude of languages and dialects, also became part of this group, while two Central American countries, Panama and Puerto Rico, also identified similar challenges to ours and joined the group.

From the beginning, the group showed different approaches to community networks, different stages of implementation, and dealt with gender challenges in different ways.

The participants in the task force are listed in Table 1. It is important to note that these participants

1 TFLAC3 was created at the 2017 Latin America and the Caribbean Regional Preparatory Meeting for the Internet Governance Forum (LACIGF) and was expected to continue until June 2018. However, since then, TFLAC3 initiatives have continued, and TFLAC3 has become a permanent forum, albeit less active. It can be thought of as a project that transformed into a forum.

2 The Internet Society Blockchain Special interest Group (ISOC-BSIG) is a group that unites global members of ISOC interested in blockchain technologies. Some of these members carry out projects in their regions. Two ISOC-BSIG members (one from Brazil and another from Guyana) founded TFLAC3.

3 National and Regional Internet Governance Forum Initiatives (NRIs) are national and regional Internet Governance Forums (IGFs) such as the Brazil IGF or the Caribbean IGF. They are all part of the network of the global IGF, promoted by the United Nations (UN).

4 The Latin America and Caribbean Network Information Centre (LACNIC) is the organisation responsible for internet address registries in the region.

5 The Latin America and Caribbean Network Operators Groups (LACNOG) is a community of network operators, organisations and professionals responsible for internet connectivity in the region.

6 The Dynamic Coalition on Community Connectivity (DC3) of the IGF is one of the intersessional bodies of the event which gathers annually. There is a group of members discussing community connectivity in a discussion list, and during the global IGF, a session on the topic is held and an outcome document is published.

TABLE 1.		
TFLAC3 members		
Name	**Chapter**	**Role in the forum**
Rodney Taylor	Barbados Chapter	Research
Talya Mohammed	Trinidad and Tobago Chapter	Secretary
Craig Nesty	Dominica Chapter	Timekeeper
Malisa Richards	Guyana Chapter / Blockchain SIG member	Research
Willis Williams	St. Vincent and the Grenadines Chapter	Research
Renata Aquino Ribeiro	Blockchain SIG member	Leader
José R. de la Cruz	Puerto Rico Chapter / Cybersecurity SIG member	Second leader
Russell Bean	Panama Chapter	Researcher

were representatives of their national chapters and not necessarily the leaders of those chapters. They would still involve their local colleagues in conducting research or implementing community networks, among other activities. Some had already studied community networks (Brazil and Guyana),[7] but others were still to begin looking for useful case studies that were relevant to the challenges they faced at the local level.

Starting to identify community network challenges in the Caribbean

The forum has held periodic webinars where participants have been able to share their findings and exchange ideas about the particular theme under discussion. Thematic aspects like the interaction of community networks with local operators, possible governmental support, and dealing with communities with different languages and cultural profiles were some of the most important topics. However, the dates of the LACIGF and Caribbean IGF to be held in 2018 were rapidly coming up, and both of these were to be important opportunities for the forum to meet and lobby for support for their activities, convincing possible partners to fund activities or provide hardware which could be helpful. Because of this, the project held periodic online meetings, presentations and debates on community networks ahead of these forums.

In our first preparatory meeting ahead of the Caribbean IGF, general observations about the local chapters were made, and some challenges identified:

Barbados – The definition of "community" itself impacted on how the participant planned research and identified community networks. Barbados has community centres and telecentres that are used for internet access. The community decided on the uses of these centres, and the educational programmes run there. However, their infrastructure was not autonomous and was dependent on government resources. Different communities in Barbados had different goals (regional commercial development, increasing quality of education, etc.), and determining who that community was and what their goals were created the profile of the network, according to researcher Rodney Taylor of ISOC Barbados. Barbados was one of the Caribbean nations heavily affected by floods and hurricanes in 2017. Much of the groundwork for community networks was destroyed or disrupted. There was a long hiatus before the networks could be rebuilt.

Trinidad and Tobago – The Trinidad and Tobago IGF has suggested new participants in TFLAC3 for the 2018 IGF. According to ISOC Trinidad and Tobago member Talya Mohammed, there are many community network initiatives being discussed in parallel in the region. For example, there is the proposal to form an IEEE SIG on Humanitarian Technology,[8] as well as an ISOC SIG on community networks.[9] Talya is one of the women involved in the ISOC SIG.

Dominica – This ISOC Chapter was founded in 2017 and Craig Nesty was the member who joined the community networks project as a researcher. Although the chapter has fewer than 100 members, the participation of women is proportionally slightly better off than in the Latin American chapters. In fact, most of the Caribbean chapters have a higher participation of women compared to the Latin American chapters, with some being board members.

7 Brazil and Guyana had studied community networks as part of governmental projects previously implemented in the region which distributed low-cost laptops to students and experimented with wireless networks.

8 The Institute of Electrical and Electronics Engineers (IEEE) is an organisation of professionals and enthusiasts on engineering topics. Its members form special interest groups, including the IEEE Special Interest Group on Humanitarian Technology (SIGHT). See: https://www.ieee.org

9 ISOC is a non-profit organisation which has groups themed under specific topics in Special Interest Groups. At the IGF 2017, the Community Networks SIG was announced as approved. See: cnsig.info

Some chapters have about 60% participation by women. Despite Craig having been involved in the project, one of the members who was going to take on and continue the research he had started was a woman, a university student. Unfortunately, Dominica was also hit heavily by hurricanes in 2017 and this member had to drop out from her volunteer work on the project to rebuild her home.

Guyana – This was the chapter most active throughout the initial duration of the project (from July 2017 to June 2018, but the project continues as a permanent forum, albeit less active and more informal) and helped spin the TFLAC3 into other projects such as a Chapterthon[10] contribution and putting the theme on the map in the first Guyana IGF. Malisa Richards, the Guyana ISOC member who participated in the forum, is an educator who became involved in the One Laptop per Child (OLPC) programme in her school.[11] She became a true believer in the importance of connectivity for regional development. Malisa was among the founders of the ISOC Guyana Chapter[12] and the Guyana IGF, and is also a member of the ISOC Blockchain SIG.

St. Vincent and the Grenadines – This chapter, formed in 2017, is also among those that have a close to equal gender distribution among their members. Willis Williams joined the TFLAC3 project in 2017, but the project was also helped greatly by Roxanne John, who is a researcher collaborating with the St. Vincent and the Grenadines ISOC Chapter, and also participated in the first IGF in the country in 2018 where the theme of community networks became an integral part of the debates. Although it is a Caribbean SIDS, St. Vincent and the Grenadines was not among the islands most hit by hurricanes and other natural disasters in 2017. This helped greatly in advancing the plans of building access infrastructure, including community networks.

Puerto Rico – ISOC chapter member José de la Cruz joined the TFLAC3 forum during the LACIGF 2017. Soon after, Puerto Rico was devastated by natural disasters which impacted specifically on technology and infrastructure. The ICANN61 meeting[13] was a few months away from happening in San Juan, the commonwealth's capital. Fortunately, despite the federal government not being helpful, the technology community gathered in several volunteer efforts to help rebuild the island. The planned meeting happened and there were several occasions when ISOC Puerto Rico could make its work known. At a civil society organisations roundtable the president of ISOC Puerto Rico, Eduardo Díaz, spoke about their efforts in working on several access initiatives, including community networks.

Panama – The LACIGF 2017 happened in Panama City, and is where TFLAC3 was born. Despite not being an island, Panama, like many other Central America countries, faces challenges similar to the SIDS and has treaties with the nearby islands. One member of ISOC Panama was particularly engaged in this project: Russell Bean. As a professional in a telecommunications company, Russell was also a representative of the technical community. He gave us hints and insights into implementing community networks. Panama also hosted LACNIC29,[14] where ISOC St. Vincent and the Grenadines was present, and ICANN62,[15] where many ISOC members in these countries met again.

The many indigenous communities and their different dialects in the Caribbean – as well as in the Amazon region – add to a composition of a varied landscape where community networks need to be mobile and can always be rebuilt whether due to migration or other reasons such as natural disasters. The relationship between mobility and migration is a dynamic typical of the Caribbean. Tribes may move to the inside of the forests if lands are devastated and may merge with other tribes. So adaptability and working with Wi-Fi routers and mobiles are key.

Sometimes the originators of these networks were specific people, a technology professor, a digital rights lawyer and even a public school teacher. The infrastructure might be funded from someone's own pocket, or through a collaborative effort at a specific instant in time. In the Amazon region, the common use of transportation by boat to forest areas made routing Wi-Fi via mobiles on the river the easiest and quickest form of communication.

Communication apps like WhatsApp were used to offer educational courses, like maths or science. The largest cities could host test periods at the end of a semester or teacher conferences, but the actual teaching, the communication between the communities, was all done via smartphones.

10 The Chapterthon project "Be a hero of your own community" introduced debates at schools on how community networks and other independent and decentralised technologies, such as blockchain, can help shape the society's future. The Chapterthon produced an online video in Portuguese: https://youtu.be/FuO5thVEIoQ

11 The OLPC programme ran in Guyana until 2016. However, many schools are still using the laptops.

12 Malisa Richards published a blog post on the founding of ISOC Guyana: www.circleid.com/posts/20171029_internet_society_guyana_chapter_officially_launched

13 https://meetings.icann.org/en/sanjuan61

14 www.lacnic.net/2386/44/evento/welcome-to-lacnic29

15 https://meetings.icann.org/en/panamacity62

Another innovative characteristic was the participation of the Blockchain SIG, which explores the impact of blockchain technologies[16] on society. The idea of identity,[17] currency and even whole governmental systems being reinvented is what brought the BSIG close to the community networks theme. Blockchain can be an ally technology, especially in underserved and less populated regions, which is the case of the Caribbean and the Amazon region.

Conclusions

The idea of community networks in the Caribbean region is still new, but the practice of sharing and enabling access for specific groups and excluded communities has long existed. What each ISOC country chapter has found is that the reasons for them implementing a community network have been similar: the need for access to knowledge/education, local communication, e-commerce, and access to health and governmental services.

Reinvention and readaptation are key in areas where major natural forces (whether hurricanes, tsunamis or climate change in deep rainforests) impact the local economy and governmental infrastructure. The impromptu build-up of community networks in a boat on the Amazon River or in a family house in the Caribbean is a "pop-up" connectivity choice, one which can be changed or abandoned if a new implementation is needed or cheaper infrastructure is found.

A lack of telecommunications infrastructure is common in many countries in the region, especially in underserved and indigenous areas. However, there is a collaborative sense of engagement with telecommunications operators. For example, Digicel, one of the largest operators in the Caribbean, joined us in a panel on community networks in the Caribbean IGF. This makes community networks not necessarily opposite to the goals of the local telecommunications service providers. In fact, community networks sometimes break ground in underserved regions and after some time the telecommunications services arrive, usually one operator at a time.

With regards to the TFLAC3 project, the group always resorts to the project webinars organised by ISOC Latin America and Caribbean (ISOC LAC)[18] and the previously recorded discussions from our online meetings.[19] These are key resources that help continue the conversation, even if informally, and help the Caribbean professionals in community networks communicate.

Action steps

We see a few scenarios for community networks in the Caribbean in the future:

- *Greater collaboration between stakeholders* – While telco operators are usually absent from the region where the networks exist, the community networks "break ground" and create a new consumer market. In the end, partnerships with telcos are possible when an operator catches up with the new demands for services. Perhaps it would be useful to have a dialogue among all stakeholders to map out areas where networks should exist and for strategic sharing of resources.
- *Learning from the other side of the world* – SIDS have commonalities which change the framework of the continent they are in or even the economic distribution of income, being the touristic spots graced with much more infrastructure than inland areas. During our project in 2017 and 2018 we learned greatly from exchanges with the Asia Pacific Network Information Centre (APNIC),[20] an organisation that conducted webinars on rebuilding networks after natural disasters. Asia Pacific islands are also hit by hurricanes and other natural phenomena. The rainforest in Southeast Asia has different characteristics but similar challenges to the Amazon region. Different geopolitical arrangements and communications are needed for the future. Civil society should engage in a multistakeholder dialogue to demonstrate that underserved regions like the Caribbean need quality access infrastructure, not only where the major tourist hotels are, but for all the people in the region.
- *More engagement:* We urgently need more actors engaged in the area of community networks in the Caribbean and solutions which are more long-term, although adaptable. This presents itself as a challenge because the populations we deal with are small in number and in largely rural or forest areas. By building community networks, we are at the same time building communities and pathways for the future of the region.

16 Blockchain is a decentralised technology platform which can serve for digital currency, digital certificates and even an alternative to the domain name system (DNS), called the Ethereum Name System, an innovation by one of many blockchain companies.

17 Blockchain can be used for digital certificates and the possibility of an online notary. Some projects have already started experimenting with a blockchain ID for refugees, which can be recognised across borders.

18 See, for example: https://nancyquiros.wixsite.com/misitio-2/inicio/community-networks-the-situation-in-the-caribbean

19 https://docs.google.com/document/d/1TLnrDpmv2umzSzYLR8xo3QCul1v27MaOsssl7sgQ90g/edit?usp=sharing

20 APNIC is the organisation responsible for local domain name registries and fosters education and technology programmes in the region. https://www.apnic.net

CATALONIA[1]

GUIFI.NET: SCALING UP A COMMUNITY NETWORK

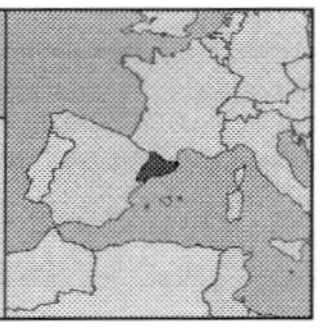

guifi.net community
Roger Baig, Leandro Navarro, Ramon Roca
and Felix Freitag
https://guifi.net

Introduction

Citizen-driven access initiatives such as community networks are often considered as the last and least "serious" option to bring connectivity to regions or sectors of the population unattended by the "serious" options – that is, private sector and publicly funded deployments (which in many cases are carried out by the same companies doing the market-driven deployments).

guifi.net is a community network with tens of thousands of working nodes, and hundreds of volunteers, professionals and public administrations involved. It proves that community networks not only can deliver "serious" services to unattended areas (e.g. fibre to rural areas), but that this can be done in a very efficient way, converting almost all investment in deployment into profitable deployments, socially and economically. This in turn refutes another globally accepted assumption imposed by "the establishment": that some regions as well as some sectors of the population will always need to be subsidised.

guifi.net is a bottom-up, citizenship-driven technological, social and economic project with the objective of creating a free, open and neutral telecommunications network based on a commons model. The development of this common-pool infrastructure eases the access to quality, fair-priced telecommunications in general and broadband internet connections in particular, for everybody. Moreover, it generates a model for collaborative economic activity based on proximity and sustainability.[2]

guifi.net started in 2004 as a telecommunications technological project in the county of Osona (Catalonia) to solve the broadband internet access difficulties in rural areas, given the lack of traditional operators to provide services there. By means of radio links built with commodity Wi-Fi routers, the neighbours deployed their own network to interconnect different locations (the so-called nodes) such as houses, offices, farms, public buildings, etc. to be able to access telecommunications and the internet wherever they needed to. A foundation was created in 2008 by the guifi.net community to give a legal identity to the guifi.net project.[3]

The guifi.net community has five main stakeholder groups according to their roles in the ecosystem and their motivations for participating in it: the volunteers, the governing bodies, the professionals, the customers, and the public administrations. These are non-profit, for-profit, and public interest groups.

As of August 2018, guifi.net accounted for more than 35,000 operating nodes. The majority of these nodes are located in Catalonia and the Valencian Community in Spain, but the network is growing in other parts of the world. The network is self-organised and operated by the users, using unlicensed wireless links and open optical fibre links.

This report shares the key factors that have enabled the scalability and the positive socioeconomic impact of guifi.net. It is based on our experience of over a decade of involvement in guifi.net, each of us in several different roles (volunteers, users, scholars, professionals). We hope this helps to establish the next steps for guifi.net, as well as to provide input to other initiatives interested in scaling up their efforts.

Policy, economic and political background

guifi.net has developed under the regulatory framework of the European Union (EU),[4] that is, a legislative body that has created a set of European guidelines that member states must fulfil, complemented by regional and local rules. As a result of the European policy, the telecommunications sector

1 Although the country names used in Global Information Society Watch are normally based on the United Nations' list of member states (https://en.wikipedia.org/wiki/List_of_sovereign_states#List_of_states), APC has adopted the decision to use the name "Catalonia" for the guifi.net report. The use of Catalonia as a country name best represents the authors and the content of the report and is consistent with values that are important to APC: community self-determination and freedom of expression.

2 https://guifi.net/en/what_is_guifinet

3 https://fundacio.guifi.net/en_US/page/aboutus

4 https://ec.europa.eu/digital-single-market/en/national-regulatory-authorities-member-states

across the region is fully liberalised, and everybody, including those running citizen initiatives, is entitled to participate in the delivery of electronic communication services.

Despite this, frequently at the state, regional and sometimes local level, guifi.net has had to deal with actors defending very particular vested interests. This has translated into lost opportunities that have damaged the project and the citizens too.

The guifi.net response to access at the local level has been about a positive and creative attitude and persistent work. The achievements include a proposal for a municipal ordinance, based on the legal and regulatory framework in Catalonia, to share telecommunications infrastructure. This ordinance has already been adopted by several municipalities and a highly successful county-level practice aimed at deploying optical fibre to all households in Garrotxa (see below).

From the socioeconomic viewpoint, guifi.net carries out its activities in a European country (where the average population has their basic needs covered) under the principles of solidarity (nobody can be excluded for social or economic reasons) and social economy (its activities are non-speculative and non-extractive). Several pre-existing community networks have merged with guifi.net and currently operate jointly.

From the tens to the tens of thousands

What are the key factors that led a local initiative to scale up to tens of thousands of nodes?

Precise definition of objectives and scope and full commitment to them

There are many initiatives and objectives that can be seen as fully aligned with the guifi.net project (e.g. those that defend freedom of speech, or promote local content). As a result, it makes sense to integrate or involve these initiatives in guifi.net. Nevertheless, a clear definition of the goals of a project, the precise delimitation of its reach and the strict observance of these goals and limits, at least (i) helps to focus the efforts, as it is clear what the project is about and is not about; (ii) broadens the community, as not necessarily everybody must share the same opinion on all issues – they must just share the common objectives; and (iii) increases certainty and reduces the likelihood of misunderstanding and conflict.

In guifi.net the objective is to build and operate a computer network that is fully inclusive (open) in terms of access and use as well as in terms of construction, operation and governance. The guifi.net leitmotif is *fent xarxa oberta, lliure i neutral* (doing an open, free and neutral network). Whichever solution is adopted or action taken at any time must be entirely compatible with these values and promote them.

The example of local content is illustrative. Obviously, a sympathy for open and local content and services can be presumed among most of the participants. Nonetheless, the rule on openness is imposed only on the content and services that are strictly necessary to run the project. Access to any other content or service is left to the criteria of the providers. (Strictly speaking, the only reason to impose a rule on openness on content and services is because openness is the only option possible when implementing an open network infrastructure, and not because of any personal choice).

The network infrastructure as an open common-pool resource

The commons is a resource management principle by which a resource is shared within a community.[5] In guifi.net the network infrastructure is held as an open common-pool resource (CPR). CPRs typically consist of a core resource which provides a limited quantity of extractable fringe units. In our case, the core resource is the network, which is created and maintained by the network segments that the participants deploy to reach the network or to improve it, and the fringe unit is the connectivity they obtain.[6] The participants can keep the ownership of the fraction of the assets they have contributed or they can donate it to the project (among other options).

The governance system

The management of CPRs is challenging because usually they are made of rivalrous and limited resources, so they are congestion prone. This is the case of computer networks, as connectivity is subtractable – it gets used – and the capacity of the links is limited.[7] The challenges are even greater when the resources are non-excludable, as in the case of guifi.net, where the non-excludability is intentionally imposed. Non-excludability in this context simply means that people cannot be excluded arbitrarily from an open network.

After in-depth studies of several CPRs, Elinor Ostrom identified a set of principles for their successful

---

5 Frischmann, B. M. (2012). *Infrastructure: The Social Value of Shared Resources*. Oxford University Press.

6 Baig, R., Roca, R., Freitag, F., & Navarro, L. (2015). guifi.net, a crowdsourced network infrastructure held in common. *Computer Networks, 90*, 150-165.

7 Well-managed fibre links have virtually unlimited capacity. The governing rules must be adapted to this circumstance.

management.[8] Ostrom's works remained unknown to the guifi.net community for quite some time. Nonetheless, the evolution of most of the concepts and governance tools developed by the guifi.net community perfectly match Ostrom's findings.

The key components (the so-called "systems") of the self-developed governance tools of the guifi.net community are:

- An investment declaration system that allows the recognition of the investments made by the participants.
- A resource monitoring system that allows the accounting of resource consumption.
- A cost compensation system that balances contribution[9] and consumption.
- A conflict resolution system with defined procedures (conciliation, mediation, arbitration).
- A gradual sanctioning system as the last resource to settle disputes. It includes temporary or permanent expulsion.

In the process of its development, the community has developed a comprehensive body of normative agreements whose components can be classified as (i) ground rules, (ii) contractual agreements, (iii) regulations, and (iv) good practices, depending on their relevance and the aspects regulated, with the licence[10] for participation in the commons being the fundamental rule. These agreements establish the objectives and scope of the project and the rights and duties of the participants, and set the basis for the development of the rest of the governance tools. Their acceptance is mandatory for participation.

Legal certainty for the socioeconomic ecosystem

Undoubtedly, the fact that most of the components of the body of normative agreements are written documents has proven to be critical in establishing the legal certainty necessary to build a competitive general-purpose infrastructure that has enabled the development of a flourishing socioeconomic ecosystem.

The guifi.net Foundation is a non-profit organisation that gives a legal entity to the project and is responsible for its core governance activities, which include the maintenance and the development of the critical components of the body of normative agreements, but also their enforcement.[11] However, the network allows for-profit activities (see below).

The different stakeholders can be classified based on their unique and non-transferable roles. Service providers sell their services (e.g. internet connectivity) over the network to the customers who pay for them according to service contracts. The volunteers are do-it-yourself non-profit participants that may organise around formal or informal non-profit organisations, associations or groups.

Public administrations must participate because they are responsible for regulating the interactions between the network deployment and the public interest (e.g. use of the public domain for network infrastructure).[12] In addition, they can also play more active roles like promotion of the project or participation in it, given the benefits to the public.

The social and economic relations within the guifi.net community are driven by the principles of non-speculation and non-discrimination, which means that prices are cost oriented and costs are shared according to the resources consumed, making sure that nobody is excluded for economic reasons. Cooperation among service providers is also well established.

Sustainability and growth

Enabling for-profit activity has two direct positive effects on the CPR. On one hand, it brings in income that makes the ecosystem economically sustainable and, on the other hand, it encourages the maintenance and upgrading of the infrastructure by professionals, as their income depends on it.[13]

guifi.net has developed a sophisticated system to ensure the sustainability and growth of the network infrastructure. The system is rooted in the obligation of certain participants, including the service providers and those who *make significant use of the network*, to participate in the economic compensation system and to fulfil the obligations resulting from it.

The economic compensation system[14] establishes who is responsible for paying for what, based on the information from traffic monitoring,

8 Ostrom, E. (1990). *Governing the Commons: The Evolution of Institutions for Collective Action*. Cambridge University Press.

9 Contributions can be in money, labour or hardware.

10 English translation: https://guifi.net/en/FONNC (outdated).

11 Its roles also include offering technical advice, creating buffer funds, providing business bailouts, etc.

12 Also called public land disposition.

13 Equivalent to "traditional" commons, such as irrigation systems. Farmers share the water and the agricultural products can be directed to self-consumption or for sale in the market.

14 Baig, R., Dalmau, L., Roca, R., Navarro, L., Freitag, F., & Sathiaseelan, A. (2016). Making community networks economically sustainable: The guifi.net experience. *Proceedings of the 2016 Workshop on Global Access to the Internet for All (GAIA 16)*. ACM. http://dsg.ac.upc.edu/sites/default/files/dsg/acm-sigcomm-gaia-guifi-econ.pdf

the investment declaration system, and the rules of participation. The rules are applied as clearing houses for each scope with its own ledger for recording and totalling economic transactions from each participant, and each participant represented. The scopes for clearing houses are defined per services or segments of infrastructure and may vary over time.[15] The clearing houses can define new rules as long as they are compatible with the rules of higher rank. The clearing houses periodically apply the calculation rules of capital expenditure (CAPEX) and operational expenses (OPEX) in the scope of the network segment or service of concern. The resulting amounts might not be settled immediately but accounted in an advanced ledger of deposits with the Foundation as intermediary. Clearing houses also have rules to decide on future investments.

XAFOGAR, a concrete success story

XAFOGAR[16] is an ongoing guifi.net project aimed at deploying fibre in all of the 21 municipalities of the county of Garrotxa in Spain. All of these did not have fibre access, except the capital (Olot). It is led by a public development agency in the county and supported by the guifi.net Foundation, four local service providers and many local businesses and small investors.

What this experience highlights is the extraordinary catalytic power of the honest involvement of a public administration. With EUR 1.5 million already invested, out of the total estimated budget of EUR 10 million, the development agency is providing irreplaceable political support and has taken the responsibility of the daily management of the project. The fact that for each euro invested by a public administration, another 12.7 have been provided by private initiatives (direct beneficiaries, service providers, investors), underlines the positive impact of these actions on the confidence of investors and the population in general.

Conclusions

guifi.net offers irrefutable evidence that large-scale competitive network infrastructures can be built and operated as an open common-pool resource. Cheaper and better quality services, higher inclusiveness, fairer salaries, local empowerment, technological sovereignty, extraordinary capacity to raise local funding, and investment coordination in a single shared infrastructure are some of the numerous benefits of this model. To realise its vision, guifi.net has developed a comprehensive governance tool set which includes mechanisms for dispute resolution, investment recognition, cost sharing, etc., and has established a non-profit institution responsible for their improvement and enforcement.

Despite this, guifi.net is still working on improving its socioeconomic ecosystem.

Many lessons can be extracted from the guifi.net case and many of the tools have already been reused in other contexts. The conception of the network infrastructure as a common-pool resource is fully in accordance with the non-speculation principle, and the relative ease with which the network can be extended and the flexibility of the uses of its capacity allow an unprecedented level of inclusiveness. To allow and promote the utilisation of the network by local commercial service providers to deliver their services has enabled a thriving economic activity.

The inclusion of measures to ensure the necessary reinvestment of a fraction of the benefits as core components of the governance has proven to be effective to prevent the depletion of the CPR and towards the redistribution of wealth. The honest cooperation with public administrations has provided evidence that the public sector has many resources to promote the open CPR model beyond providing funding.

From the theoretical perspective, the guifi.net model seems general and flexible enough to be applicable worldwide. Nevertheless, the practical implementation requires significant effort, because most of the current solutions must be redesigned to fit different contexts. To avoid bias and ensure effective implementation, these efforts must be overseen by an international organisation.

Action steps

What does guifi.net teach us?

Taking guifi.net as a reference, any community that wants to scale up in a sustainable manner should take the following steps (in this order):

- Precisely define the scope and the aim of the commons.
- Develop a governance system (as defined by Ostrom).
- Develop a resilient and inclusive ecosystem of services around the commons (the network) with special emphasis on making it inclusive and empowering the local stakeholders.

15 An area such as a town or region sharing an infrastructure, or a resource such as a set of internet uplinks.

16 www.xafogar.cat

What does it need to do to encourage actors to support it?

- Show theoretically but also by example that its proposal is serious. To do this, commitment on the above points and an action plan to achieve them must be proven. A useful response from actors interested in supporting guifi.net could come in the form of funding, technical help, making infrastructure available, etc. Good actors to engage include public administrations, service providers and local businesses, in addition to the direct beneficiaries (the communities of users).
- On the specific case of funding, we propose micro-funding strategies; that is, many iterations of small funding made available based on immediate capacities and needs.

What is the most urgent thing it needs to do right now?

- Consolidate the documentation of good practices incorporating the accumulated experience, not only on technical issues but also socioeconomic and governance issues.
- Improve communication to lower the barrier of adoption and boost collective knowledge transfer and the sharing of tools.
- Develop tools (mostly software based) to automatise the operation and assist the implementation of open governance, including using last generation technologies.
- Ensure that the community network has effective and appropriate support for developing in the real world. A key component of this support is the regulatory framework. It must effectively prevent conflicts of interest with privative models and the incumbents.

CHINA

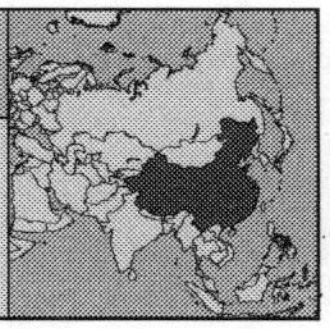

WHERE COMMUNITY NETWORKS MAY NOT GO: E-COMMERCE CLAIMS TO BRIDGE THE RURAL-URBAN GAP IN CHINA

Hudson Lockett

Introduction

Much international coverage of China has focused on policy and practices related to what happens online rather than the infrastructure that makes those connections possible. But since the mid-1990s, when the country's central government began linking up the data networks of dozens of major cities, China has brought more people online than any other country on earth.[1]

Yet as is the case in many developing countries there remains a "digital divide" between the wealthy and impoverished in China, the latter of whom are found primarily in the countryside.[2] And despite Beijing's nominal command over the state-owned companies that dominate the domestic telecoms sector, they have been slow to build out the final miles of telecommunications infrastructure needed to bring the country's most remote and impoverished citizens into the digital age.

In some countries such underserved populations might have their needs met through setting up community-owned networks, which can offer cheaper entry level prices and more transparent pricing than private companies, as was found in a report on 40 community-owned internet service providers in the United States published by Harvard University's Berkman Klein Center for Internet & Society.[3]

However, China's governance model forbids significant organisation outside the purview of the state. Thus state telecoms operators and state-supported e-commerce companies have co-opted (or fenced off) niches that might otherwise be filled by community networks. This begs the question of whether they then also provide sufficient substitutes for the access and advantages of such networks.

Policy and political background

China's telecommunications services and facilities are ultimately controlled by four state-owned companies – China Mobile, China Unicom, China Telecom and the much smaller China Broadcasting Network – on sovereignty and national security grounds.

In 2015 the State Council, headed by Premier Li Keqiang, said China would invest an estimated RMB 140 billion (USD 20.9 billion) by 2020 to expand broadband internet to 98% of the country's 500,000 administrative villages.[4] But things have not been that easy.

Reforms in the late 1990s broke China Telecom's effective monopoly under which rural service expansion was subsidised by more profitable urban branches.[5]

Now state firms supposedly tasked with rural build-out also compete with each other and are ostensibly subject to targets for returns and growth rates set by China's state-owned enterprises regulator. These contradictions in policy contribute to a persistent gap in internet penetration between cities and the countryside, as developed urban centres with existing network infrastructure offer cheaper opportunities for profit.

Although China's national internet penetration rate was 55.8% in 2017, the penetration in urban areas (71%) stood at roughly double the rural rate (35.4%). The nearly 36 percentage point gap between the two was more than double that of 14 percentage points recorded in 2005. And of the 611 million people in China who did not use the internet, 62.4% were rural residents.

1 Harwit, E. (2004). Spreading Telecommunications to Developing Areas in China: Telephones, the Internet and the Digital Divide. *The China Quarterly, 180*, 1010-1030. https://www.cambridge.org/core/journals/china-quarterly/article/spreading-telecommunications-to-developing-areas-in-china-telephones-the-internet-and-the-digital-divide/2CE1A395D9023512C10487BCEC23D16B

2 Ibid.

3 Talbot, D., Hessekiel, K, & Kehl, D. (2017). *Community-Owned Fiber Networks: Value Leaders in America.* Berkman Klein Center for Internet & Society. https://cyber.harvard.edu/publications/2018/01/communityfiber

4 Reuters. (2015, 14 October). China to invest $22 billion to extend broadband to 50,000 villages. *Reuters*. https://www.reuters.com/article/us-china-telecoms/china-to-invest-22-billion-to-extend-broadband-to-50000-villages-idUSKCN0S819L20151014

5 Peng, S. (2004). Universal Telecommunications Service in China: Trade Liberalization, Subsidy, and Technology in the Making of Information Equality in the Broadband Era. *Asian-Pacific Law & Policy Journal, 4*, 21-49.

FIGURE 1.

China internet penetration 2005-2017

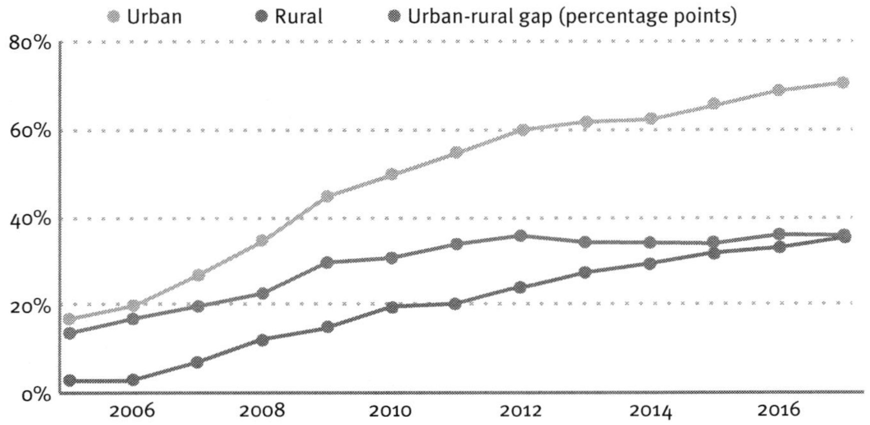

Source: China Internet Network Information Center

This would appear to highlight gaps more readily filled by community-owned networks, but multiple factors in China work against this model, including an opaque pricing mechanism for fees charged by major operators for competitors' use of their networks.[6] The central government in any case is now focused on boosting efficiency rather than introducing new competitors. In 2015 the major carriers were forced to spin off their network infrastructure assets into China Tower Corp to avoid additional reduplication of infrastructure that had resulted from competition.[7] Insofar as this research could find there are currently no national-level regulations specifically pertaining to community networks.

The development of "rural e-commerce"

With widespread creation of community-owned networks unlikely thanks to government policy, e-commerce companies have sought to cast themselves as the internet-enabled saviours of China's rural hinterlands. The most high-profile of efforts by these companies has been Alibaba's Rural Taobao programme, officially launched in 2014 to improve market penetration outside of cities for Taobao.com, its online marketplace comparable to Amazon.com.[8] (The Chinese for Taobao 淘宝 can be translated roughly as "bargain hunting".)

Alibaba has framed this drive as a programme that is equally aimed at bettering the lives of China's rural residents, and to great effect. The programme has garnered much positive attention from international press, with headlines such as "Once poverty-stricken, China's 'Taobao villages' have found a lifeline making trinkets for the internet".[9]

Chinese media descriptions of Rural Taobao frequently echo the following excerpt from a story published by Xinhua, China's state news agency:

> Rural Taobao is an ambitious effort by Alibaba to turn China's hundreds of millions of rural residents into online shoppers and sellers. It also underscores the potential of e-commerce to fuel economic activity and eliminate poverty in the country's poorer, largely agrarian regions.[10]

As of December 2017 the company said it had invested about US 1.6 billion to establish 30,000 service centres across China to enable faster deliveries and allow villagers with little or no access to

6 Wu, I. (2015, 20 September). Electricity and Telecom Regulation: China in Context. *Competition Policy International*. https://www.competitionpolicyinternational.com/electricity-and-telecom-regulation-china-in-context

7 Reuters. (2014, 11 July). China's three state-owned carriers form telecoms tower firm. *Reuters*. https://www.reuters.com/article/us-china-telecomunications/chinas-three-state-owned-carriers-form-telecoms-tower-firm-idUSKBN0FG11X20140711

8 Soo, Z. (2018, 19 April). Alibaba invests 4.5 billion yuan in online services firm to boost rural strategy in China. *South China Morning Post*. www.scmp.com/tech/china-tech/article/2142380/alibaba-invests-45-billion-yuan-online-services-firm-boost-rural

9 Freedman, J. (2018, 21 June). Once poverty-stricken, China's "Taobao villages" have found a lifeline making trinkets for the internet. *Quartz*. https://qz.com/899922/once-poverty-stricken-chinas-taobao-villages-have-found-a-lifeline-making-trinkets-for-the-internet

10 Xinhua. (2017, 11 July). UN, Alibaba cooperate to improve China's rural e-commerce infrastructure. *Xinhua*. www.xinhuanet.com/english/2017-07/11/c_136435314.htm

the internet to use the Taobao shopping platform, and had partnered with local governments to provide easier access to computers and training on how to use the internet and online payments.

Alibaba has made a special effort to depict Rural Taobao as also connecting rural producers of local goods with potential buyers in China's more affluent and populous cities – a substitute for the kind of market access that community-owned networks can provide as they link underserved communities to the broader internet within a country and beyond its borders.

Videos produced by Alibaba showcase success stories like that of one turquoise seller from a village in Hubei province. "Since starting the Taobao shop, instead of so few youths staying in their hometown now young people don't go out of town for work," she says in one video.[11] That development aligns neatly with the central government's goal of slowing migration from the countryside to major cities, which leaves behind vulnerable populations of the elderly and children.

The link between Alibaba's programme and the State Council's standing call for a more networked countryside got an official boost in 2015 when Zhang Gaoli, then vice-premier under Li, visited a "Taobao village", which Alibaba defines using the following three criteria:

1. Residents got started in e-commerce spontaneously primarily with the use of Taobao.
2. Total annual e-commerce transaction volume is at least RMB 10 million (USD 1.5 million).
3. At least 10% of village households actively engage in e-commerce or at least 100 active online shops have been opened by villagers.[12]

According to local media coverage of his visit, Zhang "praised again and again" the way villagers' incomes had risen and their quality of life increased thanks to their use of Taobao.[13]

Little wonder that in 2016, the State Council Leading Group for Poverty Alleviation and Development and a host of top government bodies jointly released guidelines that called for construction of 60,000 "e-commerce poverty relief stations" in about half of China's impoverished villages, as well as a quadrupling of e-commerce sales for villages in impoverished rural counties by 2020.[14] The year 2020 is also politically significant, as it marks the Chinese Communist Party's self-imposed deadline for total eradication of rural poverty nationwide.[15]

Also in 2016 the World Bank and Alibaba signed an agreement for "cooperation to research the Taobao village phenomenon", which the bank said was intended to produce knowledge with which "we hope to be able to help villages in poor and remote areas in China to become Taobao villages as well, and help lift themselves out of poverty."[16] Even the United Nations provided its own endorsement the following year when it allocated USD 200 million for construction of rural e-commerce infrastructure in the provinces of Sichuan, Shaanxi and Ningxia.[17]

Yet as of 2017, Alibaba reported that only 2,118 villages had enough shops and sales turnover to qualify as "Taobao villages" – about one for every 14 service centres, suggesting a less-rosy reality on the ground for most rural entrepreneurs seeking better business prospects through countryside programmes provided by e-commerce companies.[18]

The telling story of Yu Xueyi

The highest-profile critique to date of Alibaba's rural expansion programme came from Yu Xueyi, who in 2017 quit his job with Rural Taobao after two years operating a logistics outlet for Alibaba in his home village of Yongning, located in one of the poorest counties of the inland province of Anhui.

In an essay for the state-owned, English-language news website Sixth Tone, which targets foreign readers with stories that focus on personal narratives, Yu wrote that he had signed on to be a local partner for Rural Taobao in 2015 when the local government announced it had agreed to work

11 "Villager's Taobao Store Turns Turquoise into Gold", available at: https://www.youtube.com/watch?v=oFcLD-T5HmA

12 Alizila. (2016, 27 January). An introduction to Taobao Villages. www.alizila.com/an-introduction-to-taobao-villages; note that Alibaba does not give a clear definition of what qualifies as an "active" online shop.

13 Hao, G. (2015, 25 October). Zhang Gaoli visits Taobao Village in Cao County, reminds shopowners to take care of themselves. *Dazhongwang Shandong*. sd.dzwww.com/sdnews/201510/t20151025_13231521.htm

14 State Council Leading Group Office of Poverty Alleviation and Development. (2016, 23 November). Guidance on Promoting E-commerce for Targeted Poverty Alleviation. www.cpad.gov.cn/art/2016/11/23/art_624_55721.html

15 Hernández, J. (2017, 31 October). Xi Jinping Vows No Poverty in China by 2020. That Could Be Hard. *The New York Times*. https://www.nytimes.com/2017/10/31/world/asia/xi-jinping-poverty-china.html

16 Hofman, B. (2016, 29 October). The Taobao Villages as an Instrument for Poverty Reduction and Shared Prosperity. *World Bank*. https://www.worldbank.org/en/news/speech/2016/10/29/the-taobao-villages-as-an-instrument-for-poverty-reduction-and-shared-prosperity

17 Xinhua. (2017, 11 July). UN, Alibaba cooperate to improve China's rural e-commerce infrastructure. *Xinhua*. www.xinhuanet.com/english/2017-07/11/c_136435314.htm

18 Aliresearch. (2017). *China Taobao Village Research Report*. http://i.aliresearch.com/img/20171211/20171211101359.pdf

with Alibaba in hopes of boosting consumer spending and relieving local poverty.[19]

The government provided a rent-free space in the middle of town and Alibaba donated a computer, furniture and flat-screen TV to Yu, who was tasked with delivering packages to difficult-to-reach rural homes. But the programme started to lose its appeal when Yu saw it had failed to follow through on its promise of connecting rural sellers with China's prosperous cities.

"Even though the platform would advertise examples of villagers making stacks of cash by selling tea, oranges, or herbal medicines to the cities, in reality, most of these examples were mere publicity stunts concocted by local governments with the tacit approval of Rural Taobao," Yu wrote.

Yu's account also tracks with a 2017 review of academic studies published in the quarterly journal *China Perspectives*, which found that the typical experience of rural e-commerce outfits was far more cutthroat.[20] Studies cited in the review found rural online retailers in China were often rapidly emulated, since most were unable to offer products that were not easily duplicated – circumstances that often led to rapidly diminishing margins and which undermined those businesses' sustainability.

Yu also criticised the segregation of Rural Taobao from other branches of Alibaba's e-commerce network, pointing out that if rural sellers wanted to open an account with Tmall.com – the higher-profile consumer goods platform targeting China's growing middle class – they had to pay a RMB 200,000 (USD 29,830) deposit that was well beyond the means of most local shop owners.

The experience of Yu's village may well be closer to the norm than the success enjoyed by the few thousand official Taobao villages recognised in 2017. Moreover, the utility of Taobao villages in addressing the wealth gap between rural and urban China is questionable.

A study based on a list of such villages from 2013 to 2015 found that over 90% were located in more-developed eastern coastal provinces.[21] A quick calculation based on figures published by Alibaba in 2017 show those provinces' share of Taobao villages at 96%.

These same coastal provinces have been the biggest beneficiaries of China's economic development and the accompanying infrastructure build-out seen in recent decades and, perhaps unsurprisingly, enjoyed an average internet penetration rate of more than 60% at last count.

That is comfortably above the national rate, whereas the internet penetration rate for Yu's home province of Anhui stood at about 44%, ranked 26 out of 31 administrative districts.[22] And only 33 Taobao villages were located in counties recognised by the central government as priority areas for poverty alleviation and development, of which there are 933 in total.

These figures suggest that the systems being built by private companies to enable greater penetration of e-commerce into the countryside are unlikely to provide the kind of poverty panacea sought by Beijing. Yu, for his part, ultimately quit working for Rural Taobao and returned to his previous trade: rabbit farming.

Conclusion

China's transition from a centrally planned economy to a more market-oriented system in which the private sector is allowed to flourish in certain sectors has enabled hundreds of millions of people to lift themselves out of poverty. However, as recognised both within China and by outside observers, more recent years have provided ample evidence that the low-hanging fruit of economic reforms have long since been plucked.

Indeed, when the World Bank announced its cooperation agreement with Alibaba in 2016 it acknowledged that "many of the world's people are still off line and can't participate in the digital economy in any meaningful way – for China this number is still about half of the population." Its first and most emphatic recommendation for solving that problem was "making the internet universally accessible and affordable."[23]

Ultimately Beijing's embrace of e-commerce as a cure for what ails rural China ignores the role that infrastructure played in allowing urban e-commerce to flourish. Rather than deal with the lack of telecommunications infrastructure in rural areas that has resulted from government policies that privilege financial performance over public utility at state-owned providers, Beijing has opted to allow established e-commerce giants like Alibaba and

19 Yu, X. (2017, 9 March). Why I Quit Alibaba's Big Push Toward Countryside Commerce. *Sixth Tone*. https://www.sixthtone.com/news/2031/why-i-quit-alibabas-big-push-toward-countryside-commerce

20 Li, A. (2017). E-commerce and Taobao Villages: A Promise for China's Rural Development? *China Perspectives, 2017/3*. www.cefc.com.hk/issue/china-perspectives-20173

21 Xu, Z., Wang, Z., Zhou, L., & Wang, H. (2017). Spatial Distribution Characteristics and Driving Factors of "Taobao Village" in China. *Jingji Dili (Economic Geography), 37*(1), 107-114.

22 China Internet Network Information Centre. (2018). 41st Statistical Report on Internet Development in China. https://www.cnnic.net.cn/hlwfzyj/hlwxzbg/hlwtjbg/201803/t20180305_70249.htm

23 Hofman, B. (2016, 29 October). Op. cit.

JD.com to push into the countryside with initiatives that promise to finally let trade flow between remote towns and urban consumers.

But the scholarship to date on these drives, official figures and anecdotal evidence suggest that the chief beneficiaries may be the companies like Alibaba that manage to penetrate China's rural markets in order to ensure continued sales growth as urban markets become increasingly saturated. Promoting e-commerce in areas not yet fully wired to the rest of the country ignores the role foundational infrastructure played in allowing online commerce to become such an important part of China's economic growth – and thus poverty reduction – in the first place.

Action steps

The following recommendations can be made for China:

- Policy encouraging experimentation with community-owned networks that can operate within China's unique telecommunications environment should be implemented at the national level, with central government-funded subsidies made available to impoverished counties for build-out of network infrastructure where feasible.
- In the absence of community-owned networks, national-level policy should provide incentives to encourage the profitable e-commerce companies expanding their presence in the countryside to subsidise further network build-out in these areas by major state-run telecoms companies.
- To the extent that growth in rural e-commerce continues to be endorsed as a national policy priority, greater regulatory scrutiny of programmes like Rural Taobao is necessary to better gauge the extent to which they actually assist less-developed communities in the countryside in escaping poverty, as opposed to simply increasing revenues at parent corporations.

COLOMBIA

COMMUNITY MANAGEMENT FOR THE DEPLOYMENT OF COMMUNITY NETWORKS IN COLOMBIA

Colnodo
Julián Casasbuenas and Lilian Chamorro (with the collaboration of Olga Paz Martínez)
https://www.colnodo.apc.org

Introduction

Colombia has created an enabling environment for using spectrum for the deployment of wireless networks using Wi-Fi technology and more recently TV white spaces. This is an important step for the deployment of internet community networks. However, the evolution of the community networks ecosystem has been slow and it is necessary to find the tools in order to strengthen the existing processes.

In addition, other technological alternatives must be considered, such as mobile phone community networks. These have been implemented successfully in Mexico by Rhizomatica, which has set up community networks in the indigenous territories of Oaxaca.[1] This type of network is an opportunity for rural communities, since most members of the population in these areas own a mobile phone, and its implementation can be done rapidly, improving the living conditions of disconnected populations.

For Colombia it is particularly relevant to have access to community networks since they can contribute significantly to the implementation of the Peace Agreement[2] and to the social and human development in regions traditionally neglected by governments.

This report presents four community-based network experiences in Colombia.

Economic and political background

In Colombia, there is no specific policy framework for community-based networks. However, Law 1341 of 2009[3] defined the principles and concepts of the information society and the legal framework for information and communications technologies (ICTs). Among its principles is the need for non-discriminatory access to ICTs in compliance with the rights to communication, information and education, especially for the disadvantaged population and rural areas. The law highlights technological neutrality, noting that the state should allow the use of new technologies freely. In addition, it sets forth among its goals: universal service; the deployment and efficient use of infrastructure; equal opportunities for accessing resources such as spectrum and infrastructure; and the expansion of coverage in remote areas, particularly for vulnerable populations.[4]

In terms of the radio spectrum, Law 1341 establishes that the government may set aside frequency bands for free use in accordance with the recommendations of the International Telecommunication Union (ITU) as well as frequencies without licensing fees for state social programmes.[5] In Colombia, the frequency bands that are free for use include 2.4 MHz and 5 MHz[6] and TV white spaces in the 470 MHz to 698 MHz band.[7]

In line with its *Plan Vive Digital* 2014-2018,[8] the Ministry of Information and Communications Technologies (MinTIC) has been working on two relevant issues to do with internet access. MinTIC has expanded the coverage of 4G mobile networks and deployed Wi-Fi zones free of charge throughout the national territory. It has also extended the fibre optic network, which now reaches the urban areas in over 1,075 municipalities.[9] For the municipalities not covered by the fibre project, due to their geographical conditions, the ministry provides internet access through the Wi-Fi zones, *Vive Digital* centres, and some household connections using high-speed radio links.[10]

In terms of access to the internet in rural areas, the ministry has implemented a project called

1 For more information, please refer to the Mexico country report in this edition of GISWatch.

2 Final Agreement for Ending the Conflict and Building a Stable and Lasting Peace. www.altocomisionadoparalapaz.gov.co/procesos-y-conversaciones/Documentos%20compartidos/24-11-2016NuevoAcuerdoFinal.pdf

3 Law 1341 of 2009. https://www.mintic.gov.co/portal/604/w3-article-3707.html

4 Ibid, Articles 2 and 4.

5 Ibid., Article 11.

6 National Spectrum Agency, Resolution 711 of 2016.

7 National Spectrum Agency, Resolution 461 of 2017.

8 https://www.mintic.gov.co/portal/vivedigital/612/w3-propertyvalue-19438.html

9 https://www.mintic.gov.co/portal/vivedigital/612/w3-propertyvalue-647.html

10 https://www.mintic.gov.co/portal/vivedigital/612/w3-propertyvalue-7240.html

Kiokos Vive Digital,[11] which involves centres located mainly in rural schools and indigenous communities. However, while these spaces are important to provide access to the population, they are too scattered and insufficient for remote rural areas.

It should be mentioned that in the final Peace Agreement signed between the government and the FARC-EP, section 1.31.3 on electrical infrastructure and connectivity calls for the development of a national plan for rural connectivity that takes into account the provision of solutions for community access to the internet, the installation of infrastructure, and technical assistance that communities need.

Experiences of community networks in Colombia

In Colombia, the community networks movement has been active for more than 10 years, with different experiences and groups working on the issue. Below we provide an overview of some of these experiences.

Bogotá Mesh[12]

This initiative began in 2008, led by a group of "geeks" from the free software community who were interested in open networks. They set out to support communities and help them take ownership of technology using free/libre and open source software (FLOSS).

They began by installing Wi-Fi nodes close to their residences in the city of Bogotá, interconnecting them with point-to-point links. Over time they developed their own operating system (or firmware) for the routers, implementing protocols for mesh networks using 2.4 MHz and 5 MHz.

The project started to grow, particularly in Ciudad Bolívar, one of the 20 localities in Bogotá, which lacks basic infrastructure and has a low-income population. In 2014 Bogotá Mesh – the name of the network there – had 86 active nodes that connected approximately 2,500 people, with up to 1,000 concurrent users.

Among the services initially offered by the network were a local copy of Wikipedia, VoIP telephony using an Asterisk[13] server, FTP, the twister[14] microblogging platform, and even a blog with farming content. From 2008, it provided internet access to approximately 80% of the nodes via fibre from an internet service provider (ISP).

The resources for the implementation of the nodes came primarily from the management group, although they received some donations. One of the biggest challenges was the ownership of the network by the community, especially for non-technical individuals.

The group that supported the initiative did not continue. Nevertheless, some nodes still remain active using the firmware developed and some of its members are linked to other community network projects.

Red Fusa Libre[15]

This network is an initiative of the seedbed research project called Fusa Libre[16] run by the University of Cundinamarca's Engineering Faculty.

The project is implemented in rural areas in the Sumapaz region in the middle of Colombia, mainly in the rural district of Bosachoque in the municipality of Fusagasuga. In 2013 the community organised and collected money to buy the first equipment for two Wi-Fi zones. To access the internet they negotiated with people living in the nearby urban area with line of sight to the rural district. One urban household offered electricity and equipment maintenance in exchange for internet connectivity provided by an ISP and paid for by the community. In addition, they installed a content server using World Possible's RACHEL solution.[17]

In 2017 resources were secured through a research project proposal submitted to the University of Cundinamarca, which enabled the expansion of the network working in association with different programmes (Electronic Engineering and Systems Engineering) and research groups. The research project also proposed to identify educational content appropriate for rural schools.

Currently in Bosachoque there are 15 nodes that work as hotspots and are connected to the university's internet network through Wi-Fi links. When the project is finished, the community will again find a resident in the nearby urban area willing to help them with an internet connection. In the near future the network is expected to have 20 nodes and install five Raspberry Pi[18] devices with educational content.

11 https://www.mintic.gov.co/portal/vivedigital/612/w3-propertyvalue-7059.html

12 Interview with Jorge Rojas, member of the Bogotá Mesh collective. Fo more information, see: https://www.facebook.com/BogotaMesh

13 https://www.asterisk.org

14 twister.net.co

15 Interview with Wilson Daniel Gordillo, leader of the Fusa Libre seedbed research project.

16 https://es-la.facebook.com/RedFusaLibre

17 https://worldpossible.org/rachel

18 https://www.raspberrypi.org

Network Bogotá[19]

The Network Bogotá collective emerged in 2014 after meetings held at software and free culture festivals. Initially the collective met to work on different topics regarding hardware and software for community networks, as well as to share local and international experiences. Then they created a common fund for network experimentation and for deploying nodes in the homes of the participants.

They approached several communities to teach them about community networks, but it was only in 2017 that the Board for Community Action (JAC) of the Villa del Río neighbourhood in southwest Bogotá showed interest in the issue and convened meetings with the community. In these meetings it was determined that in some zones there were security problems. As a result they proceeded to install the first node in the JAC office, which was connected to other network nodes using mesh network protocols. They installed surveillance cameras, a digital video recorder and a community server. The success of this intervention boosted their confidence and the community was again convened to discuss the usefulness of the network. New nodes and services were installed, such as a captive portal, a community blog and more surveillance cameras, which are used by local residents to prevent crime and to control key points.

The network primarily uses mesh network protocols with some point-to-point links due to the difficulty of finding a router compatible with the firmware used. The resources for this implementation came primarily from the JAC and neighbourhood residents. Some organisations, such as Dongee,[20] guifi.net[21] and WirelessPT,[22] among others, provided training and equipment. The computers belong to the JAC and there is no charge for access to the network.

Weekly meetings are still held to discuss the project's progress, including the existing plans for increasing the number of nodes, installation of IP telephony services, setting up a radio station, and downloading Wikipedia and a virtual encyclopaedia to the network server.

Building a community network in Buenos Aires, Cauca

The initiative to build this network is the result of an expressed need by communities living in the rural districts of El Porvenir and El Ceral in the mountainous area of the municipality of Buenos Aires, Cauca. During the peace negotiations between the government and the FARC,[23] the communities said that one of the problems in their territory was the lack of coverage by cellular networks and poor access to the internet. Colnodo had learned about the experience of Rhizomatica[24] in Mexico and saw the opportunity to replicate this initiative in communities with low or zero coverage.

In February 2017, Colnodo, with the support of the Internet Society (ISOC) and Rhizomatica, submitted a request to MinTIC to use radio spectrum in a pilot project. Unfortunately, in Colombia there is no regulation allowing the use of spectrum for mobile communications without a public auction with more than one offer, so we began to look for alternatives with the support of the National Spectrum Agency (ANE).[25]

While conversations were held with government actors, in the municipality of Buenos Aires there have also been workshops with peasant leaders, indigenous and Afro-Colombian communities from different villages, and other communities in the territory to review the experience of Rhizomatica with Redes A.C.[26] Needs have been identified and principles established for the network. After obtaining the approval and commitment of the community, activities were proposed to make progress on the technical, legal, economic and organisational issues as well as key aspects for the installation of base stations for transmission.

Recently progress was made with an agreement between MinTIC, the ANE and Colnodo to use a portion of the spectrum in the 900 MHz band on a pilot basis in order to help develop policy recommendations aimed at the deployment of rural community-based telecommunications networks.

The pilot seeks to evaluate the technical, economic and social feasibility for implementing and operating community-based mobile phone networks in areas where there is no coverage by the commercial operators. These networks use Osmocom[27] free

19 Interview with Oscar Prieto and Leonardo Taborda, members of the Network Bogotá collective. For more information, see: https://networkbogota.org and Taborda, L. (2017). Deploying an open community wireless network: An exercise of digital sovereignty in the neighborhood of Villa del Rio (Bogotá, Colombia). *Hamut' ay*, 4(2), 80-90. revistas.uap.edu.pe/ojs/index.php/HAMUT/article/view/1486/1471

20 https://www.dongee.com

21 https://guifi.net/en

22 https://wirelesspt.net

23 After disarming, the FARC-EP changed its name to FARC.

24 https://www.rhizomatica.org

25 https://www.ane.gov.co

26 Redes A.C is a civil association from Mexico that works for indigenous and community communications and sustainable development. https://www.redesac.org.mx

27 https://osmocom.org

software to implement GSM technology using a "network-in-a-box". Furthermore, the pilot seeks to interconnect these networks to the closest municipality with a fibre optic link to the internet, in order to increase the reach of this infrastructure in the rural areas which are not connected.

The proposal is based on international recommendations aimed at facilitating the deployment of community networks, especially in rural areas without connection or those that are under-connected, such as Recommendation ITU-D 19 "Telecommunications for rural and remote areas", adopted by the International Telecommunication Union (ITU).[28] It also considers the recommendations of the ITU's Connect 2020 Agenda,[29] the Inter-American Telecommunications Commission (CITEL),[30] the Digital Agenda for Latin America and the Caribbean (eLAC2018 and eLAC2020),[31] and the global Internet Governance Forum (IGF) initiative to connect the next billion users.[32]

It is also based on national legislation, the implementation of the Peace Agreement, and the recommendations of the Colombian IGF.[33]

To strengthen the potential of the network among the community, a process to discuss ownership and governance has been implemented with a focus on gender imbalances, thanks to the support of the Association for Progressive Communications (APC).[34] For this strategy, we are carrying out ICT training workshops together with the communities and their leaders to build competencies and to visualise the potential of the network in the community before the deployment begins.

Findings

The experience of the Buenos Aires community network has demonstrated how difficult it is to access the use of spectrum in Colombia for rural community-based telecommunication networks, despite the existing laws to reach underserved communities in rural areas, the Peace Agreement, and the international recommendations to facilitate the deployment of these types of networks. For almost a year and a half we have worked together with MinTIC and the ANE, and still as we write this report, the pilot project has not been implemented because of the difficulties in accessing the use of the spectrum that we needed. As a result, we still have not been able to demonstrate the technical, social and economic feasibility of these initiatives in Colombia.

This delay has begun to frustrate the participant communities. Some leaders have withdrawn their support and, as a consequence, the managers of the initiative have lost legitimacy, given that the communities perceive this delay as a breach of their commitment to the project.

There is a critical need for a definition of regulations that facilitate the deployment of community networks in rural areas that are unconnected or under-connected in order to contribute to the connectivity of communities.

As demonstrated in the experiences presented in this report, the active participation of the community and committed leadership are key to consolidate the networks and their long-term sustainability. It is also important to build trust between the various actors involved, in order to sustain a process that demands time and dedication.

It is important to point to the work carried out in the region by ISOC, LACNIC,[35] CITEL and the Organization of American States (OAS) in the promotion of community networks.[36] This is an important factor in influencing the different international spaces – as was the case with the eLAC 2020 Digital Agenda – and for these networks to be considered as alternatives for connectivity in the region.

Action steps

- It is important to recognise the ecosystem involved in a community network. This includes the technical community, communities owning the network, local and regional governments, equipment suppliers, and facilitators of the ownership and governance processes.
- In the case of the technical community, there has been a strong investment of resources and time in local access initiatives that in many cases has not been rewarded, and this has ended up affecting the process of setting up sustainable solutions to access at the local level. Alternatives must be found to solve this problem, especially in developing countries.

28 https://www.itu.int/dms_pubrec/itu-d/rec/d/D-REC-D.19-201003-I!!PDF-E.pdf

29 https://www.itu.int/en/connect2020/Pages/default.aspx

30 https://www.citel.oas.org/en/SiteAssets/PCCI/Final-Reports/CCPI-2016-28-4000_e.pdf

31 https://www.cepal.org/es/proyectos/elac2020

32 https://www.intgovforum.org/multilingual/content/connecting-and-enabling-the-next-billion-phase-i

33 Mesa Colombiana de Gobernanza de Internet. https://www.gobernanzadeinternet.co

34 https://www.apc.org

35 https://www.lacnic.net/921/2/lacnic/lacnic-home

36 Internet Society. (2018, 12 March). Internet Society and the OAS through CITEL sign an agreement to bring the Internet closer to rural areas of the Americas. https://www.internetsociety.org/news/press-releases/2018/internet-society-and-the-oas-through-citel-sign-an-agreement-to-bring-the-internet-closer-to-rural-areas-of-the-americas

- The participation of the community is essential for the network to be consolidated. It is especially important to have leaders who are committed and who become allies of the project, as well as to create bonds of trust in the processes around the ownership and governance of the networks.
- It is also important to continue pressing the governments of the region forward in facilitating the deployment of community networks, including mobile networks that generate an immediate positive impact on the quality of life of the beneficiaries. Governments should welcome the international recommendations made on community networks and they should create an enabling environment for the implementation of these networks as an alternative to the deployment of infrastructure.
- The commitment to deploying these networks with the support of regulatory bodies is a unique experience that could encourage the implementation of community networks in other countries in the region. This could be a stepping-stone for Colombia, and would include the establishment of legislation by MinTIC that supports the development of community networks.
- In the case of internet infrastructure, it is necessary to facilitate access to fibre optic connections in the municipalities in order to increase the reach of fibre networks to rural areas that are under-connected and without access to a broadband connection.

CONGO, DEMOCRATIC REPUBLIC OF

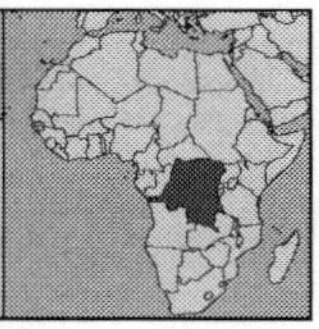

A FREE WIRELESS NETWORK IN THE DRC: AN ANSWER TO INTERNET SHUTDOWNS AND EXORBITANT ACCESS COSTS

Mesh Bukavu
Pacifique Zikomangane

Introduction

The Democratic Republic of Congo (DRC) has distinguished itself for the past four years in restricting the freedoms of its citizens and violating their rights to information and communication, which are guaranteed not only by its own constitution but also by the international law to which it has freely adhered. The Congolese authorities have repeatedly ordered the shutdown of the internet throughout the country – including the mobile messaging system – cutting 80 million Congolese off from the rest of the world, sometimes for several days.

The DRC is a vast country with an area of over 2,345,000 km², with infrastructure almost non-existent in many parts. Mobile telecoms companies such as Orange, Vodacom and Airtel are the main internet service providers (ISPs) in the country. Although the internet is of paramount importance for the Congolese population, access costs are high, which means that the majority of the poor population cannot access it.

This report discusses the Mesh Bukavu project in the DRC, an innovative community network initiative that achieves two things: it allows people in the city of Bukavu with low incomes, and especially those living in poor neighbourhoods, to have free access to a Wi-Fi intranet network, and it allows communities to bypass internet shutdowns.

Access to telecommunications and the internet in the DRC

The DRC has 46 million mobile telephony users, which corresponds to 54% of the overall population.[1] At the same time, 84% of Congolese access the internet on mobile phones, according to a study published in 2016 by Target Cabinet.[2]

In May 2016, the price of the internet packages offered by the three mobile phone companies, which are the main suppliers of the internet in the DRC, suddenly increased. The South African mobile phone group Vodacom, the country's largest telecommunications operator, started charging USD 100 instead of USD 28 for its 4 GB data bundle that expires after a month. The French group Orange, the second-largest ISP in the DRC, is now charging USD 62 for an equivalent package, as opposed to USD 35 before.[3] At Airtel, a subsidiary of the Indian group Bharti, customers have seen their bill triple – they now have to pay USD 100 for 25 GB.[4]

The emergence of community networks in the DRC – and the government's response

There is no specific policy for community networks in the DRC. However, there is Law No. 013/2002 of 16 October 2002, which regulates the telecommunications sector, including mobile telephone companies as well as the internet and all its applications.[5] On the basis of this law, the Post and Telecommunication Regulatory Authority of the Congo (ARPTC), an organ of the state in charge of granting radio frequencies and regulating the spectrum, was set up. In February 2018, the Congolese Minister of Posts, Telecommunications and New Information and Communication Technologies forced all VSAT owners, companies and ISPs to register with the ARPTC.[6] In this process the minister did not distinguish between commercial ISPs and community networks that do not aim to earn money through their activities, but which rather work for communities.

The Mesh Bukavu community network does not have VSAT and does not provide internet service, as its Wi-Fi network remains entirely local. However, it was included in the ARPTC's list of service providers, meaning it is considered a service provider by the regulator. Congolese human rights organisations suspect that the Congolese authorities have used the process of registering with the ARPTC to identify all initiatives that can enable citizens to continue to

1 https://actualite.cd/2017/03/16/45-millions-de-congolais-abonnes-a-telephonie-mobile-selon-target

2 https://www.target-sarl.cd/fr/content/rdc-84-des-congolais-accedent-internet-sur-support-mobile-selon-une-etude-de-target

3 www.information.tv5monde.com/afrique/rdc-envolee-des-prix-de-l-internet-112843

4 www.africa.airtel.com/wps/wcm/connect/africarevamp/drc/accueil/individuel/internet/offre-internet

5 www.leganet.cd/Legislation/JO/2003/JO.25.01.2003.PT.pdf

6 https://pbs.twimg.com/media/DVblcjkWkAAQ3dG.jpg:large

communicate freely with each other, and with the outside world, each time they decide to shut down the internet.

Mesh Bukavu

Bukavu is a city in eastern DRC, about 60 km^2 in size, and with a population of just over 870,000 people.

Like everywhere else in the DRC, the city experiences frequent electricity cuts, and many parts of it have no electricity at all. Insecurity is growing, and people are regularly attacked in their homes by armed bandits, which increases their need for communication. But, with the majority of the Congolese population living on less than one US dollar a day, the internet is a luxury.[7]

Because of this, the idea was to set up a wireless mesh network accessible to poor citizens in the community, allowing them to communicate and exchange information with their friends and relatives or anyone else in the community at any time.

Mesh Bukavu was set up in January 2015 by a group of journalists, bloggers and computer scientists with technical support and equipment being provided by Free Press Unlimited[8] and the Open Technology Institute.[9] It was a particularly difficult context, because it was during this time that the Congolese authorities had forced the mobile phone companies to cut both SMS and internet access across the country – a shutdown that lasted for three weeks.

Mesh Bukavu is a mesh network – a type of network in which "nodes" can connect as "peers" and dynamically route traffic across the network.[10] However, it does not give the community direct access to the internet. Rather, it functions as a kind of intranet, where people access internet resources, including websites, that have been downloaded by the Mesh Bukavu team and shared on the network. According to Benjamin Murhesa, a technician and the Mesh Bukavu Network project manager, Rocket M2s, NanoStations and TP-Link routers were used. "These devices are connected to each other remotely and can both receive and transmit information," he said.[11] They were placed on the roofs of houses to ensure wide coverage of the neighbourhood.

A central server has been set up at the local community radio station, Radio Maendeleo, a key partner in the initiative. Because there are many interruptions to the electricity supply, a generator has been installed, allowing the server to alternate between electricity supply and the generator, and to stay on permanently.

Even if people do not need the internet to use the mesh network, the team working on the network nevertheless needs the internet to regularly refresh the server and download new content to share across Mesh Bukavu. This is why we have mentioned the particularly difficult context in which it was set up. Just as two Open Technology Institute technicians arrived from the United States in the city of Bukavu in January 2015 to supervise the mesh network installation work, the Congolese authorities ordered the internet shutdown. The intelligence service (ANR) even monitored those who tried to illegally connect to the internet using their own VSATs. Since then the Congolese population has regularly had to deal with internet shutdowns, especially when the political opposition or civil society plan demonstrations. This was the case not only in January 2015 when Congolese citizens were protesting the government's initiative to change the constitution, but also in December 2016, August 2017 and recently this year in January when citizens were protesting the probable mandate of the current Congolese President Joseph Kabila.

Because of this vulnerability, the project turned to the ISPs in Rwanda, a country neighbouring the DRC, to help with the configuration of the network's equipment – Bukavu, which is right on the eastern edge of the DRC, is separated from the Rwandan city of Cyangugu by the Ruzizi River. Rwandan ISPs also use fibre optic connectivity rather than mobile. To circumvent any red tape, the project was set up under the auspices of Radio Maendeleo, which has a good reputation in the city – and across the country as a whole – especially for its editorial line oriented towards the defence of the interests of the citizens of the DRC, and of the Media Women Association (AFEM), an organisation of women journalists working for the promotion of women's rights in the country.[12] The boards of both Radio Maendeleo and AFEM were initially very supportive of the project, with AFEM organising volunteer girls from among its members to participate actively in the installation of the network.

7 Kanku Tubezele, P. (2007). *L'Afrique est à reconstruire, la responsabilité spirituelle*. Bern: Editions Scientifiques Internationales.

8 Free Press Unlimited. (2014, 30 October). People of DR Congo will communicate using their own mesh network. https://www.freepressunlimited.org/en/news/people-of-dr-congo-will-communicate-using-their-own-mesh-network

9 https://www.newamerica.org/oti

10 https://commotionwireless.net/files/cck/networking/2-Introduction_to_Mesh.pdf

11 Interview with Benjamin Murhesa, network technician and Mesh Bukavu Network project manager.

12 The city experienced extreme violence during the regional conflict following the Rwandan genocide – much of this targeted at women. In 2004, for example, about 16,000 women were reportedly raped in a single weekend by Rwandan-backed soldiers who had been given the "freedom of the city". See: https://en.wikipedia.org/wiki/Bukavu

Services available on Mesh Bukavu

Mesh Bukavu allows residents living in neighbourhoods covered by the network to communicate, but it also gives them access to online content. Chat Secure is an application that has been installed on the network and allows network users to chat instantly with each other. In addition, the network gives free access to the Wikipedia site, which is downloaded and placed on the network, and a digital library containing more than 360,000 books. Course material for online computer science and English courses is also available.

The community has participated actively in the creation of Mesh Bukavu. This has included participating in its conception, during its deployment, as well as in the management of its infrastructure. According to Murhesa, 10 boys and girls were trained on how they can repair the network when breakdowns happen.

The homeowners in neighbourhoods covered by the network have agreed that the network equipment can be installed on the roofs of their houses, and they take care of this equipment without any compensation.

During the project's installation phase, young people helped transport equipment such as antennas and climbed onto the roofs of the houses every day without asking to be paid in return.

This positive attitude to the project was in part due to a period of awareness raising prior to the installation phase, which helped orientate the community to the benefits of the network.

In 2016, Mesh Bukavu actively participated in the FASTAfrica 2016 campaign. The campaign, which involved a "week of action" with events hosted in 20 countries in Africa, aims to create an internet in Africa that is "fast, affordable, safe and transparent".[13] With support from the Web We Want initiative, the event brought together 40 students from six universities and three colleges in Bukavu for two days to discuss mesh networks: how they can be deployed and used, how content can be shared across the network, and about their right to have access to the internet. Participants also suggested the kind of content that they would like to access through the local mesh network, including courses and news.[14]

Some challenges we have faced

The city of Bukavu is in a mountainous region – a geography that does not facilitate the expansion of the network in all the targeted neighbourhoods, because it requires a lot of equipment that unfortunately the project does not have right now. Electricity cuts that go on for a long time in the neighbourhoods where the equipment is installed also affect the quality of the network. Some equipment is powered by solar energy, but not all.

Recently, changes in the boards at both Radio Maendeleo and AFEM have not been beneficial for Mesh Bakavu because the new board members are not as involved in the project. This is restricting the network's ability to extend its coverage. It is also difficult for us to go "independent" in the current context of strict regulation and internet shutdowns. It feels unlikely that the authorities will grant a legal status to Mesh Bukavu knowing that it is there as an alternative communication channel when they cut internet connectivity.

Conclusion

A mesh network is an alternative to the problem of a lack of access to the internet in the DRC, especially in poor neighbourhoods. Although it does not necessarily offer access to the internet, it can attend to many local-level communications needs and be set up so that the community has access to significant resources of information. Its maintenance is relatively inexpensive, and it does not require in-depth knowledge in network studies to keep going. It allows the community members to take responsibility for their own network. In the case of Mesh Bukavu, the community members behave like owners of the equipment that is installed on the roofs of their houses, which means that the security of equipment is provided by the community itself. As long as the community can exchange information and communicate through instant messaging, including during an internet shutdown, it is clear that they will do everything to keep the network in good condition for as long as possible.

But equipment and the will of the community are not enough: support from the authorities is also necessary so that community networks like Mesh Bukavu or any other community network can really serve the local population. This support can be provided through specific measures that promote the establishment and protection of community networks across the country. And this also means revising Law No. 13 of 16 October 2002 on telecommunications in the DRC, which allows the government to shut down the internet at will.

13 https://webwewant.org/fast-africa/toolkit/what-we-want

14 Web We Want. (2016, 27 May). FASTAfrica / Mesh Bukavu Workshop in Democratic Republic of Congo. https://webwewant.org/news/fastafrica-mesh-bukavu-workshop-democratic-republic-congo

Actions steps

The following advocacy steps are recommended for the Mesh Bukavu project:

- Speed up the process of obtaining a licence for its activities from the government so that Mesh Bakavu can be recognised as a non-profit project. Once this has been secured, the network will be freer to reach agreements and partnerships to ensure its future and growth in the community. These would include partnerships with universities, especially those that run computer science courses.
- Coordinate lobbying activities with civil society organisations, especially human rights advocates, to encourage the Congolese authorities and parliamentarians to review the telecommunication law and to develop a new law specifically on the internet and community networks. This law should formally restrict the government's use of internet shutdowns. It should also make it easier to obtain a licence for a community network, and should exempt them from paying taxes.
- Residents should be strongly encouraged to define appropriate strategies for sustaining community networks in their community. The main barriers faced by community networks in the DRC include a lack of money to pay for bandwidth, a lack of access to electricity, and the surtax on ISPs which drives up the cost of access for community networks.[15]

15 Deloitte. (2015). *Digital inclusion and mobile sector taxation in the Democratic Republic of the Congo*. GSMA. https://www.gsma.com/mobilefordevelopment/wp-content/uploads/2016/01/gsma-digital-inclusion-and-mobile-sector-taxation-in-the-democratic-republic-of-the-congo-report.pdf

COSTA RICA

THE COOPERATIVE MODEL AS AN ORGANISATIONAL MODEL FOR COMMUNITY NETWORKS IN LATIN AMERICA: THE CASE OF COOPELESCA IN NORTHERN COSTA RICA

Cooperativa Sulá Batsú
Kemly Camacho
https://www.sulabatsu.com

Introduction

This report focuses on the cooperative model used to develop community-run electricity services in Costa Rica as a possible business model for community networks in Latin America. It is based on the 53 years of experience of electricity cooperatives in Costa Rica, which, 10 years ago, started to expand their field of action to provide access to the internet and value-added digital services to the rural populations of this country.

Coopelesca[1] is one such cooperative with a long history in the country. It serves the northern zone of Costa Rica. Its history is similar to that of the other three electricity cooperatives in the country which were also started at the same time: Coopeguanacaste, Coopesantos[2] and Coopealfaroruiz, all of which have grouped into a cooperative consortium called Conеléctricas.[3] The consortium shares energy reserves and the benefits of technology transfer and also engages politically and purchases goods and services as a group.

The context in which the cooperatives started

It is important to highlight some relevant aspects of the Costa Rican context. Electricity was defined as a universal service provided by a public institution based on a solidarity model in 1949. From then until now, it is still a state service provided by the Costa Rican Electricity Institute (ICE), which develops the electricity infrastructure and provides services across the country.

However, not all of the country could be connected at the same time. Coverage had to be planned by zones. In four rural regions, the population decided to organise themselves to develop their own electricity infrastructure, administered by the communities themselves, to provide services to communities. This was so that the communities could have electricity without having to wait until their zones could be connected to the national electricity grid. As Coopelesca put it: "Here people lived off milk production and its derivatives. Families understood that electricity would generate a high added value to production, and that is why they organised."[4]

Consolidation of a technology cooperative organisation

In 1965, the Coopelesca cooperative was created by 365 members from the communities with a contribution of 25 Costa Rican colons (USD 0.0005 today) each. The total capital was around USD 80 at the beginning. They used this money to hold raffles and livestock auctions to raise money so that they could provide electricity to the communities. The cooperative has been strengthened and grown enormously. It is currently made up of 85,000 members, with 92,000 electrical connections.

Since electricity is the responsibility of the Costa Rican state, the cooperative had to negotiate a concession from the government so that 10% of the national territory would be electrified by Coopelesca. The total electrification of this entire territory was achieved in 2014, that is, 49 years later. This has been made possible by the effort of the people grouped together in the cooperative.

It is important to note that the coverage of the territory was complex due to its geography. For the electrification of each community, the cooperative had to open trails to be able to set up the infrastructure, then those trails were converted into community roads to provide maintenance to the infrastructure created. This had a knock-on positive impact in the territory: with the opening of the roads, access to health, education and markets, among other things, was increased, and development generally was accelerated. This has resulted in a rural area with greater access to opportunities.

This is another important aspect to highlight as a lesson learned: the cooperative model is a company that prioritises the development of the territory and the well-being of its members over the interests of capital accumulation. These principles are also relevant to the development of community networks.

1 www.coopelesca.com
2 www.coopesantos.com/contenido
3 www.conelectricas.com

4 Personal communication with Alvaro Chaverri, Coopelesca communication officer, 25 May 2018.

Ten years ago an important discussion was started in the cooperative on the convenience of incorporating information and communications technology (ICT) features as part of its services. This meant offering access to the internet and to digitally value-added services.

On the one hand, there were members of the cooperative who indicated that Coopelesca should focus on strengthening electrification services instead. They argued that connectivity must be left in the hands of the main telecommunications companies, Movistar,[5] Claro[6] and Kolbi.[7] On the other, there were those who proposed that the cooperative take advantage of the infrastructure it already managed to realise the community's rights to access internet services. This group also argued that it would be impossible for the telecommunications market to offer connectivity in many parts of the area that Coopelesca serves because the cost-effectiveness ratio was so poor – it implied making high investments on infrastructure for very small markets. "This is exactly why there had to be differences between technology access for the users in the central zone of the country compared to the most remote areas,"[8] Coopelesca stressed.

Through a majority vote, the members decided on the second abovementioned option. To be able to finance the connectivity, it was decided to include an additional amount of 500 Costa Rican colons (approximately 1 USD) on each electricity bill sent out to the members in the community. This surcharge was a contribution to capital – if a member withdrew from the cooperative, his/her capitalisation would be returned. This contribution to capital was accumulated to develop the necessary infrastructure for the connectivity of the most remote populations that did not have other alternatives.

The concentration of members is at the centre of the rural area. Even though they already have internet connectivity, they still contribute this monthly amount. The resources are then spent on connectivity for the more distant areas, for example, to 12 peasant settlements that could not have the connectivity service without the cooperative support. This, which we call the "solidarity model", is another of the principles that are applied in cooperatives that should be fundamental for the development of community networks. In this model, everyone pays to achieve the development of the most vulnerable.

Cooperatives that provide internet and related services are regulated at the national level by the National Superintendency of Telecommunications,[9] so the prices and costs are monitored by the state. Even cooperative members are free to choose an alternative service provider for connectivity or landline or mobile services, which means that prices should remain competitive.

It is important to mention that Coopelesca is a large cooperative whose employees are resident in the rural areas that it serves. The cooperative has been concerned to consolidate a very high technical capacity among its employees over the more than 50 years of its operation – and the same can be said for all the cooperatives that are part of the Coneléctricas consortium. Many of them, as in the case of Coopelesca, created and manage hydroelectric plants or, as in the case of CoopeSantos, they have wind power generation projects. It is the rural inhabitants themselves who have created these projects, securing credit from the government and banks, supported them with technical assistance, and run them according to solidarity business models, among other aspects. The people in these communities are trained in relevant engineering aspects, as well as in the basic management of hydroelectric plants or electrical issues such as the maintenance of wiring.

Not everyone in the community understands the technical details of setting up a connectivity project. However, any new project or investment must be approved in the general assembly of the cooperative where each member (male or female) has the right to one vote, guaranteeing that the ownership and investment are collective. It is important to mention that it is the decision of the assembly that the business model used in the cooperative is based on solidarity principles, but also that it must be profitable so that its services can be rolled out to unconnected areas. With unprofitable business models this would not be achieved – in other words, the cooperative would not be sustainable.

How is this relevant to community networks?

Three aspects are important to highlight and must be identified as good practices when talking about community networks:

- An autochthonous definition of needs by the communities, which emerges from the self-interest of their population.
- An autonomous interest on the part of the population in organising themselves to obtain the

5 www.movistar.cr
6 www.claro.cr
7 www.kolbi.cr
8 Personal communication with Alvaro Chaverri, Coopelesca communication officer, 25 May 2018.
9 https://www.sutel.go.cr

technology and the necessary infrastructure to solve this need.

- A public institution (in the case of electrification, the ICE) that supports the community network, is specialised and is willing to train the local population to allow the appropriation of technology without dependence on a third party.

In other words, community networks must be autochthonous, autonomous and independent.

With these as basic principles, the following can be highlighted about the development of community networks from considering the Coopelesca experience:

- In Costa Rica, the term "community networks" is not as well known as it is at the international level. However, the country has long-standing experiences in cooperative service provision with regards to the electricity cooperatives.
- People in rural areas can develop capacities of the highest technical level to attend to the needs of technology projects such as community networks.
- Organisations and community networks should not be conceived as small or weak organisations; they can be constituted as large, sustainable and influential organisations that are in the hands of the people they provide services to.
- The organisations that sustain the community networks must be anchored in a region, be concerned about the development of the region beyond providing digital services, and must be born from the needs of the population that inhabits the region, and not from external interests. It does not matter what sort of good intentions motivate those external interests; community networks must be rooted in the communities they serve.
- The basis of the organisational model should be the solidarity principle. Decisions should be made democratically and based on one person's vote carrying the same weight as any other. These principles should govern the collective prioritisation of actions and the technical training of the population.

Action steps

We propose the following action steps to strengthen the cause of community networks in Costa Rica:

- Our own experience as a cooperative, as well as the lessons learned in this study of Coopelesca, suggest to us that the cooperative model is one appropriate organisational model to consider for community networks in Latin America.
- At the same time, it is necessary to introduce the concept of community networks into our own country, since it is not so well known right now. We must create alliances in academia, the public sector, the private sector and civil society to strengthen the discussion on this issue in an environment where the telecommunications market was recently opened up, where there is a government fund available for connectivity projects, and where 40% of the territory is still without connectivity.
- It is critical for the public sector institutions promoting cooperatives and digital universal funds like INFOCOOP,[10] the National Telecommunications Fund (FONATEL)[11] and the Ministry of Science, Technology and Telecommunications (MICITT)[12] to develop policies for community network initiatives.
- We also must integrate the issue of community networks into the work that Sulá Batsú does with women in the digital sectors.
- Finally, the model of electricity cooperatives developed by community networks should be supported and replicated in other parts of the country, mainly in rural areas. Currently, its coverage is in four rural areas where it has been shown that the model not only works, but it also has an impact on social development generally.

10 www.infocoop.go.cr

11 https://sutel.go.cr/pagina/que-es-fonatel

12 https://www.micitt.go.cr

ECUADOR

CREATING ECUADOR'S FIRST COMMUNITY NETWORK

Anonymous[1]

Introduction

In this report we describe the process of creating a community network in our rural community of about 50 people. In order to support the creation of more community networks in Ecuador, we also look at the legal and regulatory context and the relationship between technological possibility and community, political and economic will.

In early 2017 we started a community network and now there is growing interest from various communities across the country, especially farming and indigenous communities, and those that participate in second- and third-level organisations (i.e. unions of communities and confederations of unions).

At the time of writing this report in mid-2018, we are upgrading our internet connection, expanding our physical infrastructure and beginning a more organised learning process. At the national level, the communications law is undergoing reform, and the first whispers of a community network coalition are emerging, inspired by other experiences in Latin America. People are looking to us for support about how to create networks here, since we appear to be the first such network in Ecuador.[2] This report serves as a snapshot of this moment from our perspective, which is only one of many perspectives.

Policy, economic and political background

Rapid changes in public policy, law and regulations during the past 10 years have left an uncertain field of action. The laws can be interpreted as favourable to community networks, even though community networks are absent from the regulations.[3] At the same time, the economic situation makes the creation of community networks viable though not trivial.

In Ecuador there are laws in favour of community communications but no policies specifically addressing community networks. We have also not found detailed analyses of the country's internet or spectrum regulation from the perspective of community networks, although a very useful spectrum analysis from 2011 still seems relevant even though laws have changed.[4] A government presentation from 2008 recommended four regulatory alternatives to strengthen community networks, but they were not implemented as far as we know.[5]

As mentioned in GISWatch 2017,[6] the development of the internet in Ecuador has favoured the private sector, especially multinational corporations, with some focus on the public sector, and very little functional influence from civil society, or the "community" sector as it is sometimes called in

1 The author of this report chose to remain anonymous, and APC respected this choice as it is consistent with its privacy and security standards.

2 Based on the description of community networks in the Declaration on Community Connectivity in Belli, L. (Ed.). (2017). *Community networks: the Internet by the people, for the people. Official outcome of the UN IGF Dynamic Coalition on Community Connectivity.* https://www.intgovforum.org/multilingual/index.php?q=filedepot_download/4391/1132

3 Interview with Valeria Betancourt, manager of the APC Communications and Information Policy Programme.

4 The case study notes (translation ours): "Another possibility would be to start processes of direct action based on the application of constitutional rights, and only later have these recognised by the courts. For example, start to use the spectrum without applying for a licence, [...] using the spectrum for broadband projects and then requesting protective measures from judges for these initiatives. It is important, in any case, to develop technically strong propositions, search for allies in the political push for legal reform, and fight so that the reform is debated widely and transparently. The possibilities of achieving democratic reforms are high. However, at the same time, it is possible that the debate will remain in the circles of influence of private companies, especially the big operators, and trapped in technical discourse, in which case there is also the possibility that constitutional rights will end up neutralised in the legal reforms." Navas Alvear, M. (2011). *Espectro abierto para el desarrollo – Estudio de caso: Ecuador.* Johannesburg: APC. https://www.apc.org/sites/default/files/Espectro_Ecuador_0.pdf

5 Fiallos, J. C. (2008). Redes Comunitarias. Slide show by an employee of CONATEL (former state telecommunication regulatory agency, replaced by ARCOTEL) who later became Sub-Secretary of Development of the Information Society and Online Government. It is a useful analysis of the legal and regulatory situation at that time and offers sound planning advice. It does not demonstrate an understanding of the characteristics of community networks as described by the Dynamic Coalition on Community Connectivity. Slide 25 is of interest – it seems that none of these suggestions for regulatory alternatives to strengthen community networks was implemented. The slides were presented as part of a training programme by IMAGINAR and IICD called Technical Update Seminars. https://www.imaginar.org/iicd/index_archivos/TUS12/7_Redes_FODETEL.pdf

6 Calderón, M. J. (2017). Building a national and regional internet governance agenda in Ecuador. In A. Finlay (Ed.), *Global Information Society Watch 2017: National and Regional Internet Governance Forum Initiatives (NRIs).* Johannesburg: APC. https://giswatch.org/en/country-report/internet-governance/ecuador

Ecuadorian law. A 2016 report indicates that spectrum regulation and network neutrality are at risk of political influence.[7]

In 2016, the last year for which data is published, the state reported that of individuals asked if they had used the internet in the past 12 months, the national average was 55.6%: the urban average was 63.8% and the rural average 38%.[8]

Motivation, luck, knowledge and friends enable Ecuador's first community network

Our network exists because we want it to exist; we build it, we maintain it, and we use it – and sometimes we break it, we argue about it, we insult it when it goes slower than we like or cuts off entirely, and we get frustrated about it... but mostly it works and we are thankful. Our network also works because a friend shares his internet connection with us and allows us to tie our antenna to his balcony – many thanks!

On 16 April 2016, an earthquake off the coast of Ecuador caused significant damage and loss of life.[9] As part of the response to the earthquake, a part-time community member living in the United States gathered donations from friends and colleagues and came here with a friend to help rebuild. When the reconstruction activity in our community ended, we decided to invest our attention and the remaining money in creating a community network for internet connectivity. We decided that this fitted within the scope of building community resilience to handle future disasters.

We had already started investigating the possibility of a community network. We had looked at the topography of our community and the rest of our canton; the nearby telecommunications infrastructure; the history of rural internet in the country; companies selling networking equipment; local organisations that could give social structure to the network and use the connectivity; and national organisations involved in community communications.

We asked for advice from AlterMundi,[10] contacts in Ecuador, and others. We also contacted a small internet service provider (ISP) mentioned in a 2008 article about community connectivity projects in Ecuador[11] and they visited our community. The response from everyone was more or less, "Start simple, with a single connection in your community."

Before we found a line of sight to a place with internet, an invitation arrived from AlterMundi to participate in a seven-day hands-on community network workshop with people from farming community organisations in Colombia. We spent some of the earthquake relief money and some personal money to make the trip with one community member and two young adults from a nearby farmers' union. The workshop helped us understand many things, such as how other communities organised their networks and how to configure Ubiquiti[12] Wi-Fi devices to create long-distance links.[13]

After the training, two members of AlterMundi came with us to our community, and provided the spark necessary to get our network going.

They advised us: Climb up to high places at dusk to identify potential links, and then just try the most obvious link in the fastest, least expensive way possible – fastest in terms of just buying an antenna instead of making antennas yourselves, and least expensive in terms of using a friend's internet connection instead of contracting your own. In order to take a first step, let go of the idea of building a network for five communities all at once. Maybe that will happen, but it's not the first step. Start with a single link, and that small, practical step will teach you things that enable you to grow the network later.

This turned out to work. One evening we climbed up to the top of a house being built on a hilltop and ta-da! The lights of a town twinkled in the distance! We spent the remaining earthquake money on the equipment to create the first part of our network and a few days later we had Wi-Fi internet in our community!

7 Solines Moreno, J. C. (2016). Telecomunicaciones e internet en el Ecuador del Siglo XXI: Apuntes técnicos, historia reciente y la ruta hacia el control de usuarios y contenidos. In D. Salazar & D. Viteri (Eds.), *Regulación de Internet y derechos digitales en Ecuador*. Quito: Editorial USFQ. http://libros.usfq.edu.ec/index.php/usfq/catalog/book/1 . According to the report (translation ours): "[W]ithin the process of the empowerment of society, with a growth in the flows of information and a notable influence of social networks, the Ecuadorian state demonstrates a regulatory strategy, a model of institutional design, with certain public policies oriented towards the control of users and content. Even technical aspects such as the assignment and administration of spectrum and principles such as network neutrality are at risk of political influence, which can adversely affect fundamental rights and the development of the information society in Ecuador."

8 Instituto Nacional de Estadísticas y Censos. (2017). *Tecnologías de la Información y Comunicaciones (TICs) 2016*. www.ecuadorencifras.gob.ec/documentos/web-inec/Estadisticas_Sociales/TIC/2016/170125.Presentacion_Tics_2016.pdf

9 https://en.wikipedia.org/wiki/2016_Ecuador_earthquake

10 https://www.altermundi.net

11 Carrión, H. (2008). *Conectividad rural para el desarrollo: Experiencias en Ecuador.* https://www.imaginar.org/docs/conectividad_rural_IICD.pdf

12 https://ubnt.com

13 We learned to configure devices in order to create links that are the same regardless of the frequency. Ubiquiti sells some models in 900 MHz, 2.4 GHz and 5.8 GHz, all with the same configuration interface.

A few months later we applied for a grant to expand our network to other communities, and we learned a lot in the process of applying. We did not get the grant, but we gained knowledge.

Connection and infrastructure

Our connection to the internet for the first 17 months was a DSL line with a connection of 5 Mbps and an 8:1 contention ratio from the national public ISP, *Corporación Nacional de Telecomunicaciones EP* (National Telecommunications Corporation – CNT), shared with a friend in a small town. It cost about USD 33 per month. Now, after 17 months of asking them various times, the ISP activated a fibre optic connection at our friend's house that is supposed to be 10 Mbps download and 5 Mbps upload with a 2:1 contention ratio for USD 79 per month; but in the first three weeks it has not performed better than the DSL.

Our network connects that internet connection to our community via a 22 km (line of sight) wireless link, and the network within the community is currently made up of Wi-Fi routers connected via outdoor UTP cable. We use cable instead of Wi-Fi because dense forest and hills block the line of sight between the houses.[14]

Some future connections will be wireless where there is line of sight. For those links, we want to use inexpensive antennas connected to TP-Link WDR3500 routers (or other routers compatible with LibreMesh[15] with disconnectable antennas), but if that proves too difficult or the price is close to the price of Ubiquiti devices, we will use Ubiquiti devices. We hope to acquire at least one LibreRouter[16] and backup power supplies, and to use alternative energy sources (solar, wind, micro-hydro, biogas).

We use two colours of electrical tape – red and green – for colour-coordinated markings that indicate what cable plugs into what port, so that we do not confuse what cables connect to the power-over-ethernet boxes.

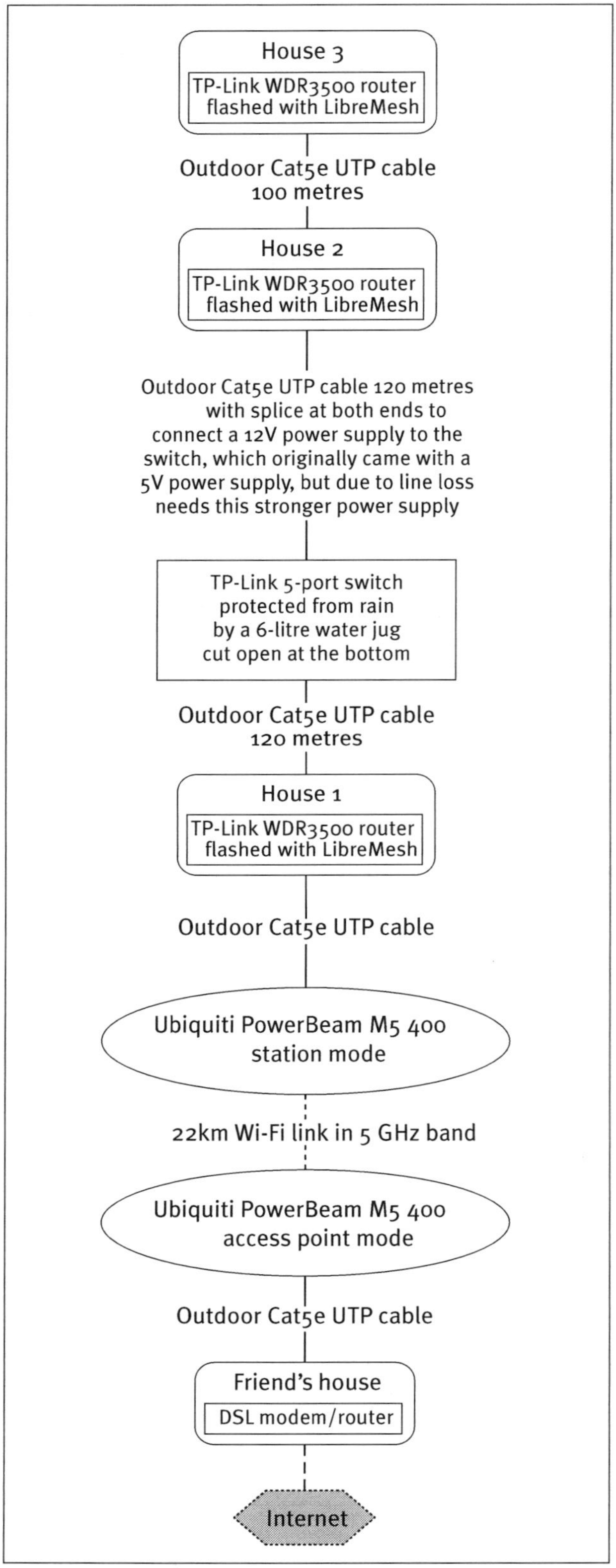

14 The Ubiquiti antennas/radios are attached to one-metre-long bamboo poles that are tied to balconies. The routers are in houses.

15 https://libremesh.org

16 https://librerouter.org

Failure and reorganisation

In December the ISP's DSL modem/router failed and we did not have internet for nearly a month, until the ISP fixed the problem. In April the connection between the antennas disappeared and no one could fix it from our community. After a month without internet we went to our friend's house and saw that a palm tree had grown in front of the antenna and blocked the signal.

The loss of internet gave us a chance to renew our way of organising the network. We started to have meetings at least once a month and soon we started contributing money and signed a contract for a fibre optic connection.

We had various challenges that we do not have space to discuss here:

- Debates about whether to connect the community centre or more homes first.
- Misunderstandings about money and motivations.
- Frustration about lack of participation.
- Lack of clarity about what routers and antennas to buy, or how to make our own antennas.
- Concerns about negative effects on human and ecosystem relationships.

Gender, age and network uses

Part of the motivation to build the network grew specifically out of the vulnerability highlighted in the April 2016 earthquake, and also from the understanding that communication enables people to interact with many processes that affect the stability of their lives.

Having internet in our community makes it much easier to coordinate tourism, an educational programme in the community, or volunteers, and internet also attracts more visitors. This benefits the members of the community tourism association, most of whom are women. When people visit the community, women earn money by cooking for them and cleaning the community centre where they stay or have meetings. Men are occasionally hired as tour guides to take people on hikes in the forest. Many community members also appreciate the cultural experience of meeting people from other places.

Men tend to use the internet to communicate about work, some women use the network for academic study, and everyone uses the network for social communication with family and friends and entertainment. Women have participated in almost all the community network activities, such as meetings, installing the gateway node and antennas, and extending the network to more houses.

We have given specific attention to including women, people from all the families in our community and people of all ages in the conversations that guide our network and in the considerations of the impact of the network. The conversations happen in community meetings and in people's homes. The core group that has implemented the network so far makes an effort to include people in the construction of the network, to explain how it works to everyone interested, and awaken interest and a sense of capability in everyone in the community. We think that with encouragement, everyone is capable of understanding how the network functions and participating in the network as they choose.

The list below gives an indication of the gender breakdown in participation in various network activities:

- First months of preparation and coordination – one male, in conversation with many people.
- Installation of primary link – two females, three males (of these, one female and one male from AlterMundi providing support).
- First meeting – participants not recorded.
- Second meeting – four females, five males (after the meeting one male taught one female how to put RJ45 connectors on a UTP cable).
- Third meeting – seven females, nine males.
- Fourth meeting – six females, six males.
- Reposition of source antenna – two females, four males.
- Communicating with ISP to contract fibre optic connection – one male (contract signed by one female).
- Visit to town with internet connection for fibre optic link activation – one female, two males.

All installation and maintenance activities have involved people between the ages of 19 and 35. At meetings the age range has been about 10 to 65. The network offers teenagers and young adults in our community a chance to take responsibility for a community system, to learn and to demonstrate to their parents and themselves that they are capable of managing the community network. The adults already manage the water cooperative, the community bank, and the community tourism association.

State policies and laws

We have not dealt with national internet policy directly, other than interacting with its effects: the lack of internet access in our community.

Our community and the surrounding area have very little state presence, so we have not found it necessary to approach the state (public ISP, regulatory agencies, legislators, elected officials) and we have not yet read all the laws and regulations – we just maintain our network and move along with our lives.

However, for our network to survive in the long term and to support the creation of new community networks throughout Ecuador, we consider it wise to understand the law and to influence it and interpret it in ways that defend and support our community networks. As far as we know, the legal system, the regulations and the people tasked with enforcing them have not interacted with community networks, since we are the first such network here.

Due to 10 years of tight state control in many areas of society (2007 to 2017), we have so far chosen to remain unnoticed, rather than risk state attempts to co-opt, regulate, or shut down our network. Recently an ally in contact with state telecom institutions told us that the current national government is favourable towards the idea of community networks and community spectrum use. We have also had some contact with a rural mayor who wants to support a community network in his canton even though we explained that the legal aspects are unknown.

We serve as an early experience of the potential of community networks in Ecuador, and we plan to use our experience adapting this model to our context as a basis for conversation with other communities, organisations, ISPs and the state about how to support this new way of co-creating internet infrastructure. We hope that the state and ISPs of all sizes adapt to this reality, rather than trying to stop or control the formation of community networks.

Conclusions

We draw the following conclusions from our experience of setting up a community network in Ecuador:

- The conditions in Ecuador are ripe for community networks, but a spark and organising initiative such as a national coalition has been lacking until now.
- In our community, communicating clearly and organising ourselves is very important. Even if no one is charging us money for our internet connection, we still benefit from organising ourselves, and it makes sense to collect money for future expenses.
- Achieving and maintaining connectivity requires attention, time, money, understanding, confidence and perseverance. Otherwise people give up and say "I don't know how" or "We don't have time and money to go wait for the technicians to install the connection."
- Communication with external actors is a key part of organisation.
- In many communities we (humans) get things working just enough, and then shift our focus to the next urgent issue. In our case we created a precarious connection and did not focus on stabilising it until it broke down for a month, and when we started to help set up other networks in other parts of the country and wanted our network to serve as an example of a well-run community network.
- Extending the physical infrastructure starts with deciding who pays for equipment, who performs maintenance and management, what the technical design is and what hardware will be used. In order to create clear understanding among network members, it helps to make these decisions before heading out into the field.
- Interact with other communities that have experience in or the desire for community communications and self-managed infrastructure. Connect community networks (the organised people, the infrastructure, and the concepts) with the global and local movements towards well-being and freedom based on peer-to-peer cooperation. Participating in movements helps us to learn, build community relationships, and improve our ability to influence state and commercial processes.
- Support communities to create their own processes of appropriating information and communication technologies, recognising that our current concept of "community network" is shaped by our cultural perspective and our history of appropriating technology, and can look different in different communities and change over time.

Action steps

Action steps for our network:

- Improve skills in meeting facilitation and conflict transformation.
- Learn how to use the internet for useful and creative things.
- Decide on ways for neighbours and visitors to participate in our network.
- Improve and extend the physical infrastructure.
- Participate in the global community network movement.

Action steps to advance community networks in Ecuador:

- Advocate for:
  - Community network access to infrastructure and resources including towers, power sources, TV white space frequencies and the idle bandwidth of public institutions.
  - Exempting community networks from licensing or registration requirements that require payment or advanced technical studies.
- Grow and strengthen the emerging network of people and organisations working to advance community networks in Ecuador via training, financing, advocacy and movement building.
- Participate in the Internet Society (ISOC) Community Networks Special Interest Group.[17]
- Conduct a market scan to clarify what networking hardware is available in Ecuador.
- Acquire LibreRouters in order to build community networks with a lower cost and more efficient technical design. The LibreRouter includes three radios in a single device that is projected to cost USD 150.
- Get the LibreRouter through the state approval process (so-called homologation or type certification) so that we can use it in state-funded or state-regulated networks.
- Consider ways to access the state's Universal Access Fund to train communities and buy network hardware.
- Implement the ITU-certified Diploma in Community Networks.[18]
- Promote the Inter-American Telecommunication Commission (CITEL) Draft Resolution on "Connecting the Next Billion, Boosting New Communication Patterns for Unserved Areas".[19] This promotes community networks and could serve as another instrument in advocating for state support of community networks.

17 cnsig.info

18 https://techiocomunitario.net/diplomado

19 https://www.itu.int/net4/proposals/WTDC17/Detail/Index?idProposal=48505

GEORGIA

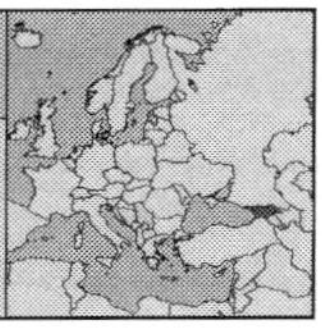

SAY "NO" TO ISOLATION: THE TUSHETI WI-FI COMMUNITY NETWORK IN GEORGIA

Small and Medium Telecom Operators Association of Georgia and Tusheti Development Fund
Ucha Seturi and Rati Kochlamazashvili
toa.ge/en/?p=974;

Introduction

This report is about the Tusheti Wi-Fi community network project in Georgia. Tusheti is the name of the remote northeastern region of the country, located on the northern slopes of the Greater Caucasus Mountains. It is part of the Tusheti Nature Reserve and National Park, which is the largest protected region in Europe.[1] At 2,345 metres, the village of Bochorna is the highest settlement in Europe.

Tusheti can only be reached by an unpaved road that is considered one of the world's most dangerous,[2] and which crosses the 2,850-metre-high Abano Pass. The population of Tusheti is physically isolated from the rest of Georgia, in part due to the heavy snow in the region.

The community Wi-Fi project has brought high-speed internet to more than 33 villages in Tusheti, as well as connectivity to the Abano Pass and for the roughly 14,000 tourists visiting the region every year. It is a good example of installing a community Wi-Fi network in a high mountainous region. At its peak, the network reaches elevations of 3,500 metres.

The project has opened new development opportunities for tourism and agriculture, and helps to preserve the unique local culture. It also provides an essential communication channel for healthcare and other emergency sectors. It supports the economic sustainability of the region and creates business opportunities for the local community, benefiting trade in products and access to services.

The project was implemented by a local community organisation, the Tusheti Development Fund, with funding and support from the Internet Society (ISOC).[3] It also received mentorship from the ISOC local chapter,[4] and in-kind support, including co-ordination, from the Small and Medium Telecom Operators Association of Georgia.[5]

It is a successful example of a private-public partnership that includes the participation of the local community, the local and central governments of Georgia,[6] the Georgian National Communications Commission,[7] private businesses and international donors that helped with equipment and the training of members of the community in using the internet, and in e-business skills.[8] It demonstrates a sustainable business model that is the result of community-led development.[9]

We also believe our experience will be helpful to other isolated communities in high mountainous areas. All technical information is open and ready for sharing.

Policy, economic and political background

Georgia does not have specific policies on community networks. You do not need a licence to use the 2.4 MHz and 5 MHz spectrum, and you do not need special permission to set up a community Wi-Fi network or to operate as an internet service provider (ISP) – you just need authorisation, which can be done online. ISPs, however, have to pay a regulation fee.[10]

The special tax regime of 0% value-added tax for small and medium-sized enterprises (SMEs) applies to community Wi-Fi networks. Settlements in high mountainous regions also receive other special tax cuts, such as being exempt from income tax. As it is a protected area, legislation also allows for some benefits for the residents in Tusheti.

While Georgia has several community radio stations, the Tusheti community network is the first and only one in Georgia. There are no specific political issues inhibiting the roll-out of community

1 Patterson, K. (2011, 12 October). Special Program Highlights Largest Protected Area in Europe. *National Park Service*. https://www.nps.gov/romo/learn/news/pr_news_release_special_program_october_20_2011.htm

2 See the BBC documentary at: https://www.dailymotion.com/video/x1tpe2k

3 https://www.internetsociety.org

4 https://www.facebook.com/groups/isoc.georgia

5 toa.ge/en

6 www.economy.ge/?lang=en

7 gncc.ge/en

8 projects.worldbank.org/P152441?lang=en

9 Palovirta, M. (2017, 9 October). A Community Network in the Remote Georgian Region of Tusheti. *Internet Society*. https://www.internetsociety.org/blog/2017/10/community-network-remote-georgian-region-tusheti

10 The fee is 0.75% of monthly income from the activity.

networks in the country. Legislation is aligned with European Union (EU) laws, and therefore can be considered supportive for community networks overall. More competition is nevertheless needed to improve access to the internet in the country. The population of the country is 3.7 million, and half of the people live in the cities. Mobile penetration is nearly 130%,[11] and the number of fixed-broadband subscribers is 774,000 (at end of 2017).[12]

Setting up the Tusheti community network

In September 2016, the ISOC Georgia Chapter, the Small and Medium Telecom Operators Association of Georgia, the Freenet Ltd. Association, and the Tusheti Development Fund – a non-profit organisation set up for the project – signed a memorandum of understanding (MoU)[13] with the ISOC EU bureau to build and develop a new wireless network for the Tusheti community. This MoU creates the multistakeholder group that guides the development of the community network.

Equipment was procured over the next few months, and the network itself was set up in just 60 days – between the end of June and the end of August 2017.

It was built using only solar energy, six masts up to six metres in height, and a few gap fillers for "white" areas. We also bought and installed a few end-user computers for the most vulnerable villages and used them for site and network checking and testing. This high-speed wireless internet service can deliver a 10 Mbps connection to each user.

The network connects to the internet in the Kakheti region (the village of Ruispiri is the closest point for internet connection) via a 42 km radio link with a 120 Mbps connection speed to a mast on the Abano Pass. The Abano Pass is then connected to the Diklo mountain mast (this is a 20 km link). From this point a connection is made to three masts that cover all three of the gorges in the Tusheti region. The network provides access to 85% of the villages in Tusheti, and covers more than 260 square kilometres. Freenet provided technical training and helped with on-site installations.

The parallel activities by the government, such as those by the Ministry of Economy and Sustainable Development's Innovation and Technology Agency (GITA), and the World Bank-funded GENIE Project[14], have also been very important for getting the network going. The GENIE Project trains disadvantaged citizens in internet and e-commerce skills. Participants also receive a voucher (USD 70) as a contribution towards purchasing equipment and paying for bandwidth.

11 https://www.budde.com.au/Research/Georgia-Telecoms-Mobile-and-Broadband-Statistics-and-Analyses

12 GNCC Annual Report: gncc.ge/uploads/other/3/3117.pdf

13 https://www.google.com/url?sa=t&rct=j&q=&esrc=s&source=web&cd=1&ved=2ahUKEwjd243w_J_dAhVECpAKHT6qBAYQFjAAegQIABAC&url=https%3A%2F%2Fwww.internetsociety.org%2Fwp-content%2Fuploads%2F2017%2F08%2FPress20Release20MoU20Tusheti20Georgia.docx&usg=AOvVaw2mDG7-vkCtA_-bvoztXd8Q

14 documents.worldbank.org/curated/en/320681467987898858/pdf/PAD1543-PAD-P152441-R2016-0029-1-OUO-9.pdf

Network sustainability

We want to highlight the sustainability and business model of the community network. Firstly, it is important to involve the local community in the project – in conceptualising the project, in installing the network, and in managing the project's finances. The community needs to have a good understanding of the idea of the social ownership of the network. Right now the Tusheti Development Fund works well as a mechanism to manage the network. Through this the community is able to manage the project on its own, whether it comes to technical or managerial skills.

The network provides cost-effective internet connectivity to dispersed communities located in a challenging terrain, and this point is very important for us.[15] The profit generated from the network is not allowed to exceed 10% of all operational and capital expenses.[16] Any profit made is used to develop the network further, repair or replace equipment, or for other activities that promote and develop the network.

Tourism is fundamental to the financial sustainability of the network. Currently Tusheti is in high demand from local and international hikers and adventure tourists. As a result, hostels and guesthouses contribute the most to the income of the Wi-Fi network. Good quality Wi-Fi is also important to begin thinking about the development of meetings, incentives, conferences and exhibitions (MICE) tourism.

Most importantly for many permanent residents, the internet is free and running costs are covered by the Tusheti Development Fund during the winter time, when tourism is low – although it is hoped that the young tourists will find other reasons to stay in the area when poor weather rules out hiking or adventure sport.

A clear understanding of the importance of the "bottom-up" approach is crucial. Each community also has to understand the importance of volunteer activities and how this can contribute to the development of their community or region. The voluntary activities provide in-kind support from local community members and local stakeholders. This is the type of spirit necessary not only for running the Wi-Fi network, but also to meet the needs of the community, such as roads, electricity, and water supply. The state and donors create opportunities for micro and small businesses, but often these lack "community soul". As a result, they do not bring the desired results and benefits for the whole community and for individuals inside of the community.

Conclusion

The benefit of the Wi-Fi network to the community is clear.

Irakli Khvedaguridze is a 76-year-old doctor living in the village of Bochorna. During winter he is the only resident in the village. When local technicians arrived to install the antenna and get the network connection running, he was overjoyed, even though he did not yet have a smartphone or even a computer. He recounted the following story: One winter a tree fell on a man in Tusheti, hurting his back. The man had to walk with his injured back for three days through waist-deep snow to find someone who could then contact the doctor to help. Even then, helicopters could not reach the man and a team of seven men had to walk from another village

15 Quarmyne, N., & Granville, K. (2018, 5 January). Hauling the Internet to an Ex-Soviet Outpost High in the Caucasus Mountains. *The New York Times*. https://www.nytimes.com/interactive/2018/01/05/technology/caucuses-mountains-internet.html

16 Maka. (2018, 19 March). Project for Tusheti Community Internet Network, Georgia. *Small and Medium Telecom Operators Association of Georgia*. toa.ge/en/?p=974

and carry him out on a stretcher. The doctor's first thought was how much of this could be avoided if there was an internet connection.

Just a year's experience shows that the demand for local tourism facilities has increased rapidly. The Tusheti network allows tourists to easily connect directly with the locals, and those in the tourism industry can now focus on offering their services without an intermediary. Local guesthouses need the internet to support their hospitality businesses and stay competitive in the market – a high-speed internet connection gives them an advantage. A guesthouse owner in Darkizanidze has expanded her guesthouse to meet the demand of an increase in the number of tourists since the Wi-Fi was installed.

The farmers and guesthouse owners deep in the Gometsari gorge, who previously did not have the internet, are now are able to boost their businesses and have an "easier life". The farmers are also able to reach out more to buyers of their products, besides enjoying connecting to their relatives in faraway places. Archil Elizbaridze, a sheep farmer, says he enjoys the conversation with his family and friends every evening after he takes care of his sheep.

The hard winter in the region undermines the network, and brings us new challenges, such as those related to electricity and solar energy supply in the long winter nights. Two high-altitude sites (the masts in Abano Pass and Diklo) need their batteries replaced and additional solar panels installed because of a lack of sunlight in the wintertime, and over periods of heavy snowfall. Other than this, the radio equipment and other devices are working without any issues, network performance is good, and technical troubleshooting has been carried out in an appropriate way.

The Small and Medium Telecom Operators Association of Georgia is now working with local partners on a different community network project in the neighbouring Pshavi and Khevsureti areas and we hope to find donors and supporters for this network. This will be connected to the Tusheti community network. It is important for both of these areas to get access to the internet, because they have two schools with more than 30 children, and more than 50 SMEs.

Action steps

Firstly, our plans relate to the Tusheti project. This year – and for next two years – we have to monitor the system we have built and keep it running, offering help wherever it is needed. We will check all sites and will install additional filler masts so that there is full coverage in the Tusheti area.

One challenge is the lack of qualified employees in the local community. However, we have managed to find a decent technician and we will train and prepare him for on-site activities.

Tusheti also faces the important opportunity of deserted villages. The internet could allow these to be resettled by people in the areas near tourist trails. Secondly, ISOC is working in cooperation with the Kyrgyzstan government, the regulator and the ISP community in that country, as well as the ISOC Kyrgyz Chapter, among other stakeholders, in order to replicate the Tusheti model there. We will support this in any way we can.

GERMANY

FROM THE SHELTER TO THE CLASSROOM: TWO CASES OF CIVIC PARTICIPATION THROUGH FREIFUNK BERLIN

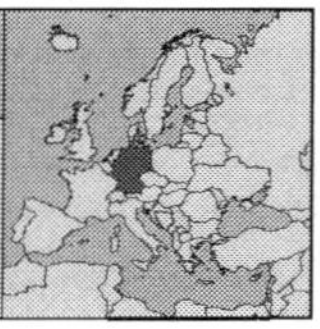

Freifunk
Tim Schütz and Monic Meisel
https://freifunk.net

Introduction

Whether campaigning for net neutrality, internet access as a human right, or civic engagement during the so-called refugee crisis, the "Freifunk" ("Free Wireless") initiative continues to receive attention from journalists, politicians and activists alike. By connecting over 300 refugee shelters and reception centres,[1] Freifunk could underscore the significance of its hacker practices for digital infrastructure politics in Germany. Yet what could be framed as an exceptional "humanitarian media intervention"[2] can be understood within an ongoing move to establish Freifunk as a legitimate form of "digital volunteering", which includes a sustained engagement with public institutions and a struggle in legally backing up its own emerging practices.

In our report we use two case studies to investigate the question of what "doing" free wireless network activism means in different contexts, drawing on Freifunk's engagement with youth centres and refugee shelters and reception centres[3] in Berlin. We discuss key problem areas, ranging from the influence of "humanitarian" logics on Freifunk activism to the challenge of involving teachers in passing on Freifunk practices in the classroom. We also consider the overarching restrictive legal regulations, broadband policies and the developments towards specific forms of participation and citizenship that currently affect Freifunk.

Policy, economic and political background

Founded in the early 2000s, the emergence of Freifunk as an initiative and socio-technical practice is often framed as a response to the "market failure" of telecommunication companies to provide internet access in a recently unified East Berlin and rural Germany.[4] Early workshops included tinkering with wireless devices, free software, organisational forms and routing protocols. The proliferation of affordable broadband access in the mid-2000s then decreased the initiative to a set of core participants in Berlin's underground hacking scene. Surprisingly, there are now more than 100 active Freifunk communities in cities and towns all over Germany and other German-speaking countries.

Situated in close proximity to the growth of hackerspaces, fab labs[5] and other do-it-yourself (DIY) practices, the initiative emphasises its heightened political awareness, both through its activism and network policy advocacy. Under the slogan "Freifunk Against Fear", communities have sought to challenge the so-called "network liability law" in Germany which puts the legal responsibility for online activities on the clients of internet service providers (ISPs). After several years of challenging and subverting the law in written form and via devices that reroute internet traffic to countries outside of Germany,[6] the law was abolished in 2017. Although this opens new possibilities for free wireless networks in Germany, there are commercial practices and national and regional regulations and policies that still concern Freifunk:

- The overuse of licence-free spectrum by commercial players, coupled with a lack of frequencies for non-commercial public use.

1 Schröder, I. (2017). Freifunk Hilft. In W. Schiffauer, A. Eilert, & M. Rudloff (Eds.), *So schaffen wir das – eine Zivilgesellschaft im Aufbruch: 90 wegweisende Projekte mit Geflüchteten*. Bielefeld: transcript Verlag.

2 Kubitschko, S., & Schütz, T. (2017). Humanitarian Media Intervention: Infrastructuring in Times of Forced Migration. *Spheres: Journal for Digital Culture #3*. spheres-journal.org/humanitarian-media-intervention-infrastructuring-in-times-of-forced-migration

3 Reception and emergency centres are the first institutions where refugees have to register, before they are eventually relocated to long-term shelters or private housing.

4 Petersen, G. (2014). Freifunk: When Technology and Politics Assemble into Subversion. In J. Leach & L. Wilson (Eds.), *Subversion, Conversion, Development: Cross-Cultural Knowledge Exchange and the Politics of Design*. Cambridge, MA: MIT Press.

5 https://en.wikipedia.org/wiki/Fab_lab

6 To bypass the regulation, all traffic within the Freifunk network is rerouted through virtual private networks (VPNs). This extends the connection to countries where the liability law does not apply or to a collectively run server in Berlin. Additionally, the Freifunk initiative appeared as experts before the "Digital Agenda" federal committee and advocated for a reformation of the law on secondary liability of open Wi-Fi networks. They also used negative declaratory actions to reconsider the law. See also the Freifunk statt Angst blog at: freifunkstattangst.de

- Unused local frequencies which should be taken back for dedicated non-commercial use (e.g. TV white space or LTE).[7]
- The continuing need to make access to communication networks a basic human right, particularly for minority communities.[8]
- The problematic European Union (EU) radio and data retention directives that create hurdles for community networks.
- The EU funding regulations for community networks (such as Wifi4EU), which demand a central registration that conflicts with the newly enforced General Data Protection Regulation (GDPR) and Freifunk values of abstaining from collecting data.
- The pending implementation of the public benefit status of Freifunk communities, which was already included in the latest government coalition contract.[9]

While these struggles certainly impact Freifunk, the following cases mostly respond to the last point, namely, Freifunk's legal recognition as a civic entity, rather than an open and spontaneous collective of individuals. Our two examples render visible the limits and creative workarounds of doing Freifunk that emerged over the last few years.

Case 1: Humanitarian interventions in refugee shelters

The "long summer of migration"[10] marked a turning point that entangled the struggles of the Freifunk initiative with the realities of people on the move. It became among the most visible "tech" projects in an upsurge of volunteer activism labelled as "Willkommenskultur" (a "Culture of Welcoming"). Yet a commitment to support refugees was itself not novel, since Freifunk participants in Berlin had already provided internet access to migrant camps in the Kreuzberg district.[11] In doing so, they amplified a stance shared by other non-governmental organisations, that digital devices are more than "luxury" items,[12] but are crucial for protecting (digital) communications as a basic human right. In parallel to the traditions of established hacker organisations like the "Chaos Computer Club",[13] the Freifunk initiative provided the socio-material practices to problematise the infrastructural politics of refugee shelters and reception centres, but also sought to actively reconfigure them.[14]

To shed light on this, we conducted an interview with Philipp, a 31-year-old master's student in computer science who, since the very beginning of the refugee crisis, was involved in refugee tech activism in Berlin's Neukölln district. His engagement began when an emergency camp was opened in his university's gym. As he recalls, the focus of most volunteers was to provide items for basic hygiene, clothes and social support. Philipp was interested in the digital infrastructure, but his idea to reroute access via the university's eduroam[15] network proved difficult for legal reasons. Nevertheless, the experience led him to bring his idea to a refugee support collective located in his neighbourhood. Together with two other friends, he contacted companies or individuals that would be willing to reroute their private internet uplink to refugee shelters and reception centres in the neighbourhood.

With improvised housing facilities for the refugees mushrooming all over Berlin, Philipp then began to assemble a public wiki to keep an overview on the "status" of different installations.[16] These were now organised collectively through a regular meeting at the "c-base", Berlin's most well-known hackerspace. At its peak, up to 30 people would gather for planning possible installations. While this included several supporters and managers of shelters, he acknowledged that refugees and asylum seekers only occasionally found their way to the crowded seminar room.

7 High-speed wireless for mobile devices. See: https://en.wikipedia.org/wiki/LTE_(telecommunication)

8 Kettemann, M. C. (2015, 16 October). Zugang zum Internet: Ein Grundrecht auch für Geflüchtete. *iRights.Info*. https://irights.info/artikel/internetzugang-fluechtlinge-voelkerrecht-verfassung/26266

9 Freifunk Darmstadt. (2018). Digitales Ehrenamt – Jetzt! www.digitales-ehrenamt.jetzt

10 Kasparek, B., & Speer, M. (2015). Of Hope. Hungary and the Long Summer of Migration. *Bordermonitoring.eu*. bordermonitoring.eu/ungarn/2015/09/of-hope-en

11 Andre. (2013, 22 October). Embassy of Hope. *hamburg.freifunk.net*. https://hamburg.freifunk.net/2013/10/532.html, accessed 28 May 2018.

12 Amnesty International (2017, 15 December). When smartphones are a lifeline, not a luxury. *Amnesty International New Zealand*. https://www.amnesty.org.nz/when-smartphones-are-lifeline-not-luxury

13 Kubitschko, S. (2018). Chaos Computer Club: The Communicative Construction of Media Technologies and Infrastructures as a Political Category. In A. Hepp, A. Breiter, & U. Hasebrink (Eds.), *Communicative Figurations: Transforming Communications in Times of Deep Mediatization*. https://link.springer.com/book/10.1007%2F978-3-319-65584-0

14 In a now offline document by the Berlin County Office for Refugee Issues, a provision of free wireless network access in public areas as well as two desktop computers (per 100 people) are designated as minimum quality criteria. See also: Landesamt für Flüchtlingsangelegenheiten. (2018). Qualitätssicherung. https://berlin.de/laf/wohnen/informationen-zum-betrieb-von-unterkuenften/qualitaetssicherung

15 https://www.eduroam.org

16 https://wiki.freifunk.net/Berlin:Refugees

According to Philipp, problems with doing installations in shelters are less legal than financial in nature. At first, many social workers and managers were sceptical about the legal status and technical details of Freifunk, particularly in relation to network liability. In order to persuade them he then needed to show how – through the use of VPNs mentioned above – liability laws can be circumvented. Concern then turned to budget issues, including what scale of the installation was possible (e.g. should it only be available in certain spaces or the entire building) and how running and maintenance costs would be covered.

Philipp found that there were only a few social service providers who fully integrated the digital infrastructure into the thinking behind newly built accommodation for refugees. Though some managers agreed to cover the full costs for the routers and antennas that needed to be used, others had to rely on the Freifunk initiative to provide donated equipment and time, as they did not have designated funding for media and communication infrastructures. Furthermore, improvised emergency shelters like the university gym are repurposed buildings that are expected to only be used for short periods of time.

Phillip says the initial rapid growth of "Freifunk for Refugees" was only experienced in the first stage of the project. Now, he says, only a handful of people show up to the meetings, while shelters have closed down, and migrants have been allocated to individual housing projects or deported. Though he still gets asked via word-of-mouth to help with installations, he founded a one-man enterprise through which he can now negotiate contracts and small reimbursements for installation and maintenance. He is aware that this is not an uncontroversial practice among Freifunk communities, since it violates traditions based on reciprocity, non-commerciality and passing on skills to others. To this criticism he responds that many shelters and reception centres simply demand a more "professional" service with clear responsibilities. Furthermore, he thinks that it would be unfair for them to "simply lie back and relax". Rather, he says, the management should be held responsible by at least covering the ongoing financial costs of the infrastructure.

The debate on how to make internet access possible by "any means", but in doing so having to negate some of Freifunk's principles, indicates how the DIY practices of Freifunk can become entangled with what anthropologist and medical doctor Didier Fassin calls "humanitarian reason".[17] For some Freifunk participants, this break with shifting established practices was only questioned at certain points. Frequently discussed examples were when social service providers urge them to install content filter software on the network or set an automatic shutdown of the network at certain times of the day. In this situation, many participants have resisted the request or stopped the installation altogether. By the same token, Freifunk routers and practices are turned into "humanitarian goods"[18] ready for philanthropic fundraising campaigns (with a peak in fundraising experienced in 2016). This also puts Freifunk projects in competition with a larger market focused on devices and funding opportunities for other "good projects".[19]

Case 2: Bringing Freifunk to the classroom

With forced migration resonating as a major issue throughout wireless communities in 2015, many were in parallel looking to expand Freifunk practices to educational and social institutions. To explore this potential, Freifunk participants collaborated with the Media Institute Berlin-Brandenburg (mabb) to fund a joint project.[20] In its first round, the project focused on 10 youth centres in Berlin which already qualified for teaching "media competency" to kids and adolescents. The plan consisted of holding several empowerment workshops through which staff and teenagers would learn how to "flash" (install modified firmware) and set up routers locally. It was further assumed that the workshops would lead to the formation of working groups that would continue to expand the network into the adjacent neighbourhood.

One of the participants was Holger, a 47-year-old IT specialist, who we interviewed about his experience. Now employed as a system administrator for a company that provides IT solutions in social and educational institutions, he previously used to work as a media instructor in one of the designated centres. Sympathetic to the idea of community networks, he felt strongly about the specific notions of freedom that underpin Freifunk: not only should access to the internet be without costs, but it should also come without a content filter. Instead of tabooing "harmful content", the openness provided by Freifunk serves as a way to promote responsibility among the youth centre's clients.

---

17 Fassin, D. (2011). *Humanitarian Reason: A Moral History of the Present*. University of California Press.

18 Collier, S. J., Cross, J., Redfield, P., & Street, A. (2017). Preface: Little Development Devices / Humanitarian Goods. *Limn, Issue 9*. https://limn.it/articles/precis-little-development-devices-humanitarian-goods

19 Krause, M. (2014). *The Good Project: Humanitarian Relief NGOs and the Fragmentation of Reason*. Chicago: University of Chicago Press.

20 www.mabb.de/information/digitale-welt/freifunk.html

Even though Freifunk networks were established at eight out of 10 youth centres, running the workshops according to Freifunk values proved difficult for both the centres' staff as well as the attending kids. Due to his background in free software, Holger explained, it was easier for him to engage in Freifunk practices, which he classifies as rather "high level" compared to other activities at the centre. Still, several kids around the age of 12 actively helped him to set up the equipment.

Holger pointed out that compared to the cosmopolitan inner city, the centre is located in an infrastructurally marginalised district and that fostering participation is not an easy task. Yet he remains optimistic, especially when it comes to promoting gender equality through engagement with technology. Though the majority of participants in the Freifunk workshop were young boys, Holger has carried out coding classes at the centre where the gender distribution tended to be equal. This, he argues, can be linked to broader shifts in computing and the gaming scene, a trend that might also reach Freifunk.

To foster this development, Holger emphasised that Freifunk's DIY approach should already be part of the school curriculum to enable an in-depth learning experience using digital technologies. It is precisely this question that will be taken up in the second round of the project: to try to get the Freifunk model taught in schools in an interdisciplinary course covering computer science, physics, maths and ethics. The project is further driven by the fact that many schools in Germany are rather poorly equipped when it comes to digital media infrastructure. It is not uncommon that maintenance of IT systems heavily depends on the commitment of individual teachers. It is here that the second phase of the project resonates most closely with the first, given that it takes intensive training for teachers to become apt at "doing" Freifunk. One response is to develop open educational resource materials that can be used for teaching purposes.[21] For this, role models can be found in projects like the Junge Tüftler ("Young Tinkerers").[22]

Ultimately, the aim of the Freifunk schools initiative would be to offer an inroad for kids and young adults to question the commodified and privatised nature of contemporary digital infrastructures. In his article "There Is No Free Software", anthropologist Christopher Kelty points out the intense commercial extraction of "open source" practices to fit the needs of both large software companies and (precarious) software workers. According to him, the political significance of free software was derived precisely via its hybrid position "between the corporate forms of intellectual property-saturated IT industries and the cultural uptake of software and tools."[23] Therefore, the schools project is another fresh attempt to expand Freifunk in new contexts through workshops, educational materials and dialogue with key political players.

Conclusions

Juxtaposing the two different cases above renders visible the multiple natures of the Freifunk initiative and how its traditional practices play out in different political and educational contexts. On the one hand, participants managed to challenge the precarious media infrastructures encountered in crumbling public institutions and facilities guarded by an oppressive German border regime. On the other, it presents room for the concerns of Freifunk participants about adjusting to humanitarian logics, facing unexpected "professionalisation" and translating their practices to differently situated communities. Some more than others might heavily limit what it means for Freifunk to assemble people around a shared concern

It is important to keep in mind the limits of our report, focusing predominantly on the experiences of relatively well-situated and educated middle-class activists. While indeed representing a significant part of the Freifunk community, the two cases can be linked to a transnational cultural form based on solving socio-political problems with means developed in Silicon Valley-influenced tech communities.

This phenomenon is rendered visible by science and technology studies scholar Lilly Irani in a beautiful article investigating a design event in Delhi, India, that centres on "hackathons". Encountering similarly well-situated middle-class Indians, she shows that the hackathon does not necessarily produce any functioning products but rather encourages the "entrepreneurial citizenship" of participating subjects.[24] Freifunk participants need to be aware of this when encountering middle-class imaginaries of the present and future, and engaging with differently situated adolescents or migrants who either struggle for citizenship or the means to overcome its limitations.

21 https://freifunkoer.github.io/Freifunk-OER

22 https://junge-tueftler.de/fuer-umdenker

23 Kelty, C. (2013). There is no free software. *The Journal of Peer Production #3*. peerproduction.net/issues/issue-3-free-software-epistemics/debate/there-is-no-free-software

24 Irani, L. (2015). Hackathons and the Making of Entrepreneurial Citizenship. *Science, Technology, & Human Values*, *40*(5), 799-824.

Action steps

In sum, it becomes clear that Freifunk communities in Berlin and beyond are advocating for emerging forms of "digital volunteering" presented in the two cases. It is the unruly potential of Freifunk as an initiative to foster decentralised organisation, an infrastructural commons and public engagement without surrendering to the dominant Silicon Valley startup model. To keep this experiment going, we propose the following steps for policy makers, Freifunk communities and future wireless activists to consider:

Policy makers (EU and Germany)

- Increase public funding for free wireless networks.
- Consider a separation of "network" and "service" to protect net neutrality (i.e. the Swedish model).
- Expand EU-based funding schemes to go beyond acquisition of hardware for installations, to also cover the running costs for maintaining networks, and their general sustainability.
- Acknowledge the role of Freifunk and other community networks in public participation and public education.
- Foster dialogue between Freifunk members and social/public institutions.

Freifunk communities and future wireless activists

- Work towards reflecting the political dimensions of practices framed as "political" or "humanitarian" so as to question asymmetries between "givers" and "receivers".
- Leverage their role as political advisors and explore new means of fostering basic rights to access, particularly for minority communities.
- Interrogate the forms of subjectivity, citizenship and exclusion they produce in their institutionalised/commercialised practices.
- Strive for further transnational engagement.

GHANA

TRIUMPHING OVER DISTANCE AND TIME: THE CASE OF WIRELESS GHANA

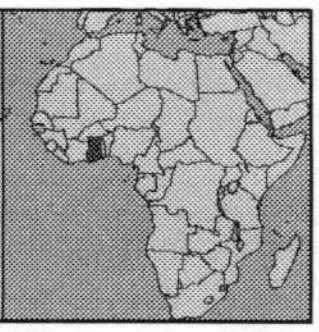

Afrotribune
Emmanuel Elolo Agbenonwossi
https://icannwiki.org/Afrotribune

Introduction

The internet has become an effective tool for communication at different levels in the life of Ghanaians since its introduction in the country in the late 1990s.

Ghana is considered one of the more stable countries in West Africa since its transition to multiparty democracy in 1992, and consistently ranks among the top countries in Africa when it comes to freedom of speech and press freedom.[1]

Although Ghana was among the first countries in sub-Saharan Africa to have internet access, the penetration rate remains worryingly low. As of December 2017, only about 10 million people, or roughly 35% of its population of about 29.6 million, used the internet.[2]

While still a "young" technology, the internet has had an impact on the life of Ghanaians, with many of them trying to challenge the digital divide by connecting the unconnected in their communities. This report discusses one such attempt: Wireless Ghana.

Policy environment

Ghana's telecom sector was one of the first on the continent to be liberalised and deregulated, allowing universities and communities to build their own networks. The privatisation of Ghana Telecom in 1996 was the catalyst for an extraordinary growth in market competition across the mobile, internet and fixed-line sectors. The reforms yielded mixed results. Landline telephone penetration increased dramatically (78,900 to 130,000 as of December 1997), while the number of mobile subscribers surpassed even this higher level of fixed-line subscribers.[3]

In Ghana, the National Communications Authority (NCA) is the government agency responsible for licensing and authorisation for the operation of communication systems and services. It was established by the National Communications Authority Act, 1996 (Act 524).

The NCA assigns, allocates and regulates the use of frequencies and develops strategies for the communications industry in the country. The NCA is responsible for managing civilian access to radio spectrum. Licensed spectrum in Ghana includes licensed "exclusive" spectrum (traditional analogue TV, mobile cellular) and licensed "shared" spectrum (LTE/LTE A, 2.3 GHz and 3.5 GHz Bands), while unlicensed spectrum includes "unlicensed exclusive" spectrum (the extremely successful Wi-Fi bands of 2.4 GHz and 5.8 GHz) and "unlicensed shared" spectrum (TV white space coexisting with licensed TV).

Currently there is no legislation or policy governing community networks. Most of the community networks operate using free 2.4 GHz and 5 GHz wireless frequencies.

Wireless Ghana

One of the most interesting chapters in internet development in the country has been the founding of Wireless Ghana,[4] formerly known as the Akwapim Community Wireless Network. The name Akwapim Community Wireless Network was changed to Wireless Ghana after the pioneers of the project decided to expand to other regions of the country. Today Wireless Ghana provides a large number of professional services in the fields of wireless technologies, software solutions and IT support in different communities in Ghana.

Wireless Ghana is a non-profit project implemented by Community-Based Libraries and Information Technology (CBLit), a non-governmental organisation based in an isolated community in the mountainous eastern region of the gold-rich nation. This Akwapim North District has 17 towns and several villages with a total population of about two million people.

The Wireless Ghana project was started in 2005 at the Apirede Resource Centre (CBLit's first

1 Reporters Without Borders. (2018). Ghana: Media pluralism but not enough independence. https://rsf.org/en/ghana

2 Daily Graphic. (2018, 22 February). Over 10 million Ghanaians use the internet – Report. *GBC*. www.gbcghana.com/1.11795371

3 Haggarty, L., Shirley, M. M., & Wallsten, S. (2002). *Telecommunication Reform in Ghana*. World Bank Policy Research Working Paper No. 2983. https://ssrn.com/abstract=636345

4 wirelessghana.com

Testing of the internet on the street after installation.

community resource centre) in Akwapim in response to the local community's requests for internet connectivity to help them break their isolation and also connect their children to the digital revolution.

The project and its resource centre have become the game changer with respect to accessing information and triumphing over distance and time. It cuts down barriers with the click of a mouse, helping schools, teachers, churches, farmers and youth groups in the community communicate with the rest of the world.

The network was developed and is managed and maintained by the community members themselves with the help of volunteers from the United States Peace Corps programme.[5] The management committee together with the chief and elders of the community monitor the daily operations of the network to help create a professional environment, and ensure that services are of quality and delivered on time. They also manage the accounts and bills from internet service providers (ISPs).

The network currently has over 20 nodes, and spreads out over a 10 to 15 km range, offering connectivity to Koforidua Technical University, secondary schools, churches, non-profit organisations, businesses and community activity centres throughout six towns in the mountainous region.

Since its introduction in the community 13 years ago, the initiative has also built local digital libraries which have become hubs for free and open access to information and documentation for students and teachers. The community libraries are equipped with computers with internet access and allow students to browse and do their research. The volunteers of Wireless Ghana also help train the students on computer literacy, internet use and coding.

In 2014 the initiative was extended to other regions. Wireless Ghana has helped establish a wireless network at the University of Cape Coast in the western region. The project is called Campus Wi-Fi Project, helping students and lecturers to share resources and data among themselves. Another network was launched at Sakumono in the Tema Metropolitan District, a district in the Greater Accra region of Ghana.

Wireless Ghana does not operate as an ISP. The organisation only builds and maintains the infrastructure for the communities. The internet services are managed by the communities themselves, contracting with normal ISPs. There is fibre and VSAT

5 https://www.peacecorps.gov

Testing of the local antenna after construction.

in the communities where the community network operates.

Technologies involved

According to the chief technology officer, Wireless Ghana is an open-architecture mesh network that uses low-cost Wi-Fi hardware and open-source software, with most installations being point-to-multipoint networks with routers constructed from old computers and new wireless cards.

Most of the nodes were made from Dell OptiPlex units, D-Link DWL-G520 Wi-Fi cards and Cisco Meraki routers. However, with the rapid growth of the project, the founders have changed technologies to be able to cover long distances, and today, most of the equipment has been purchased from the US supplier Ubiquiti. This includes the NanoStation NSM2 (2.4GHz), Bullet M2 and PicoStation M2. The nodes on the Akwapim network are connected to the rest of the internet via a shared 128/32 kbps VSAT connection.

Funding and challenges

The initial phase of the project involved grassroots engagement with the community, together with Peace Corps Ghana volunteers. Funding and skills and technical support were also received from the Champaign-Urbana Community Wireless Network (CUWiN),[6] Ghana Educational Services,[7] the Akwapim North District Council, TakingITGlobal,[8] and the Student World Assembly,[9] among other partners. The initial funding was between USD 10,000 and USD 15,000.

The cost of a typical node installation for individuals and institutions in the community is about USD 500. This covers all of the equipment and installation expenses necessary to mount an antenna on a rooftop, run cables indoors, set up an indoor router and connect personal computers to the router. Nodes are guaranteed to be installed within two weeks, depending on the availability of equipment that must be imported.

Before the installation of a node, the Wireless Ghana team writes a contract for potential node users that allows them to obtain access to the network. The start-up contract informs the node user of the cost of the equipment, the timeline for

6 transmissionproject.org/projects/champaign-urbana-community-wireless-network-cuwin
7 moe.gov.gh
8 https://www.tigweb.org
9 https://orgs.tigweb.org/student-world-assembly

installation, and the cost of internet services on the network.

The nodes are built and installed by the volunteers working on Wireless Ghana. Following the installation, the beneficiaries (communities or universities) are trained on how to use and maintain the nodes. The training is usually conducted by the volunteers.

Community members who use the network pay fees every month to the volunteers to enable the sustainability of the network and also to pay the ISP bills for internet access. The two universities that have been connected to Wireless Ghana self-manage the equipment and internet bills.

Other challenges faced by Wireless Ghana are radio-frequency interference and lack of electricity in some of the villages. Wi-Fi devices and non-Wi-Fi devices that occupy the same shared radio frequency spectrum usually cause interference on the network, and sometimes there are electricity blackouts.

The electricity grid in the region reaches only 60% of homes. This means about 40% of households lack access to any means of conventional power. Even in the areas where the national grid is present, the quality of the service is an issue due to incessant power outages, sometimes for a considerable length of time.

To solve the power problem, the Wireless Ghana team sometimes builds alternative power supply systems with recycled car batteries and other locally available equipment. In one case, for example, a broken uninterruptible power supply (UPS) was rebuilt with car batteries. This was innovative because the team took broken computer parts and turned them into a device that is capable of powering a wireless network node for up to 24 hours during a power outage.

It is, however, important to note that it is difficult to provide internet access and other communication services to areas not connected to the electrical grid, and this constitutes a major challenge for Wireless Ghana. This is a significant limitation to the development and spread of community networks in rural Ghana.

Action steps

While growth in the internet market has increased with higher penetration in Ghana in recent years, there is enormous potential in the coming years for community networks. Communities are capitalising on consumer demand for services by upgrading existing infrastructure. But funding, capacity building and electricity remain the main challenges. It is therefore important for donors and the government to make funding and capacity-building programmes available for communities who wish to set up community networks.

TV white space (TVWS), otherwise known as dynamic spectrum allocation, is also seen as a promising form of connectivity for extending broadband networks to rural areas across Ghana. The NCA authorised Spectra Wireless to run a six-month trial on 8 January 2014.[10] The Ghanaian regulator granted the first licence to operate a commercial TVWS network on the continent on 8 December 2014.[11]

Using the new service, students at Koforidua Polytechnic and in some surrounding dormitories in the eastern region can get a Wi-Fi connection starting from 2 Ghanaian cedi (USD 0.61) a day for unlimited bandwidth. Students can choose from symmetrical services with speeds ranging from 0.5 Mbps to 2 Mbps.

The project is considered a promising alternative to broadband access and the government of Ghana and all the stakeholder groups must work together to take advantage of it to expand community networks.

Lastly, it is particularly important that the power problem confronting rural communities in Ghana be addressed. Pilot experiments in the use of solar energy and photovoltaic cells should be supported.

10 BBC. (2014, 28 May). TV White Space in Ghana. https://www.bbc.co.uk/programmes/p01zdr24

11 Okpamen, E. (2015, 26 January). First Commercial TV White Spaces Services Launched In Ghana. *Ventures Africa*. venturesafrica.com/first-commercial-tv-white-spaces-services-launched-in-ghana

GREECE

CONNECTING THE UNCONNECTED IN RURAL GREECE: THE CASE OF SARANTAPORO.GR

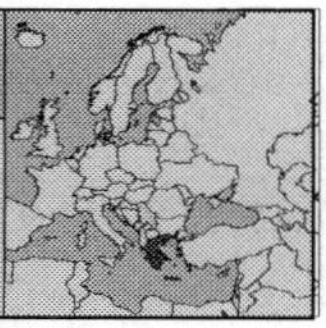

Sarantaporo.gr
Vassilis Chryssos, Aimilia Voulvouli, Alexandros Papageorgiou, Giorgos Klisiaris and Nicholas Kourtzis
www.sarantaporo.gr

Introduction

In 2010 a group of friends from the village of Sarantaporo in the municipality of Elassona in Greece built a website, Sarantaporo.gr, to promote their village. However, they could not show the website to people in the village because the villagers did not have internet access. Due to the remote location and small population density of the area, the telecommunication providers found little or no commercial interest in deploying infrastructure to bring internet connectivity to the region, leaving the local population with no alternative.

With no previous knowledge and experience in building a wireless network, the team turned to examples of community networks in other parts of the world to learn from their experience and practices, such as the Athens Wireless Metropolitan Network,[1] guifi.net[2] in Catalonia, and others.

Volunteering their time and effort, and collaborating with the community, they soon managed to build their first wireless mesh network in the village of Sarantaporo. In doing so they provided open internet access to the inhabitants and visitors to the area.

Soon community members from neighbouring villages approached the project's core team – called the "Sarantaporo.gr" core team after their village website – and asked them to build similar networks in their villages. Over the following three years, until 2013, the network was deployed in 15 villages in the area.

In 2013 we founded Sarantaporo.gr as non-profit organisation (NPO) and soon after, at the beginning of 2014, we managed to secure funding from a European Union (EU) FP7[3] programme to build our backbone network. The wireless network connected the villages to each other and extended as far as the nearest city, covering a line-of-sight distance of 50 km. These wireless links provided connectivity via the local university, the University of Applied Sciences of Thessaly,[4] which offered a 1 Gbps backhaul as a social responsibility service to the local unserved communities of the region. Today 11 villages are connected to the backhaul, served by an average of more than 30 Mbps symmetrical internet connectivity.

The agreement signed with the university was part of a collaboration strategy which involved a diverse array of stakeholders, such as the Greek Free/Open Source Software Society (GFOSS),[5] Athens' Hackerspace.gr,[6] P2P Lab,[7] the Alliance of the Commons,[8] the Social and Solidarity Economy (SSE)[9] movement in Greece, third-party service providers, and others. These collaborations outlined the character of the Sarantaporo.gr wireless community network as a holistic approach to supporting the local communities, and not strictly as a technological endeavour.

Our mission is to bridge the digital divide in the region. Sharing this mission with local communities has been perhaps the single most significant challenge we have faced until today. As none of the Sarantaporo.gr core team members lives in the region (the closest member lives in Larisa, 80 km away, while most of us live in Athens, the capital of Greece, 450 km from Sarantaporo), it has been quite challenging to maintain communication with locals and even more so to share our vision and mission and align these with local communities.

To tackle this challenge we spent a great deal of effort and resources organising events and training workshops and tried diverse communication tools. Eventually we managed to identify local champions in every village. These are the most engaged and active members of the local community. Their contribution has been invaluable in maintaining and

1 awmn.net

2 https://guifi.net/en/node/38392

3 Seventh Framework Programme, an EU research and development funding programme. https://ec.europa.eu/research/fp7/index_en.cfm

4 https://www.teilar.gr/index_en.php

5 https://gfoss.eu

6 https://www.hackerspace.gr

7 www.p2plab.gr/en

8 https://commons.gr/about_en /

9 An umbrella term for various social and solidarity initiatives in Greece.

expanding the local community network and also in creating ties between the villages, empowering the community which is built around the wireless network. Today we count more than 70 community members in our network, 20 of whom are women.

Sarantaporo.gr has managed to drastically change the lives of local people. Visiting doctors in the villages are now able to prescribe medicines using the health portal, the elderly inhabitants get to have their grandchildren visiting for longer periods since they now can enjoy internet connectivity in the village, local farmers can discover new markets for their products and raw materials, and locals can communicate with friends and family abroad, which makes them feel less isolated.

Organisational sustainability

As an NPO, Sarantaporo.gr consists of a group of 10 people (the core team). The NPO was deemed the most appropriate at the time of registering it in 2013, due to the social and voluntary nature of the endeavour. The NPO is run by an executive committee, comprised of up to three administrators who are elected every two years by an assembly. Important decisions about the course of the endeavour are reached via consensus in the assembly, which takes place bi-monthly, or on an ad hoc basis when the need arises. The basic mode of operation is "do-ocracy": members suggest actions and, if not vetoed by other members, they go ahead and try to fulfil them.

Every three months we organise an open assembly in one of the villages. All of the local community members can participate and voice their proposals. Decisions reached in these assemblies are advisory and not binding for the NPO.

In order to empower locals to engage more actively in the life of the community networks, we organise and deliver workshops and seminars. Apart from digital skills building, these seminars are also about communication and community building. We secure sustainability by including train-the-trainer sessions for the more keen community members.

In terms of funding we try to diversify our streams to ensure financial resilience. Currently our revenue mix comprises private donations, yearly member subscriptions, local services fees, grants and programme funding. The Internet Society (ISOC), an international organisation, and the EU-funded programmes netCommons[10] and CONFINE[11] are a few of the funding sources for our organisation.

10 https://netcommons.eu

11 https://confine-project.eu

Part of our revenue is used to participate in international meetings, conferences and networking events with other community networks around the world. In November 2017 we co-founded the ISOC Community Network Special Interest Group (CN-SIG)[12] along with other community networks from all over the world during the Internet Governance Forum (IGF) 2017 in Geneva. The CNSIG serves as a vehicle through which community networks from around the world can develop, strengthen and promote the community network model, draft common strategies, share experiences and expertise, debate policy and regulatory issues, and present their views through their direct involvement and participation.

Technical sustainability

Our community network comprises two layers: the backbone and the access layer. The backbone network is a tree and mesh topology based on 802.11 and other networking standards. It utilises redundant routing per backbone node with two or more links with failover (not load-balancing). The access network is a mesh topology, based on 802.11 and proprietary standards. The available throughput (~60 Mbps symmetrical on average currently) is not purposefully limited per device or user. It is provided on a best-effort basis and shared among members on a good neighbour principle.

The main considerations concerning the backbone are capacity, reliability, resilience, and central monitoring and administration. Resilience, for example, is a critical factor for the given region. Power outages are quite regular in the area and when they happen, not even phone landlines work (legacy phone lines had been converted to VoIP). This, combined with the isolation of the area due to extremely bad weather, results in quite severe conditions for local inhabitants. Adding resilience to our network by installing UPS devices in each node ensures that a communication channel will almost always be available for locals to use.

Access network considerations concern ease of installation, admission control, central monitoring, end-user support, and administration. Of great importance to our community network is end-user support. Considering that our network users are not mere customers, but members of our local community, it is critical that we build a trusting relationship and that we do not just provide customer support, but also empower local community members. We partly achieve this via our instant messaging channel, which we use daily as a communication tool.

12 cnsig.info

A statement by one of our community members is indicative: "We couldn't have dreamt of this level of support from incumbents."

Sarantaporo.gr: Social actors and a community that is a "work in progress"

In the case of Sarantaporo.gr we have learned that community networks are not only networks of a certain community but also networks that enable community. Community networks constitute infrastructure that, in a first instance, aims at covering the tangible needs of a local population by filling, bridging and closing gaps, whether they be communicative, social, technical, institutional or other. However, this reading often presupposes that there is "community", that its members will want to own and manage the network, that the gaps are acknowledged by all, and that their filling or bridging is consciously pursued.

The villages in the area west of Mount Olympus do not form a cohesive community, either in administrative, cultural or economic terms. Nor were locals familiar with the existence of the technology in question and the possibilities it could offer. Consequently, the initiative and mediation by a group of experts/activists (the core team) was essential for the community network to be created, developed and sustained, and for locals to be informed and involved. We should not fall into the trap of viewing rural communities as the passive recipient of a one-way beneficial offering. Neither should we assume that the communities we work in have a strong sense of social cohesion. Instead, based on our experience in the wider area around Sarantaporo, we need to emphasise the enabling dimension of community networks, the potentiality and openness that they activate, the serendipitous, dynamic and collaborative manner in which needs are traced, problems are addressed, solutions are devised and community is moulded, in ways that could not have been anticipated or planned.

Having said that, we need to be aware of history and its implications in trying to discuss "the sort of people whose names are usually unknown to anyone except their family and neighbors but who nonetheless are major historical actors when they act collectively."[13] According to Hobsbawm, such people, widely known as "the common people", are actually far from common given that when they have acted collectively they have made a difference and can again shape history. So Hobsbawm advocates for a "history from below", the history of committed men and women not as passive subjects of macro-history, but as progressive forces of society, which is something else we have learned in the field working on Sarantaporo.gr.

Drawing on participant observation in the area where the infrastructure of Sarantaporo.gr is located as well as social media ethnography and multi-sited ethnography – that is, ethnography pursued in more than one geographical location – to include not just the local users/node-holders of the network but also the core team members that do not reside in the area, we became aware of the importance both of personal histories of the individual participants and of the historical context of Thessaly, the administrative region in which the villages of the network are located.

The rural movement in Thessaly, which sprang up at the beginning of the 20th century and extended into the first years of the century, was one of the strongest movements within the then newly established Greek state. Similarly, the cooperative movement was born in 1900 when 24 peasants established the first agricultural cooperative in Almyros, a town in Thessaly.[14] The contemporary collective action in Elassona that is the result of setting up the community network is seen in this light.

The life trajectories both of the local node-holders and of the core team members who were interviewed shed a similar light on the communities we worked in. Sarantaporo.gr is a "work in progress", a community of people whose participants are rational social actors rather than docile consumers. Given that the state was unable and the market unwilling to provide connectivity, they were in a position to identify the crack in the system that gave life to Sarantaporo.gr. Following Hobsbawm's train of thought of a "history from below", we argue that communities have the potential to muster collective power that can bring about social change.

Legal and regulatory issues

As far as the law in Greece is concerned, non-profit community networks are considered equal to any other internet service provider, with the same licensing requirements for spectrum. This, for example, makes it very difficult to license channels in an interference-free band such as 11 GHz, which would be great for point-to-point backbone links. Yet the Wi-Fi band which we utilise is unregulated, and this allows us to operate our community network without the requirement for any special licence.

13 Hobsbawm, E. (1998). *Uncommon People: Resistance, Rebellion and Jazz*. New York: The New York Press.

14 Patronis, V., & Mavreas, K. (2004). Agricultural Cooperative Organizations in Greece throughout the 20th Century: A Critical Overview. *Journal of Rural Cooperation, 32*(1), 51-62.

The owner of the equipment is the Sarantaporo.gr NPO. Each piece of equipment is provided to the community members as a loan. The maintenance cost for the access layer is covered by the village community, while the cost for the backbone layer is covered by the NPO.

Action steps: The future

When we first deployed our community network, many locals felt that there was no use for it, since they were used to living an isolated life in the rural countryside. Soon the profound impact became evident in every aspect of their lives: social cohesion, economic development, citizenship empowerment, health services. The local population realised the possibilities of how digital inclusion could improve their lives. This was a stepping stone for the locals to pursue even better connectivity and more active engagement. We continue to nurture our relationship by organising local events and assemblies, providing training via workshops and promoting transparency about the operation and management of the community network.

Sharing and an open culture lie at the heart of our endeavour. We strive to collaborate with other community networks around the world and seek to share our knowledge and experience so that others can benefit from them, just like we benefited from the stories of other community networks before us. The three components that are *sine qua non* for creating a successful community network are: infrastructure, local community and training/education. Deploying infrastructure, building a local community to run and manage it, and educating locals to acquire digital skills are necessary pillars towards a sustainable community network.

Over the next few months we are organising our first mission to Northern Tzoumerka, another mountainous area in Greece, in the Epirus region, in an attempt to work with locals towards building the first node of their community network. Given that it is important to cultivate human relationships in parallel to the network, we are taking members of our local communities to the region to share their experiences and tell their stories.

Among the many challenges we face for the future, perhaps the single greatest is to discover ways in which the community network can serve inhabitants as a local infrastructure. In other words: what is the added value of the infrastructure for locals if it is disconnected from the internet? The answer to this question can be a defining one for the future of our community network and others around the world. Currently and in the near future we will be exploring technologies and experimenting with peer-to-peer approaches and the local deployment of services. Data retention and management, safeguarding privacy, information sharing in the local context, e-health and agricultural internet-of-things technologies are fields of interest we wish to explore.

HONDURAS

THE AZACUALPA COMMUNITY NETWORK: EMPOWERING THE LENCA PEOPLE THROUGH TECHNOLOGY

Red de Desarrollo Sostenible-Honduras and Internet Society Honduras Chapter
Eduardo Tome
https://rds.hn

Introduction

The Azacualpa Valley is located in the village of Yamaranguila, Intibucá, right in the heart of the Lenca corridor in Honduras. Surrounded by mountains and completely isolated from the world, the people of Azacualpa have no paved roads, no telephone lines, no internet access, no power supply, no running water and no health services; the nearest town is 45 minutes away by bus.

Red de Desarrollo Sostenible (RDS), the organisation I represent, is a Honduran NGO that has been working in capacity building with a focus on information and communications technology (ICT) since 1994, with a vast experience working with communities all throughout the country dealing with subjects such as disease prevention, food security, community radio and internet governance. We started working with the community in 2017 through the help of Olivia Zúñiga Cáceres, who is currently in the Honduran Congress, and Salvador Zúñiga, an RDS board member; they are also the daughter and former husband of late indigenous leader and environmental activist Berta Cáceres.

In that year we were able to launch – through a Cultural Survival Community Media Grant[1] and with the help of the "Impactos" programme – the first community radio fully directed and operated by women in the country, *Radio Azacualpa – La Voz de las Mujeres* (Radio Azacualpa, the Voice of Women). By empowering the local women in becoming leaders in their communities, a strong community structure has flourished and developed in Azacualpa, generating an environment in which community members have been seeking opportunities to overcome the limitations they currently face.

Last year we approached them with the idea of starting their own community network; they immediately jumped on board and though they do not fully grasp the concept of internet, they understand it to be a tool that can help them unlock the great potential that their community holds. This is the story of the Lenca people of Azacualpa, and their experiences and their challenges as they become the first community network in Honduras.

Socio-political context

The population in Azacualpa is 1,070 inhabitants. Each family has an estimated six to eight members, and there are around 300 families in the entire valley. The main economic activity in the Azacualpa Valley is agriculture.

The Lencas, like most indigenous communities in Honduras, endure hardships to protect their lands and resources and most have no access to basic services. They have no education in their native language, and their communities generally do not have properly equipped education centres; in the words of a local Azacualpa teacher: "We chose not to give the students homework because we know they don't have anywhere they can do research." Their access to health is also limited; they have a high mortality rate amongst children with diarrhea being the leading cause.

Community radios have helped with this situation. They have empowered local leadership and helped create consciousness amongst the Lencas regarding their rights, and knowledge about the historical struggles of their people. This includes the life of environmentalist Berta Cáceres, murdered in 2016 while in the middle of protests to protect the Gualcarque River from a power company, a river that lies in historical Lenca grounds and that serves as a water supply to several of their communities.

In Honduras, the digital divide is progressively being reduced. According to a 2017 report by the National Telecommunications Commission, an estimated 2,240,400 people have internet access, which is 25% of the general population. This number amounts to approximately 29 out of 100 people who are internet users. In the urban area, around 42.6%

1 The Community Media Grants Project provides opportunities for international indigenous community radios to strengthen their broadcast infrastructure and systems while providing training opportunities to their community journalists through a participatory and dynamic grants programme. Overall, the initiative enhances the community's efforts to establish and ensure sustainability of indigenous community-controlled media. https://www.culturalsurvival.org/current-projects/community-media-grants-project

of the population has an internet connection, while in the rural area the digital divide is much bigger, with only 14.1% of the population with internet access.

Policy-wise, community networks have a long way to go in Honduras. The fact that there is no legislation mentioning community networks is a double-edged sword for potential efforts. Without regulation, communities can explore a vast array of possibilities to get connected to the internet – however, this comes at a risk as it leaves community networks without legal recognition and without any protection from the state.

The start of the community network in Azacualpa

The idea of the Azacualpa community network was first raised in 2017 at the launch of Radio Azacualpa. The aim was to submit a proposal to the Internet Society's Beyond the Net Large Grants[2] for USD 30,000, which would be used to set up the network's infrastructure and secure connectivity for at least 18 months. A month before submitting the proposal, we wanted to be sure that people in the community were still on board, so we decided to conduct a survey in the households of community members we had met previously (around 40 households) to determine whether they were still interested and if the sustainability of the network was viable. The response was overwhelming: we discovered that they were already spending an average of USD 30 per month per household on phone services that they did not have access to locally, and that they often had to travel long distances on foot just to be able to make a phone call or send a text message.

They were interested – to say the least – although not entirely aware of how the internet works or what it is. Presenting the idea to the whole community was a bit tougher. Though the community is organised, geographically it is dispersed. The terrain is rough, and the distances people have to walk to get anywhere are very long. We knew the only chance of presenting the idea to them all at once was to do it in the only place the community congregates as a whole on a regular basis: church.

Deciding on responsibilities and tasks

With the help of the local priest, we were able to pitch the idea to the whole community with positive results, including the formation of three working groups to strengthen the social structure of the community network: the technical working group, administrative working group and governance working group. These working groups are formed by both men and women and have 17 members all together. Their tasks vary, though they complement each other in a very organic way. While the technical working group is in charge of the maintenance of the equipment and infrastructure of the network, it is also responsible for teaching peers in the community how to take full advantage of the network's technology. The administrative working group is in charge of managing the community network's resources, such as the computer lab established for such purposes, collecting the necessary financial contributions from the community to sustain the network, and making sure these contributions have a positive effect on the network to ensure its sustainability. The governance working group, which is made up of "house leaders",[3] serves the goal of consolidating the network from an organisational point of view; the social structure behind the network must be strong in order for it to survive. It is also important for the community to make alliances with key stakeholders in the local, national and global context.

Keeping the technology manageable

The success of these working groups will rely on how well they are able to communicate with each other and work together to create a community consciousness of the importance of the network and – most of all – to ensure that the community is able to appropriate the technology to suit their own purposes.

The technical aspects of the network are very simple and straightforward. This was done in order to keep the technology manageable for a community that had very little technical training and experience. The network consists of four nodes connecting Azacualpa via four radio towers to the nearby city of La Esperanza. Wi-Fi will cover an area of 70% of the Azacualpa Valley, reaching most of the community and some parts of neighbouring communities.

As mentioned, current regulation does not deal with community networks. The closest thing in the local legislation that resembles community

2 Beyond the Net Medium and Large Grants look for innovative ideas that advance the Internet Society's mission and contribute to the empowerment of people through projects focused on local impact. The initiative supports mainly projects from Internet Society Chapters and Special Interest Group members.

3 The Lenca people are very devoted to Catholicism. The Azacualpa community has divided itself into what they call "houses" which are largely family groups devoted to one specific saint, and there are eight of these houses in the community. These houses often organise religious celebrations together, such as pilgrimages within the community in which they parade the saint's images around and daily prayer sessions in certain months of the year to bless their crops and pray for rain.

networks are private networks which require no permits to operate as long as they do not sell their internet to third parties.

Setting up an association

The social structure that constitutes the network needed a legal structure in order to consolidate the network; the community decided that the best way to do so was to create a social economy association.[4] This allows them to negotiate better prices for their products, and reinvest profits in communal projects that will allow the community to continue their economic and social development on their own terms.

While the community's main income comes from agriculture, and though they grow high-quality produce, they often face hardships since they have no way of moving it to a nearby market, and intermediaries often underpay them for their products. They barely break even at times. The idea of the association has motivated the community to think of new ways to market their products, and to form groups that can negotiate better deals as they intend to wholesale their produce.

With the help of the internet, they will both earn more money and secure the economic sustainability of the community network.

Empowering women

The people of Azacualpa are very humble, hard-working people. When we first approached them, we found a community that was very united but in which women had a very restricted role in their households. With little to no sexual health education, the average Azacualpa household consists of eight family members. This means that men generally dedicate themselves to manual labour such as agriculture and tending livestock, while women generally stay home and raise their kids, tending to crops near their houses.

When they started the community radio, the women – who were often shy to speak – seemed to transform. As they progressed through their preparation before the official launch of the station, leaders were identified among community members such as María Guadalupe Gutiérrez and María Santos Sánchez, and now, in the community network, María Lázara Rodríguez. Seeing this has motivated more women to join the process.

This has had positive results in the Azacualpa community overall. Before the arrival of the community radio there were three women's groups identified in the village; now that number has effectively tripled, resulting in greater involvement of women in local organisations. Today women's groups have started tackling issues like sexual education that were once taboo in the community. We hope that these types of organisations keep thriving in the local context to help young Lenca women envision a better future for themselves and their people.

The community network has kept this spark alive with a large number of women joining the working groups established to support the network. The empowerment of women in technology will be key in the deployment of this network. Both the men and women of the community see eye-to-eye on gender equality issues and understand that unless they work together as equals, the community will not be able to fully reach its potential.

Local content and recovering the Lenca language

Another key aspect which the network wishes to address is the revival of the Lenca language, a language that has been lost not just in Azacualpa but in Honduras in general. A few years ago, anthropologists Alan King and Jan Morrow published *Kotik Molka Niwamal*, a dictionary of Lenca terms they were able to recover by speaking to Lenca elders in various communities throughout the country. We hope that one day the community will be able to publish content that allows the people of Azacualpa and other Lenca communities to recover their native language. Although the funding for this is not there yet, we believe the governance working group will be able to achieve this milestone for their culture.

Conclusion

While it is too early to determine the impact that the community network is having on the Azacualpa community, in the process of setting up the network we have witnessed a lot of enthusiasm, and the growth of great vision that the community has for its own development. The internet will not only allow them to sell products online and secure a better livelihood for their families, but will show them a new world filled with possibilities and opportunities for them to grow. They have shown great interest in the internet's potential for education, since the educational centre they have only reaches the ninth grade. They know that while the internet means more options for education, it also entails

4 In 1993, the Honduran Congress approved the Social Economy Development Law, which allows communities to organise themselves into Social Economy Associations. It is a business model that allows small producers to sell their products collectively to achieve a better quality of life as a community.

capacity-building processes that can strengthen the whole community.

The process has taught the community that having a clear view of the future and addressing these issues as a whole and not as individuals gives them a far greater power than they have ever imagined. They have embraced equality, and women in the community have a far greater voice than they ever had at any point in their history – this gives us hope that the Lenca people will be able to break a long tradition of subjugation in which the role of women was limited to childbearing and tending the house.

Women now stand arm-in-arm with their husbands and their sons in the process of revitalising their community, and are leading the process of teaching everyone an important lesson: a community will never be able to thrive if both men and women do not have equal opportunities to grow and express themselves.

Action steps

The following steps are suggested for strengthening the Azacualpa community network, and community networks generally in Honduras:

- The governance working group of the Azacualpa community network has to approach local government authorities and local associations and establish networks that will allow it to consolidate and prosper and ensure its sustainability through time.
- Funding for the production of local content in the Lenca language must be sought by the community. In this way they will be able to fully appropriate the technology they are using and make the internet a space for the Lenca people to protect their heritage and their traditions.
- It is hoped that the story of the Lenca people of the Azacualpa Valley will motivate other indigenous communities to replicate the experience. Once their story is out there, and local governments understand the benefits of community networks and the positive impact they can have on the development and empowerment of indigenous communities, policies will have to be created to ensure that community networks can prosper in Honduras.

INDIA

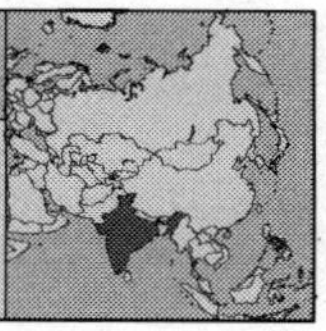

WOMEN CREATING SOCIALLY SUSTAINABLE WIRELESS COMMUNITY NETWORKS IN INDIA

Digital Empowerment Foundation (DEF)
Ritu Srivastava
https://defindia.org

Introduction

At one time, all roads were supposed to lead to Rome; in today's world, the "information highway" leads Rome right to our homes. Using a laptop, mobile phone or any other internet-enabled device, we can access data about Rome or virtually any other area under the sun on the internet. But how many have used the power of the internet?

The number of internet users in India is likely to cross 500 million this year; however, the percentage of women internet users in India is only 29% (approximately 143 million).[1] The telecom sector has one of the lowest percentages of women in the workforce – an average of between 8% and 15%, far less than the 26.6% female participation rate in the total workforce, according to the Gender Diversity Benchmark for Asia 2017 study.[2] The number of women in the science and technology workforce is 12.5%.[3] In rural areas, the gender divide is also felt. Few women engage in technical projects or have technical skills, largely due to cultural and social restrictions, but also because there are few opportunities for work in information and communications technologies (ICTs) at the grassroots level.

The Wireless for Communities (W4C)[4] project of Digital Empowerment Foundation (DEF) was initiated in 2010 with the aim of connecting rural and remote locations of India where mainstream internet service providers (ISPs) are unwilling to provide internet connectivity. In the last eight years, the programme has adopted various models of engagement – "hub and spoke", "wireless on a wheel", "Zero Connect box" and "women wireless entrepreneurs" being some of them – and connected 35 districts (in 18 states)[5] by establishing 178 access nodes, engaging men and women equally. By engaging women, DEF's community networks have created safe spaces for other women but also made them socially viable. This report discusses how women's participation and engagement has transformed our wireless community networks into socially sustainable networks. In particular we highlight how creating so-called women "barefoot engineers" in the community means that learning and knowledge are transferred, while women are empowered to act in a field that they were previously restricted from entering.

Policy challenges and opportunities for community networks in India

Last year, the global community reaffirmed the principle of "digital equality"[6] – or equal access to and use of ICTs for all people. The principle is seen as critical for socioeconomic growth and creating opportunities to achieve the Sustainable Development Goals (SDGs) by 2020. One important way to achieve digital equality is to roll out broadband to rural and other unconnected regions. However, taking connectivity to the rural regions of India has been a tough task for several reasons.

While the launch of the National Optic Fibre Network (NOFN) in 2011 aimed to connect 250,000 villages in India and the government's 2012 National Telecom Policy[7] called for "broadband access for all", fixed broadband penetration in India is only 1.4%.[8] Limited existing infrastructure to deliver broadband, including the availability of backbone fibre, network towers, a lack of backhaul connectivity, and the high cost of providing fixed-line

1 Agarwal, S. (2018, 20 February). Internet users in India expected to reach 500 million by June: IAMAI. *Economic Times*. https://economictimes.indiatimes.com/tech/internet/internet-users-in-india-expected-to-reach-500-million-by-june-iamai/articleshow/63000198.cms

2 Deloitte. (2017). *Women in the Boardroom: A global perspective.* https://www2.deloitte.com/au/en/pages/risk/articles/women-in-the-boardroom.html

3 Huyer, S., & Halfkin, N. (2013, 29 October). Study reports India's slow progress in advancing women in science and technology. *Elsevier*. https://www.elsevier.com/connect/study-reports-indias-slow-progress-in-advancing-women-in-science-and-technology

4 wforc.in

5 Andhra Pradesh, Assam, Bihar, Delhi, Karnataka, Tamil Nadu, Telangana, Uttar Pradesh, Madhya Pradesh, Punjab, Orissa, Rajasthan, Puducherry, West Bengal, Uttarakhand, Delhi, Jharkhand and Gujarat.

6 Commission on Science and Technology for Development. (2018). *Implementing World Summit on the Information Society outcomes, 2017*. unctad.org/meetings/en/SessionalDocuments/ecn162018crp2_en.pdf

7 meity.gov.in/writereaddata/files/National%20Telecom%20Policy%20(2012)%20(480%20KB).pdf

8 International Telecommunication Union 2017 data.

broadband, are some of the inhibitors to roll-out. Traditional ISPs are unwilling to go to rural areas as they are uncertain about the return on their investment. This is mainly because rural areas are either not as densely populated as urban areas or they are geographically isolated. Since rural people are less aware about digital tools and their usage, service providers are also unsure that they will be willing to pay for a service or afford their internet plans. Without raising digital awareness, it is difficult to increase the demand for connectivity.

The lack of penetration in rural areas due to these barriers further discourages already marginalised people from becoming digitally literate. This leaves us in a deadlocked situation, with rural India trapped in a vicious cycle of being disconnected. There are relatively few working models which provide connectivity to communities excluded from access as a result of income levels, size and geographic isolation. One of these models is the community wireless network. Realising the importance of Wi-Fi networks, the Telecom Regulatory Authority of India (TRAI) identified community wireless networks as a model to promote public access in its 2016 paper, "Consultation on Proliferation of Broadband through Public Wi-Fi Networks".[9] Based on this paper, the Department of Telecommunications introduced the designation of Virtual Network Operator (VNO), enabling telecom service providers to utilise their networks and spectrum efficiently by sharing active and passive infrastructure to facilitate services at the resale level.[10] This has opened up opportunities for small enterprises and non-profit organisations to become small ISPs, which was difficult to do before due to regulatory challenges.

In India, there are very few social enterprises that are actively designing, deploying and setting up wireless networks specifically catering for rural communities. AirJaldi[11] in Dharamshala, Gram Marg[12] in Mumbai and DEF (based in Delhi) are among the few organisations that are providing basic connectivity and enabling access to information for citizens outside urban areas.

AirJaldi started as a social, non-profit enterprise established in Dharamshala, Himachal Pradesh and provides affordable and reliable internet connectivity using unlicensed spectrum and wireless networks in rural communities. Gram Marg, an incubation of the Indian Institute of Technology (IIT) Mumbai, uses TV white space and now Wi-Fi to provide internet connectivity in 13 villages in Maharashtra. DEF also uses low-cost wireless technology, unlicensed spectrum bands – 2.4GHz and 5.8 GHz – and line of sight to support the provision of affordable, low-cost and reliable internet services in 38 districts of the country.

Wireless community networks and women's entrepreneurship

Wireless community networks – or community networks – are "bottom-up" network models defined by various academicians and institutions as networks owned and managed by communities. Community networks are also defined by some as "crowdsourced networks" that are structured as free, open and neutral; they are built by a community and managed and operated as a common resource.[13] Elkin-Koren offers a more technical definition of community networks as distributed network architectures in which users can implement a physically decentralised network through the decentralisation of hardware.[14] According to the European Commission, community networks are a "private initiative by the local residents of the community using a so-called bottom-up approach." In general, community networks offer an alternative and complementary approach to the traditional commercial model where internet connectivity is sold to the user.

DEF, founded in 2002, aims to connect unreached and underserved regions of India in an effort to bring them out of the digital darkness and empower them with access to information. With a firm belief that access to information and community engagement can reduce information poverty in communities, it became essential for DEF to support first-mile connectivity solutions and to develop digital literacy to understand why and how relevant information can improve the socioeconomic conditions of people living in underdeveloped contexts.

In 2010, DEF and the Internet Society (ISOC) initiated the W4C project to help provide internet connectivity where traditional ISPs were unwilling to operate. The first pilot project was started in a handloom community in Chanderi, Madhya

9 https://www.trai.gov.in/sites/default/files/Wi-Fi_consultation%20Paper_13_july_2016.pdf

10 Guidelines for Grant of Unified License (Virtual Network Operator). www.dot.gov.in/circulars/guidelines-grant-unified-license-virtual-network-operators

11 https://airjaldi.com

12 grammarg.in

13 Baig, R., Roca, R., Freitag, F., & Navarro, L. (2015). guifi.net, a crowdsourced network infrastructure held in common. *Computer Networks, 90*, 150-165. https://doi.org/10.1016/j.comnet.2015.07.009

14 Elkin-Koren, N. (2006). Making Technology Visible: Liability of Internet Service Providers for Peer-to-Peer Traffic. *New York University Journal of Legislation and Public Policy, 9*. https://ssrn.com/abstract=924316

Pradesh, with the aim of providing the community with information about weaving (while also allowing them to produce their own content) and the tools to archive their designs online, and connecting them to a wider market for them to sell their products. Gradually the project transformed into the first community network project in India established in a weavers' community. Using low-cost, line-of-sight wireless technology, and the unlicensed 2.4 GHz and 5.8 GHz spectrum bands, W4C now creates community-owned and community-operated wireless networks.

The W4C programme has four main components: 1) train the community members on using wireless networks and its components to create "barefoot network engineers"; 2) use open source practices, low-cost technology and frugal methods[15] to set up community wireless networks (such as using line of sight to find the tallest building for setting up the antennas); 3) create a platform to develop local content using the wireless network; and 4) improve the socioeconomic lives of community members.

Along with establishing the wireless networks, information hubs known as Community Information Resource Centres (CIRCs) or "wireless centres" are set up to provide digital literacy training to the community. These centres work using DEF's "AHEAD Agenda",[16] which stands for:

- A – building *awareness* on social rights and services through online avenues, and on laws and issues such as the Right to Information Act and women's empowerment.
- H – *health*, such as telehealth services to connect primary health centres to district hospitals and enable local communities to access health-related information through the internet.
- E – *education* for young people who have not completed their schooling, and access to online learning materials.
- A – *activating* entrepreneurship by enabling community members, particularly women, to set up e-commerce sites and businesses.
- D – *delivery* of governance services and enhancing state transparency and accountability.

In eight years, the W4C programme has reached 38 districts across 18 states by deploying about 200 hub-based access points. These access points currently connect more than 4,000 people and user numbers are still growing. Most of these networks are located in tribal and telecom dark areas, where people have seen a computer and experienced the internet for the first time.

There are two aspects to sustaining the networks. The first is establishing, managing and operating the network and the second is creating and managing the content. While developing these two aspects, DEF realised that most of the networks we set up are managed by young men, who often leave for cities to find better work and other opportunities. Women mostly stay at home or do odd jobs. Even if women visited the CIRCs, their role was limited to searching for information. Their participation in managing and operating the networks was negligible.

At the same time, managing wireless networks in rural regions is difficult, as wireless network engineers from urban areas are not willing to travel to rural regions.

Realising this gap, we saw a need to have a person who could operate and manage these centres and interact with users or community members on a regular basis, providing wireless training to women and enabling them as what we called "women barefoot engineers" and "wireless women entrepreneurs". This became a priority for W4C. We wanted to achieve two things: 1) create a network of women barefoot engineers who could manage the network, and who could also develop local content; and 2) by doing this, improve the presence of women in the telecom and wireless sector.

Training the first woman barefoot engineer

In 2014, Kainat Ansari became the first woman appointed by DEF to set up a wireless network. Kainat received three months of wireless networking training in Guna, Madhya Pradesh.[17] Initially, she was engaged in the daily management of the centre. Then she started managing the wireless network, including configuring devices and setting client access levels, as well as interacting with community members on their tech needs. Her tasks also included live network set-up, DVR configuration, point-to-point configuration and even troubleshooting at the server or client level. Kainat also helped us set up other wireless networks including in Aron, Madhya Pradesh and Saiddanpur, Uttar Pradesh.

Kainat acted as an influencer for other girls living in Guna. Fauzia was one of them. She was visiting the centre to access YouTube and other

15 https://en.wikipedia.org/wiki/Frugal_innovation

16 Srivastava, R. (2016). A Network by the Community and for the Community. In L. Belli, *Community Connectivity: Building the Internet from Scratch. Annual Report of the UN IGF Dynamic Coalition on Community Connectivity*. bibliotecadigital.fgv.br/dspace/handle/10438/17528

17 Interview with Kainat Ansari.

entertainment websites. Following in Kainat's footsteps, she joined the wireless networking training programme. She is now able to make PCB circuit boards, climb network towers, do TP-Link router configuration, and troubleshoot at the centre and client level.[18]

Empowerment through digital literacy

Another local content intervention was our Wireless Women for Entrepreneurship and Empowerment (W2E2) project, which was part of the W4C programme, and initiated in partnership with ISOC in 2014.[19] It aimed to help women in grassroots businesses make use of digital tools and the potential of e-commerce. Bidyawati Mehar, a girl from Orissa who had not finished her schooling, is part of a weaver family living in a small village called Barpali. Barpali is famous for hand-woven ikat[20] sarees, and the village is home to about 20,000 handloom weavers. Bidya was the first of a few students who joined our wireless network centre in the village and soon became among the fastest-learning students at the centre. The weavers in Bidya's village lacked knowledge about new designs, business practices and modern technologies. Most of the designs made by the weavers had not been set to paper. This restricted experimentation and innovation in designs. Being a digitally literate weaver gave Bidya opportunities to introduce various digital interventions into the weaving process. Bidya is now part of a "mobilisation team" that encourages more weavers to adopt digital tools, and trains scores of youth from her community in digital literacy and design.[21]

Last year, Bidya's efforts were recognised by ISOC's 25 Under 25[22] programme. She was one of the 25 individuals, all under the age of 25, who travelled to Los Angeles in the United States on 18 September 2017 to receive their awards. She says, "Today, I'm the only woman master weaver in our cluster. People who criticised me earlier for choosing this occupation now call me 'Lady Master Weaver'."[23]

The engagement with women has created new opportunities for rural women who want to learn technical skills but do not get the opportunity due to a lack of projects, programmes or institutions that offer training, or because they face social and patriarchal challenges which exist in the society. As part of W4C, over 30 rural women have been transformed into barefoot wireless engineers. These women barefoot engineers act as influencers of others and motivate them by showing how this sort of training can improve their livelihoods.

Conclusions

Community networks are recognised as a catalyst for development, especially for women, who have fewer opportunities to access digital tools compared to men. But when the opportunity is given to them, we find that they manage our wireless networks effectively and efficiently – just how they manage their homes. These barefoot women wireless engineers, who can barely read and write, have demystified technology and transferred the control, management and ownership of the technologies to the community.

Engaging in experiential learning and using the technology in a way that improves lives and skills has changed the definition of "literacy" and what it means to "be educated". This approach has enabled rural women to impart their skills to others in a sort of multiplier effect. Engaging women has not only given them an opportunity to learn about technology, but also enabled them to find a comfortable place in an area mainly dominated by men. This is a small effort that helps to bridge the gender divide in the telecom sector. In the last three years, it has led DEF to engage more women in the wireless sector, as they not only make the network economically sustainable, but also make it socially viable.

Action steps

Over the years, the W4C programme has adopted various kinds of models as interventions, depending upon the needs of the community and the social environment. One thing that was common in all models was the need to capitalise on the social and human value already present in the community, and transforming this into a "socially sustainable wireless network model". Here are a few action points that those involved in community networks should consider:

- Increase gender sensitivity: Community networks should be designed in a more gender-sensitive manner. For example, it should be made mandatory that at least 50% of the people participating in activities should be women and girls.
- Promote different kinds of community network models: Presently most of the community network models in India are only talking about leveraging technology and innovation, but

18 Interview with Fauzia.

19 wforc.in/w2e2-2

20 https://en.wikipedia.org/wiki/Ikat

21 One or two weavers also received training to maintain the wireless network.

22 https://www.internetsociety.org/25th/25-under-25

23 Interview with Bidyawati Mehar.

there is need to sustain these networks socially – including creating a conducive environment for women to engage with the community. Stakeholders should promote and support community network models that do this.

- Make funds available: Universal service funds and other funding mechanisms should be made available to increase participation in community networks.
- Create opportunities for learning and employment. For example, community networks can be used to create new opportunities for women and girls who want to learn technical skills but do not get the opportunity because of patriarchal structures or because training programmes are not available.
- Create self-learning tools that can be used by others who have an interest in learning wireless networking so that they are able to set up their own wireless networks.
- Create a network of "barefoot wireless engineers" who not only manage the network but also develop local content and make it accessible for everyone.

In 2017, DEF initiated the concept of an "Internet in a Box" or, as we call it, a Zero Connect box,[24] which is a solar-powered box with an antenna. The box is a plug-and-play configurable networking solution for deploying a wireless network for people in predefined small-range coverage areas. It is solar enabled and the size of a suitcase, and is fixed with a trolley mechanism for easy mobility. The lightweight unit (7 kg) is made of aluminium and is divided into four sections to hold the tripod and hotspot printer (to print receipts for internet vouchers); the charger controller; the batteries; and the router, antenna (5.8 GHz and 2.4 GHz) and other equipment. A solar panel on the top of the box is fitted for the purpose of charging the batteries. The batteries can provide power for six to eight hours. The unit has the power to connect as many as 200 individuals at a time within a radius of 500 metres to 5 km. To make this solar-enabled box, we will be using low-cost wireless technology such as MikroTik routers and omni-directional antennas, and using both 5.8 GHz and 2.4 GHz unlicensed spectrum to provide the connectivity.

The prototype model was implemented in Anantapur district in Andhra Pradesh. Presently, DEF is scaling these Zero Connect boxes in other wireless locations. This year DEF also initiated the Solar Women Wireless Engineers for Entrepreneurship and Empowerment (SW2E3) programme to provide solar and wireless training to women. It aims to create a network of women wireless engineers who can not only set up and deploy the Zero Connect box, but are also able to operate and manage the wireless networks.

24 internetinabox.in

INDIA

ADDRESSING SUSTAINABILITY IN RURAL CONNECTIVITY: A CASE STUDY OF GRAM MARG COMMUNITY-LED NETWORKS

Gram Marg rural broadband project, Department of Electrical Engineering, Indian Institute of Technology Bombay
Sarbani Banerjee Belur[1]
http://grammarg.in

Introduction

India's digital landscape has rapidly evolved with the introduction of the government's "Digital India" initiative.[2] This has led to an unprecedented growth in communication technologies. Internet connectivity, with the aim of achieving "accessibility for all", has penetrated at a rapid pace. This has impacted and revolutionised lives in a manner that perhaps no other technology has achieved so far.

In spite of this growth, approximately 49.5% of the population of India is still unconnected.[3] Wide rural-urban connectivity gaps exist that evidence a stark digital divide. While urban India is almost completely covered with voice and data service, rural India still suffers from inadequate connectivity. In a country with a population of 1.34 billion, there are only 325 million broadband subscribers.[4] Internet penetration in rural areas stands at 21% compared to 65% in urban areas.[5]

The Indian government, through the BharatNet initiative, aims at digitally connecting 250,000 Gram Panchayats[6] (local self-government offices at the village level) by 2019,[7] extending this to villages and households in villages through local internet service providers (ISPs), run by both government and private telecom operators. Such large-scale deployments involve significant cost to the government and, as a result, there is a need to enable this connectivity to be permanent and sustainable.

The sustainability of rural internet connectivity is based on the basic premise of supply and demand. The demand is by and large dynamic in nature due to a lack of digital awareness and an unstable customer base (primarily farmers) with no fixed monthly income. On the supply side, a lack of digital infrastructure and services, the ineffective use of available connectivity, and low profitability of investment play major roles. Due to this unevenness of demand and supply, rural areas are underserved and unreached.

In order to address this, partnership models have been identified as a suitable method for enabling sustainable connectivity.[8] There have been several types of such partnership models, such as the BOT (build operate transfer) model, BTO (build transfer operate) model, and joint venture model.[9] Globally, the private and public sectors are the

1 I thank the Gram Marg team members, in particular M. Khaturia, N. P. Rao, J. Singh, T. Ghadge, A. Patil and V. Kavale. Special thanks to the principal investigators of this project, Prof. A. Karandikar and Prof. P. Chaporkar, for their valuable insights. This study is also a part of IEEE-SA's Digital Inclusion Through Trust and Agency (DITA) programme (the affordability and accessibility work streams).

2 The Digital India initiative was launched by Prime Minister Narendra Modi on 1 July 2015 with the objective of connecting rural areas with high-speed internet connectivity and improving digital literacy. The vision of this programme is centred on three key areas: digital infrastructure as a utility to every citizen, governance and services on demand, and the digital empowerment of citizens.

3 Broadband Commission for Sustainable Development. (2017). *The State of Broadband: Broadband catalyzing sustainable development*. https://www.itu.int/dms_pub/itu-s/opb/pol/S-POL-BROADBAND.18-2017-PDF-E.pdf

4 Telecom Regulatory Authority of India. (2017). The Indian Telecom Services Performance Indicators: July-September, 2017. https://www.trai.gov.in/sites/default/files/PIR_July_Sept_28122017.pdf

5 Indo-Asian News Service. (2018, 20 February). Acute Urban-Rural Divide in Internet Penetration in India: Report. *Business Television in India*. www.btvi.in/news/acute-urban-rural-divide-in-internet-penetration-in-india--report/57421

6 All villages in India are either Gram Panchayats or only "villages", depending on the population size of the villages. Villages that are small will fall under the same Gram Panchayat. So a Gram Panchayat may have three to four villages under it.

7 Indian Institute of Technology Bombay. (2016). *Planning for BharatNet Phase 2*. Bharat Broadband Nigam Ltd. bbnl.nic.in/WriteReadData/LINKS/Rpt_Nw_Plg_tool6cad24a7-eea8-4adc-91ff-468d7119eeb0.pdf

8 Nungu, A., & Pehrson, B. (2011). Towards Sustainable Broadband Communication in Rural Areas. In R. Szabó et al. (Eds.), *AccessNets 2010: Lecture Notes of the Institute for Computer Sciences, Social Informatics and Telecommunications Engineering, 63*, 168-175. https://pdfs.semanticscholar.org/13a6/88fc3d27f7f9b507424203c6acb5659b-1fe6.pdf; Belur, S. B., Khaturia, M., & Rao, N. (2017). Community-led Networks for Sustainable Rural Broadband in India: The Case of Gram Marg. In L. Belli (Ed.), *Community Networks: The Internet by the People, for the People. Official outcome of the UN IGF Dynamic Coalition on Community Connectivity*. https://bibliotecadigital.fgv.br/dspace/bitstream/handle/10438/19401/Community%20networks%20-%20the%20Internet%20by%20the%20people%2C%20for%20the%20people.pdf#page=194

9 In the BOT model, the private service provider provides the capital required to build, maintain and operate the service for a contract period under a concession from the government and then returns the service to the government after its contract has expired. In the BTO model, the private service provider builds the infrastructure and then transfers it to the public owner. The public owner is given the right to operate and a return on investment. This could also be called a private investment and public facilitation model. In a joint venture model, the investment is shared between the private service provider and public sector.

two key players that collaborate on initiatives to bring voice and data connectivity in rural areas and in most cases they work in partnership with each other. However, planning and maintenance delays, inadequate monitoring, funding gaps and improper management are some of the shortcomings. Because of this, these models are unable to sustain themselves in the field for very long. Apart from this, a key feature that these models lacked was involvement of local people for whom the network has been facilitated. In India, these models largely follow a top-down approach, concentrating on a customer base and a suitable return on investment. The local and regional needs are sidelined and unattended to in such partnership models, making them unviable in rural areas. As a result, in order to make connectivity sustainable in rural India, there is a need to develop a suitable partnership model with village involvement as the crux.

This report[10] discusses the Public-Private-Panchayat Partnership (4-P) model developed and validated on the ground for its sustainability in villages where connectivity has been enabled through our project. We also discuss how the 4-P model is implemented in Gram Marg villages and generates revenue, thereby making it a sustainable model. Currently this model is working successfully in the villages in the Palghar district of Maharashtra, where internet connectivity has been enabled by Gram Marg. The first part of the report gives a short overview of Gram Marg community-led networks. In the second part of the report, the development and validation of the sustainable 4-P model will be discussed. The final section highlights relevant policy measures needed for the adoption of a sustainable model in rural connectivity.

Towards rural connectivity: Gram Marg

The Gram Marg rural broadband project[11] at the Department of Electrical Engineering, Indian Institute of Technology Bombay (IIT Bombay), aims at connecting the unconnected by overcoming the barriers and challenges to connect rural India. In order to provide ubiquitous connectivity to rural, remote areas, research and development at the Gram Marg lab has suggested that a shift is required from traditional technologies to a more affordable, efficient and robust technology. To this effect, Gram Marg has deployed two large-scale test beds since its inception in 2012.

The first test bed that was set up by Gram Marg was based solely on TV white space (TVWS) technology, covering seven villages in Palghar, spanning an area of 25 km², in 2013-2014. The purpose of setting up a TV UHF band test bed was to check the feasibility of using TV UHF for middle-mile connectivity to provide high-speed broadband access to the villages. The test bed consisted of a single base station which connected to 12 clients situated at selected locations at varying distances. A 20 Mbps leased line was provisioned at the base station. A total of 10 Wi-Fi access points were deployed as well as three community kiosks, which were backhauled using the TV UHF band. Detailed results from this test bed have been published in papers.[12]

The second test bed is a scaled-up version of the earlier test bed which experimented with the feasibility of unlicensed bands such as 5.8 GHz for middle-mile connectivity covering 25 villages in Palghar, Maharashtra. The test bed spans an area of about 300 km², a network diagram of which is shown in Figure 1. Villages are divided into clusters of four to five villages, resulting in six clusters. Each cluster has one base station that has a fibre point of presence. In one cluster group of 15 villages, access is offered at Gram Panchayat offices only. In a second cluster group of 10 villages, Wi-Fi access points are deployed at strategic locations that can be accessed in and around those locations. In all, a total of 60 Wi-Fi access points have been deployed in the 10 villages, with six access points per village. These access points are set up at the Gram Panchayat office, primary health care centres, at least one school and one community centre. A total of 106 Mbps bandwidth has been procured to serve the 25 villages.

Establishing a community-led network

Seeding the growth of a community-led network has been an important achievement of the test bed. An internet needs assessment survey suggested that enabling connectivity was not enough for the villages. There needs to be ownership of the network by the village authorities so that local and regional needs will be prioritised. Community involvement in the connectivity can help in the maintenance of the network and take care of the security of the devices. Local youth from the village community can also be effectively engaged through skills development and training.

10 This report has been an outcome of a research project partly funded by Tata Trusts.

11 www.grammarg.in

12 Kumar, A., Karandikar, A., Naik, G., Khaturia, M., Saha, S., Arora, M., & Singh, J. (2016). Toward enabling broadband for a billion plus population with TV white spaces. *IEEE Communications Magazine, 54*(7); Khaturia, M., Belur, S. B., & Karandikar, A. (2017). TV White Space Technology for Affordable Internet Connectivity. In R. Stewart, D. Crawford, & A. Stirling (Eds.), *TV White Space Communications and Networks*. Woodhead Publishing.

FIGURE 1.

Network diagram of 25 villages in the Palghar project

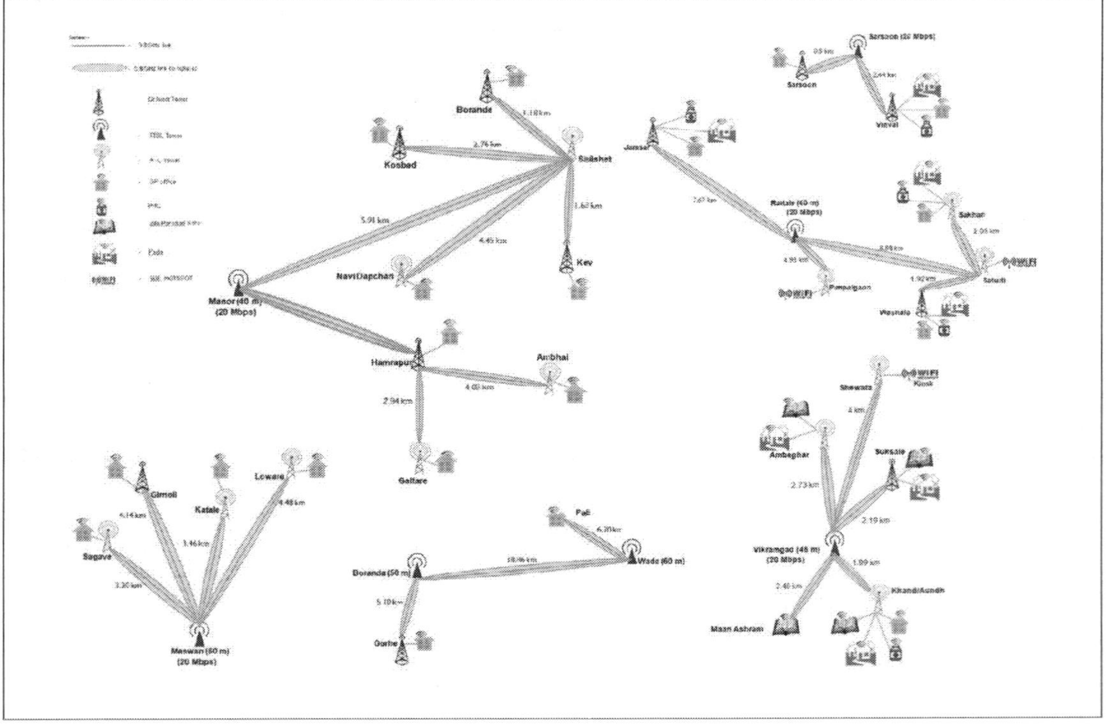

Out of the 25 villages covered in the second test bed, community networks were set up in 10 villages only. They were selected for two reasons: i) the location of these villages was in remote pockets and hilly areas, and ii) there was no mobile coverage in these villages. Figure 2 depicts the network diagram of the 10 villages where the community network is currently operational. As seen in the figure, a point-to-point 5.8 GHz link has been set up from the tower at the base station to the Gram Panchayat office. Connectivity is further distributed inside the village in a mesh network through six access points that are placed strategically inside the village. As a result, the entire expanse of the village is Wi-Fi enabled.

The network in the 10 villages is maintained through community participation. A village youth is nominated as the Village Level Entrepreneur (VLE) who undergoes training and skills development to maintain and operate the network in the village. The VLE is registered with the government's Common Service Centre (CSC) programme to ensure citizen service centres operate in all Gram Panchayat offices in India. CSC e-Governance Services India Limited, a special purpose vehicle, has been created by the Ministry of Electronics and Information Technology (MeitY) in order to extend digital services in rural areas. Wi-Fi Choupal[13] was set up alongside this to extend connectivity using Wi-Fi-based network infrastructure.

As an ancillary to the CSC, Wi-Fi Choupal has partnered with Gram Marg in the 25 villages Palghar project. Wi-Fi Choupal, as an ISP licence holder, uses the VLE to sell bandwidth inside the village through fixed-pricing plans. In these villages the connectivity is currently being used both on smartphones as well as in the village Gram Panchayat office for accessing and enabling e-governance services.

Sustainable economic partnership model

As mentioned, in order to make the connectivity sustainable, the 4-P model was developed. This partnership model was developed for all the 25 villages in Palghar. The main reason behind developing the partnership model has been to throw light on how different partnerships can facilitate connectivity to remote, rural villages. The model also lays stress on the fact that no single entity (i.e.

13 www.wifichoupal.in

FIGURE 2.

Community network villages in Palghar

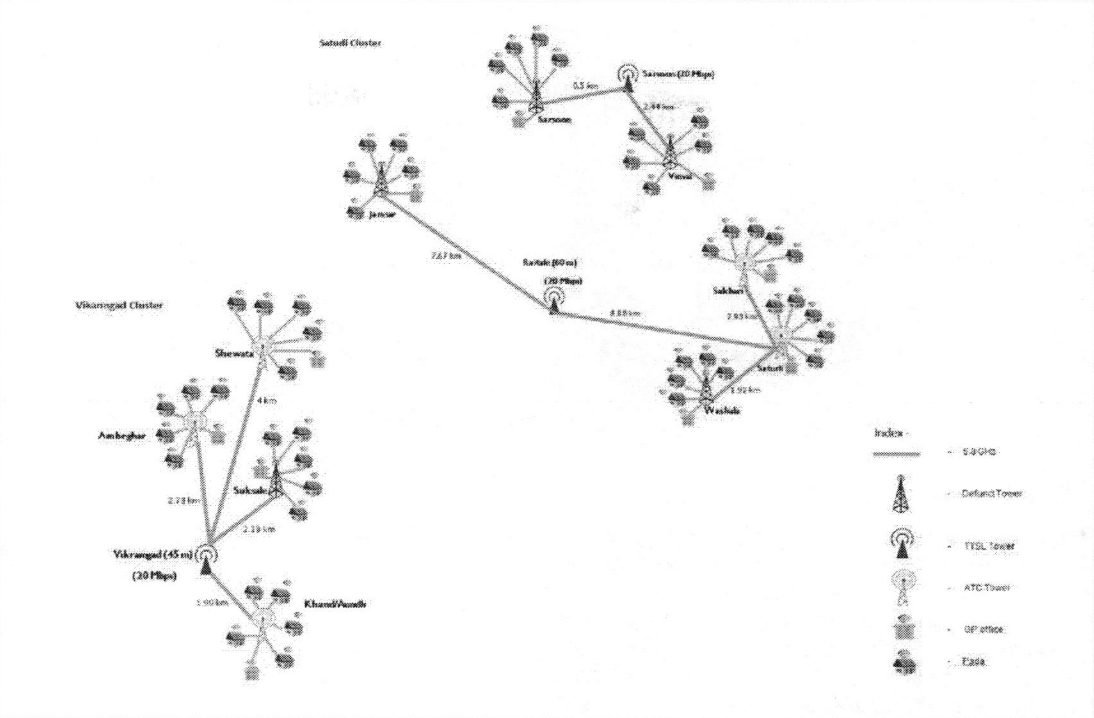

private telecom operators) or the government is responsible for enabling connectivity to the unserved. Important aspects of the model are: i) the Panchayat (i.e. the village administration) has been introduced into the partnership model alongside the public and private partners, and ii) the partnership model adopts a bottom-up approach with the involvement of the villagers, focusing on the local and regional needs with regard to connectivity. As the Panchayat represents the village administration, and is elected by the people of the village and backed by the district and state government, its participation adds value to the partnership in terms of authority, ownership and financial disbursement.[14]

The role of each of the partners in this partnership model is as follows:

- The Panchayat owns the network at the village level. It plays a major role in defining priorities for the local digital needs of the villagers. In one revenue model it purchases the bandwidth and enables revenue generation by reselling the bandwidth to the villagers. In another revenue model, local youth from the villages are appointed as VLEs by the Gram Panchayat and operate and maintain the network in the village, also selling bandwidth to the villagers.
- The private sector partner provided the bandwidth which enabled connectivity to the 25 villages.
- The public sector partner plays a vital role in technology innovation, deploying the network and providing the capital expenditure (CAPEX) funding for setting up the network infrastructure in the villages.

Validation of the 4-P model on the ground

A partnership model is not sufficient in itself if it does not have revenue generation as an important part of it. The revenue generation aspect of the partnership model addresses the sustainability of the connectivity after the unconnected villages are connected. For the validation of the 4-P model on ground, the first step has been to identify the two important cost indicators, i.e. CAPEX and operational expenditure (OPEX). For setting up a network,

14 Further details about the 4-P model can be found in Belur, S. B., Khaturia, M., & Rao, N. (2017). Op. cit.

FIGURE 3.

Data usage and number of users from traffic data

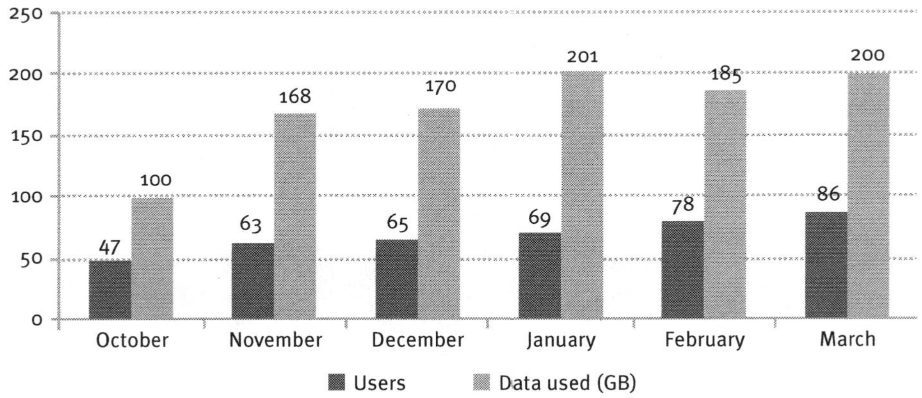

there is a CAPEX investment that is needed and the OPEX needs to be recovered. Revenue generation has, as a result, been identified as an important aspect of the model, without which the model cannot be sustained. This incentivises the Gram Panchayat and the VLE to maintain the network in the village. Usage of the connectivity was also analysed through monitoring traffic data, which helped in understanding the bandwidth requirement at each Gram Panchayat.

In order to test the sustainability of the 4-P model, the 25 villages have been divided into two groups each with a different revenue-generation method. The villages were separated into groups based on their proximity to the highway (Mumbai-Ahmedabad Highway): 15 villages that are located close to the highway comprised one group and 10 villages that are located in the remote tribal areas comprised the second group. The 4-P model is being validated in the two groups to test the model's viability, scalability, replicability and adoption in different village contexts.

The first step in the validation process has been to look at the traffic data of all 25 villages. The network went live in October 2017. Initial analysis of the traffic data from all the villages suggests that connectivity is being used by the people. Data from October 2017 to March 2018 (as illustrated in Figure 3) show an increase in number of users and in data usage. This can be viewed as a satisfactory service in terms of internet availability and speed.

Data on monthly uploads and downloads by the internet users in these villages shows that there are

FIGURE 4.

Data usage and number of users from traffic data

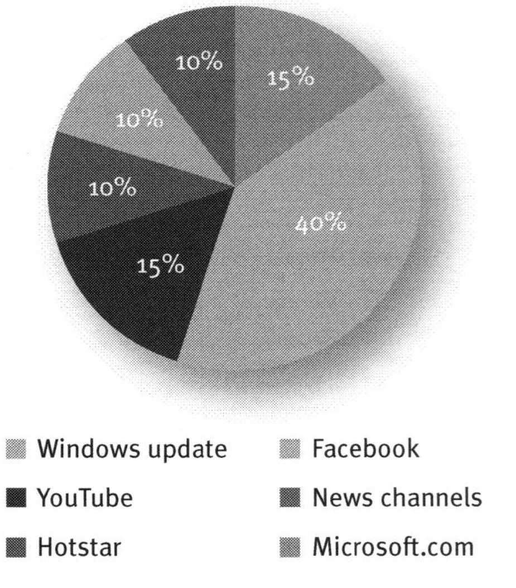

more downloads than uploads. From October 2017 to March 2018, downloads increased from 96 GB to 154 GB. While uploads account for less data usage than downloads they are also increasing, from 6 GB to 15 GB. This is in line with consumer patterns where downloads are more than uploads.

The traffic data of the 25 villages also gives a broad overview of the utilisation of bandwidth by the users in the form of websites most visited. As shown in Figure 4, 40% of the bandwidth has been used for Windows updates, followed by YouTube, Hotstar[15] and visits to Microsoft.com. In the case of Windows updates, this is likely to be the result of automatic updates by the software vendor. Due to the high illiteracy levels (60% in these villages), users say that they are unable to do keyword searches. As a result they prefer sites like YouTube and Hotstar.

Revenue generation as part of the 4-P model

In the 15 villages cluster (the first group), the revenue model is through the local ISP. The local ISP has enabled 2 Mbps bandwidth at each Gram Panchayat office and gets directly paid by the Gram Panchayat. The local ISP further sells the bandwidth inside the village as part of its marketing strategy and generates revenue from the connectivity. The Gram Panchayat office pays a fixed price of INR 1,000 (USD 14) for 2 Mbps of bandwidth. This cost includes the bandwidth cost, operation and maintenance of the link and device cost if the device needs replacement due to damage. However, as the Gram Panchayat office does not use the entire bandwidth, the unused bandwidth is sold to the villagers in the form of "pay as you use" daily coupons of a duration of one hour each costing INR 10 (USD 0.14). This connectivity is accessed at the Gram Panchayat office. It has been observed that an average of five to 10 people use the internet at the Gram Panchayat office per day, which totals INR 50 to INR 100 (USD 0.70 to USD 1.40) per day and results in a monthly income of INR 1,500 to INR 3,000 (USD 21 to USD 42). This contributes to the monthly revenue of the Gram Panchayat. Out of this amount, INR 1,000 (USD 14) is paid to the local ISP, as noted above. The Gram Panchayat office plans to use the accumulated amount for development activities within the village.

The second set of 10 villages has a VLE-focused revenue model, where CSC Wi-Fi Choupal has acquired 30 Mbps bandwidth from a local ISP[16] and distributes the same to different villages depending on internet use and number of customers in each village. The VLEs maintain the network in these villages and sell bandwidth to the villagers in the form of coupons based on the fixed pricing plan. The monthly customer base of the VLEs includes new customers as well as returning customers. The revenue plan of the VLEs has been devised in a way that it maximises profit for the VLE, thereby providing incentive to perform. It has been observed that of the coupons sold per month, 40% of the coupons are of INR 10 (USD 0.14) in value, which gives 500 Mb of data for 10 days. The next most popular coupon amount, accounting for 22% of purchases, is INR 100 (USD 1.40), which is valid for 28 days and gives 12 GB of data.

Revenue information suggests that in those villages where there is a substantial use of internet data and a large customer base, the monthly revenue generated by the VLE is in the range of INR 5,000 to INR 6,000 (USD 70 to USD 84). In other villages, the monthly revenue generated is INR 3,000 to INR 4,000 (USD 42 to USD 56) on average.

In the months to follow, it is expected that the number of broadband subscribers will increase, which will directly have an effect on the revenue generated by the VLEs. A steady growth in revenue generation by the VLEs suggests that the model will perform well and also offers lucrative value for the investment made. A nominal pricing plan for data will facilitate greater usage by rural villagers.

Conclusions and the road ahead

In this report, we address the importance of sustainability of broadband connectivity in rural areas of India. We have discussed Gram Marg's 25 village Palghar test bed and how it has attempted to answer various issues regarding rural connectivity through test bed deployments and setting up community networks in 10 villages in Palghar, Maharashtra. We also discuss the 4-P model which has been developed and successfully validated in the field with regard to sustainability. As the 4-P model takes a bottom-up approach, it is a robust and scalable model which generates revenue, enabling the internet to thrive and grow sustainably.

Even though the 4-P model has been successfully validated in the Gram Marg villages, we need to replicate the model in different village contexts, and with different CAPEX contribution scenarios, to understand the model in its entirety. As of now, the model is based on actual expenditure and real field data on usage, revenue generation and recovery of OPEX. In time we shall come up with a projection model taking into account the growth rate of internet subscribers, expected growth in internet demand, and yearly increases in the monthly cost of internet usage. This will enable predictions of revenue accrued by the VLEs and the Gram Panchayat over a period of time.

The district administration also needs to have a voice about their need to be connected and take ownership of the network as soon as connectivity has

15 An Indian digital and mobile entertainment platform. https://www.hotstar.com

16 Wi-Fi Choupal is an ISP licence holder only. It acquires bandwidth from private telecom operators and then distributes the bandwidth and generates revenue.

been enabled. With the Panchayat at the core of the 4-P model, the model becomes very localised, which will eventually help in the betterment of the model.

This sustainable economic model can be the foundation for making rural broadband become financially self-dependent at the village level. As the model focuses on broadband usage, supply and demand, it can effectively formulate the cost effectiveness of technologies used. Unless such a self-sustainable model is implemented, it will be difficult for the internet to penetrate rural areas.

Action steps

We would like to emphasise the following policy recommendations in support of community networks in India:

- Community networks are allowed to operate in India but there are no specific policies that support such networks. They should be promoted and encouraged by the government.
- Mere internet connectivity should not be the only agenda when connecting rural villages. Seeding the growth of community networks, developing community technologies and encouraging the meaningful use of connectivity should be set as priorities by the government.
- Sustainability of connectivity is a serious issue that needs to be addressed. In most cases, this is thought about only after the network deployment has taken place and the project funds are running dry. However, if the sustainability plan is chalked out and conceptualised during the planning phase of network deployment, all stakeholders in the model can take equal responsibility.
- The village administration (i.e. Gram Panchayat) should be given the authority to address connectivity challenges in their own villages. Government policy measures should help Gram Panchayats to form cooperatives or groups with a legal status to own the network. There can be a provision for such groups to acquire a special ISP licence, making them eligible to procure and sell bandwidth.
- Internet connectivity is still not a part of the development plan of Gram Panchayats in India. As each Panchayat is expected to develop an annual development plan, internet for development can be included in it. In this way, the money for enabling connectivity to the Gram Panchayats can be financed by the state government.
- It is necessary for a sustainable partnership model to grow within a community network. This will enable better community participation and involvement for the sustainable model to thrive in rural areas.

INDONESIA AND MALAYSIA

SMART PARTNERSHIPS AND PARTICIPATORY ENGAGEMENT: A COMMUNITY NETWORK FOR INDIGENOUS BORNEO

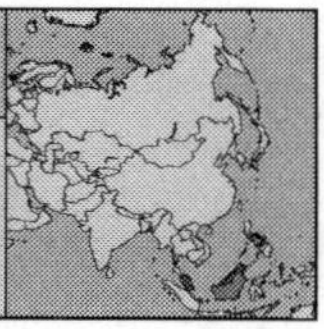

Advisor to Forum Masyarakat Adat Dataran Tinggi Borneo (FORMADAT)
Roger W. Harris
www.formadat.com

Introduction

Orthodox state-sponsored development often inadvertently damages the cultures and lifestyles of indigenous peoples, as well as plundering their natural resources, despoiling their environment and sometimes moving them into new forms of poverty where none previously existed. Opposition to such initiatives can be misinterpreted as opposition to development, which indigenous representatives counter with calls for development that is indigenised – addressing their needs as they express them and fulfilling their aspirations rather than those of outsiders.

Information and communications technologies (ICTs) have been convincingly demonstrated to offer opportunities for indigenised development when introduced within processes that take full account of the local socioeconomic context. The outcome is to empower indigenous communities to devise and implement development activities of their own design. Furthermore, many of the core characteristics of ICTs are conducive to the promotion and implementation of indigenised development. This report describes how a cash-strapped but resourceful grassroots indigenous organisation mobilised a diverse set of smart partnerships to create a cascading set of community networks and other information solutions at the local level.[1] It discusses how the organisation went on to deploy innovative participatory methods of public engagement to devise uses for the networks that address the specifics of the communities' development needs as people with unique cultures, traditions and lifestyles and as guardians of Asia's last great rainforest.

Policy, economic and political background

Borneo is divided into 73% Indonesia and 26% Malaysia, with the remainder comprising Brunei. While both Malaysia and Indonesia have ICT policies that target wider access to the internet – with Malaysia achieving 77% of households with internet access and Indonesia 25% – rural access lags well behind national averages and in the remote and isolated areas of Borneo it lags even further behind.[2] However, the government of Malaysia is promoting high-speed internet, with particular attention to rural access through universal service provision initiatives. These include the establishment of "1Malaysia Internet Centres". Internet centres are common across Indonesia, and in rural villages are typically government or donor-funded. Mobile phone usage has expanded rapidly across Borneo and is now reaching even the most remote communities.

Against the background of state sponsorship of rural telecommunications, local ICT networks consisting of infrastructure that is owned by the community are rare. Among the barriers, especially in rural areas, are affordability, insufficient skills and awareness and cultural acceptance, as well as the lack of content and services in local languages. By promoting ICTs as fundamental tools for national development, both governments open themselves to questions about the implications for their rural, indigenous and other underserved populations. While the response has been to slowly roll out some form of shared subsidised access, this has not always been accompanied by robust measures to ensure that such access translates into socioeconomic inclusion or anything else that recipient communities would regard as development that is relevant to their needs.

FORMADAT and its community ICT networks

FORMADAT is a trans-boundary grassroots initiative that works to increase awareness and understanding of the communities of the highlands of Borneo – to maintain their cultural traditions, build local capacity, and encourage sustainable development in the Heart of Borneo without risking the degradation of the quality of the social and natural environment. The Heart of Borneo is a conservation area designated under an international agreement initiated by

1 The term "community ICT network" is used in this report to distinguish it from other (non-physical) forms of network.

2 ITU. (2017). *Measuring the Information Society Report 2017.* https://www.itu.int/en/ITU-D/Statistics/Documents/publications/misr2017/MISR2017_Volume2.pdf

the World Wide Fund for Nature (WWF) to protect a 220,000 km^2 forested region on Borneo island that is known as Asia's last great rainforest (see Figure 1). FORMADAT is a formally registered society in Indonesia and Malaysia, made up of the rural indigenous communities that occupy part of the Heart of Borneo.

FORMADAT comprises members of the following marginalised ethnic groups: Lundayeh/Lun Bawang, Kelabit and Sa'ban residing in Bario, Ba'kelalan, Long Semadoh and Long Pasia in Malaysia, and the Krayan and Krayan Selatan districts of Indonesia. This population amounts to approximately 12,500 individuals. FORMADAT won the United Nations Development Programme's (UNDP's) 2015 Equator Prize for reducing poverty through sustaining biodiversity, emphasising community empowerment and cultivating innovative partnerships for local climate solutions.[3]

FORMADAT's community ICT networks have grown over the years with the growth of the organisation and with the expansion of its set of supportive partners (see Table 1). They began in 2005 with the multi-award-winning eBario telecentre project, which later expanded to incorporate a telecentre in Ba'kelalan and Malaysia's first community radio station in Bario. eBario project staff, supported by WWF, later assisted with the establishment of a telecentre in Long Bawan, Indonesia. The next expansion included a telecentre in Long Lamai, Malaysia, the home of a remote Penan community. Although not formally a member of FORMADAT, this telecentre and the community participate in a range of joint activities with FORMADAT that are supported by ICTs. Most recently, the government of Malaysia has established public telecentres in both Bario and Long Pasia which also provide access to ICTs for the resident members of FORMADAT.

FORMADAT is an indigenous community organisation that draws support from several partners. As a founding member, the Kelabit community of Bario promoted the use of ICTs towards the advancement of its and the organisation's goals based on their many years of experience with the eBario project. This was instituted by researchers at the Institute of Social Informatics and Technological Innovations (ISITI) at Universiti Malaysia Sarawak (UNIMAS), who also helped with the establishment of community-operated telecentres in Long Lamai and Ba'kelalan before they were handed to the community. Ultimately, FORMADAT and UNIMAS agreed on a memorandum to promote joint action research

FIGURE 1.

The Heart of Borneo and FORMADAT communities

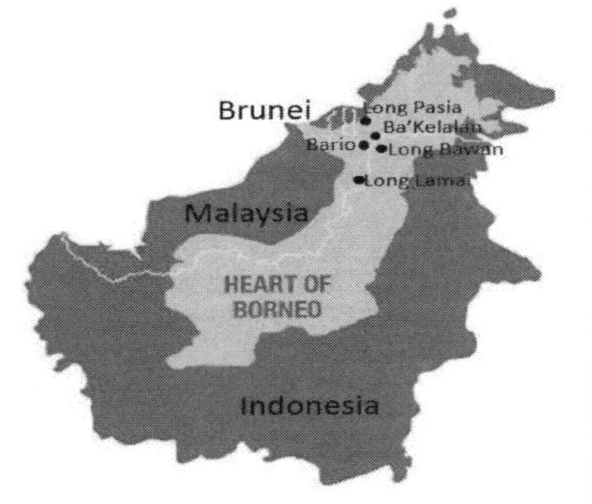

and project development that utilises the ICT networks to advance the goals of sustainable development and environmental protection in the Heart of Borneo. Alongside the growth of FORMADAT's ICT networks, the organisation conducts networking and knowledge-sharing activities to promote their use towards endogenous indigenised development – the animation of development along a bottom-up trajectory, involving the search for development resources and mechanisms that focus on the local territory and which address the specific needs of the indigenous residents.

The early components of what became the FORMADAT ICT networks were established before the organisation was formed but the initiative grew as the various stakeholders collaborated with each other in pursuit of the same goal of endogenous development in the highlands of the Heart of Borneo. The assertion by WWF that the lifestyles and knowledge of the indigenous residents of areas that are the most vulnerable to climate change – such as rainforests – are critical for a better understanding of both its impacts as well as possible mitigation measures increased the importance of efforts to preserve the lifestyles of the FORMADAT members and added impetus to their use of their ICT networks to achieve such outcomes.

As a result of the widely publicised successes of the eBario project, the other FORMADAT communities were alerted to the potential for positive outcomes from similar projects for themselves. Accordingly, they were highly motivated towards participating in the establishment and operation of their own ICT installations, which facilitated

3 WWF-Malaysia. (2015, 15 October). FORMADAT wins UNDP's Equator Prize 2015. www.wwf.org.my/?20465/FORMADAT-wins-UNDPs-Equator-Prize-2015

the recruitment of volunteers and also helped in mobilising community engagement towards the identification of suitable ICT applications that would be capable of contributing towards their well-being. Nevertheless, there were still significant challenges, not the least being the logistical aspects relating to the remoteness of the locations, but more significant was the imperative for establishing trust between the implementation teams and the communities.

TABLE 1.

The FORMADAT ICT community networks

Location	ICT components	Community	Ownership and management	Partners	Status and impact
eBario, Sarawak, Malaysia	Telecentre: computers, VSAT[1] internet access, telephone, solar power	1,000 Kelabit residents	eBario Sdn Bhd, a social enterprise promoting ICTs for community development. Run by volunteers.	UNIMAS[2] IDRC[3] MIMOS[4] MCMC[5]	Opened by UNIMAS in 2001 and handed over to the community in 2005. Having achieved its objectives of raising awareness and skills and stimulating local development, rebranded in 2016 as a knowledge centre and community museum.
Bario 1Malaysia Internet Centre	Computers, internet access and Wi-Fi	1,000 Kelabit residents	MCMC with a locally employed manager.	MCMC	Opened 2016. Raising ICT skills for e-learning, social media, entrepreneurship, multimedia, and life-long learning.
Radio Bario, Sarawak, Malaysia	Low-power 50-watt FM transmitter	1,000 Kelabit residents	eBario Sdn Bhd. Run by female volunteers.	IFAD[6] RadioActive (UK)	Went on-air October 2011 and still operational, broadcasting twice daily to residents living within a range of 15-25 km. Delivers local news in the Kelabit language.
eBario, Sarawak, Malaysia	VSAT for internet access and Wi-Fi	1,000 Kelabit residents and 5,000+ diaspora	eBario Sdn Bhd	OmniAccess S.L. (Spain)	Provides internet access for ground-station testing to be used for data streaming and digital broadcasting of Radio Bario.
Ba'kelalan, Sarawak, Malaysia	Telecentre: computers, VSAT internet access, Wi-Fi, telephone, solar power	2,500 Lun Bawang residents	UNIMAS and the community. Run by volunteers.	UNIMAS MIMOS MCMC	Opened by UNIMAS in December 2010. Raising awareness and skills and stimulating local development.
E-Krayan Telecentre, Long Bawan, East Kalimantan, Indonesia	Computers, VSAT internet access	3,800 Lundaye residents	Managed by FORMADAT Krayan.	eBario WWF-Indonesia German government (BMZ-FiT)	Opened in April, 2011. Provides ICT access for residents, local schools, government officials, visitors, and the health centre. Supports promotion of local products and ecotourism.
eLamai, Long Lamai, Sarawak, Malaysia	Computers, VSAT internet access, Wi-Fi, telephone, solar power	600 Penan residents	UNIMAS and the community. Run by volunteers.	UNIMAS Information Society Innovation Fund Asia	Opened in 2009. Raises awareness and skills promoting health, ecotourism, highlighting local development issues, and documenting aspects of local culture.
Long Pasia, 1Malaysia Internet Centre, Sabah, Malaysia	Computers, internet access and Wi-Fi	500 Lun Bawang/ Lun Dayeh residents	MCMC with a locally employed manager.	MCMC	Opened in 2017. Raising ICT skills for e-learning, social media, entrepreneurship, multimedia, and life-long learning.

1) Very small aperture terminal. 2) Universiti Malaysia Sarawak. 3)International Development Research Centre (Canada). 4) Agency of Malaysia's Ministry of Science, Technology and Innovation for Applied Research and Development in ICTs. 5) Malaysian Communications and Multimedia Commission. 6) International Fund for Agricultural Development.

The eBario telecentre research project has been widely acclaimed, winning multiple awards for its innovative approach to localised development for rural and remote communities. As a result, it is possible to claim considerable influence over Malaysian government policy for accelerating the provision of subsidised internet access to its underserved rural population. The application for a licence for Radio Bario lead directly to the government liberalising its policy to allow community broadcasting. With the formation of FORMADAT, WWF became aware of the impact of these initiatives and championed the establishment of the telecentre in Long Bawan, Indonesia. Subsequently, seeing the convergence of each other's goals, WWF-Malaysia and Indonesia have entered into a formal partnership with UNIMAS and FORMADAT to conduct joint action research around the theme of ICTs for endogenous and indigenised development among the communities in the Heart of Borneo. Simultaneously, the government has extended its 1Malaysia programme for public internet centres into the area, stimulated – it can be argued – by the achievements of the previous implementations. Additionally, Malaysian telecommunications operators have extended their mobile phone networks into the area, recognising the potential market among the now technology-aware communities and their desire for improved communications.

Against this background of technology diffusion, and arising from their studies of the impact it could have, UNIMAS researchers quickly understood that desirable results stem from how the technology is used and that communities would benefit from facilitated processes to define uses of technology that would contribute to their specific needs and aspirations.[4] By adopting a process of participatory engagement for mutual learning, they were able to jointly develop an agenda that defined applications capable of achieving this. One important component of this is the eBorneo Knowledge Fair (eBKF), which began in 2007 in Bario and has been held every other year since then, moving to Ba'kelalan in 2015. The fair is an immersive experience in rural telecommunications for local development – for researchers, government officials, policy makers, private sector representatives and development professionals. The event features the innovative use of technologies for localised development and it offers opportunities for knowledge exchange and sharing between community members and various other stakeholders. eBKF is organised as an "unconference" – structured and led by the people attending it. Instead of passive listening, all attendees are encouraged to participate. eBKF features open discussions, workshops and "walk-shops" (in which participants are taken on a walk around the village to illustrate the topic they are discussing) rather than individual speakers giving talks.

The result has been a range of implementations that demonstrate improvements in education, health, enterprise development, social interactions and cultural preservation, all of which are greatly welcomed by the participating communities. Women have benefited particularly from additional incomes through the promotion of community-based ecotourism, which has expanded considerably and which generates jobs that they typically perform.[5]

Conclusion

Overall, research findings suggest that the community ICT networks and their associated activities for engagement with their users have empowered communities towards development that they welcome by giving them voice, stimulating involvement in public debates that affect them, and enabling more equal participation in the information age and digital economy policies that the respective governments are advancing. The major lessons are that i) smart partnerships can create win-win opportunities for communities to acquire otherwise unaffordable technology resources, and ii) while the technology is essential, alone it is insufficient, requiring associated activities for public engagement in order to generate positive outcomes that are relevant to local needs. Insofar as the networks have been able to stimulate locally derived development, they can be seen to have achieved positive results.

However, with regard to the wider context of regional and national development, the participating communities are still experiencing tension between what is regarded as development by them and how it is defined by national authorities and public bodies. Accordingly, there is now a pressing need to influence national policy making in both countries towards the concept of endogenous and indigenised development that can be fostered with ICTs

4 Bala, P., Harris, R. W., & Songan, P. (2004). eBario Project: In Search of a Methodology to Provide Access to Information Communications Technologies for Rural Communities in Malaysia. In S. Marshall, W. Taylor, & Y. Xinghuo (Eds.), *Using Community Informatics to Transform Regions.* IGI Publishing. https://www.igi-global.com/book/using-community-informatics-transform-regions/1022

5 Harris, R. W. (2009). Tourism in Bario, Sarawak, Malaysia: A Case Study of Pro-poor Community-based Tourism Integrated into Community Development. *Asia Pacific Journal of Tourism Research, 14*(2), 125-135. https://doi.org/10.1080/10941660902847179

in accordance with the principles of environmental protection and sustainable development. In this regard, despite the commitments to preserving the Heart of Borneo, it remains unclear that such policies have been internalised within national development programmes, and it is this that makes up the greatest challenge to ensuring that ICTs achieve their optimum contribution to development.

Action steps

The following recommendations can be made when setting up a community network:

- Find partners who are willing to learn and share their knowledge and resources.
- Focus on turning access to technology into relevant use by communities by seeking answers from them.
- Accelerate the diffusion of access to technologies and acknowledge the long-term necessity for subsidies.
- Extend universal service provision fund eligibility to qualified community-based organisations.
- Promote development policies that acknowledge the imperative for indigenised development.

ITALY

BUILDING COMMUNITY NETWORKS IN ITALY: HACKER-LED EXPERIMENTS TO BRIDGE THE DIGITAL DIVIDE

University of Trento, CNIT and ninux.org
Leonardo Maccari and Claudio Pisa
ninux.org

Introduction

This report describes community networks on the Italian scene, with particular attention to ninux.org, which has the largest coverage in the country and the longest history. Ninux started as a "geek experiment", and maintained this approach throughout its evolution.

This gave it a specific ethical and ideological purpose, and allowed it to actively contribute to the spirit and development of the European community network movement. Its approach, although not focused on internet access, was successful, especially in urban areas, in a period in which wireless technologies were expanding, and the Italian hacker scene was very active. Today, however, we are seeing a decrease in interest and energy compared to other European initiatives.

The report describes ninux.org's trajectory, while also considering other internet-based initiatives in Italy that are expanding their user base. It describes two possible futures for ninux, which may be emblematic of the hard decisions that many involved in the early community network movement worldwide might face.

Policy, economic and political background

Three features of Italy are worth describing to introduce the context:

- Italy is one of the European countries with the largest digital divide (in 2017 only 69.5% of Italian families had access to the internet through fixed broadband, according to the Italian National Statistical Institute).[1] This is probably due to the fact that the Italian population is scattered over a large area: 55% of its people live in cities and towns with fewer than 50,000 inhabitants, and about 18% in towns with fewer than 5,000.[2] The country also has an extremely variegated geography, made up of flatlands and many mountainous and hilly zones. While this diversity is culturally astonishing, it is a nightmare from the point of view of developing infrastructures.
- Italy is one of the countries in the European Union that was hit most severely by the economic crisis in the last decade. According to the National Statistical Institute, in 2006 the number of people living in absolute poverty was about 1.9 million, while by 2016 this had grown to 4.7 million.
- The Italian population is ageing, and declining in number. Italy has one of the highest rates of people (especially young people) emigrating to foreign countries and one of the lowest number of people with a university degree.

These statistics are important because ninux.org emerged in the early 2000s as a community where hackers (primarily young and educated males) engaged in the creation of an alternative internet, with internal rules derived from their own ethical and political vision. Today this approach faces the challenges of a society that is more unequal, precarious, uncertain, and less educated, especially when it comes to young people. Can an advanced, progressive hacker experiment thrive in a declining society?

The history of ninux.org

Ninux.org was started in Rome in the early 2000s and was the initiative of a computer science engineering student, Nino Ciurleo. Nino had grown technically in the ham radio community as well as the Italian hacker scene and was influenced by the punk do-it-yourself attitude. One day he read about the Seattle Wireless community network in a magazine, liked the idea, and decided to use his personal web page – ninux.org (a pun on "Nino" and "Linux") – which was hosted on a server in his room, to search for other enthusiasts to help him build a wireless community network in Rome. To help spread the word, stickers were printed and placed around the city. After a couple of years, the ninux network was bootstrapped, and the core of the network, composed of three nodes, was up and running. Many people with different (but still technical) backgrounds were then joining the ninux mailing

1 https://www.istat.it/it/archivio/207825

2 From the document *Strategia italiana per la banda larga*, available at www.sviluppoeconomico.gov.it/images/stories/documenti/ITALIA_Strategia_BUL-Piano_di_investimenti_fin.pdf

The sticker promoting the early ninux community network.

list and meetings. The motivations for joining the community ranged from socio-political reasons, to helping to bridge the digital divide, a desire to learn by doing, down to pure curiosity.

In spring 2006, a handful of ninux members participated in the Wireless Community Weekend in Berlin, getting a grasp of the philosophy of the Freifunk community network[3] and acquiring skills in mesh networking and open source firmware operation and development. Back in Rome, these skills were developed by the core members of the community and put into practice. However, some obstacles were in the way: the hilly topography of Rome and the unclear legal framework for outdoor wireless networks.

In 2009 the ninux community organised the first "Ninux Day", a two-day event to which several community network members were invited: community network members from Freifunk, guifi.net and Athens Wireless Metropolitan Network (AWMN), developers from the OpenWrt community,[4] and other enthusiasts from all over Europe gathered in Rome to hack together and give and attend talks. The most interesting outcomes from the ninux perspective were the enthusiastic response from other European community networks and the understanding that Europe has a common legal framework, which potentially allowed ninux to circumvent what appeared to be the legal limitations for outdoor wireless in Italy.

Just some weeks before Ninux Day 2009, some ninux members attended the Wireless Battle Mesh v2 in Brussels,[5] an event to build a wireless mesh network and test the performance of different wireless mesh network routing protocols. This led to ninux organising the Wireless Battle Mesh v3 at a campsite next to a lake near Rome in 2010, replicating the success of the Ninux Day event and also involving a range of different people with different skills.

Since then, many things have changed. Ninux is now a community with about 350 nodes scattered around Italy. It is an integral part of the European community network movement: it hosts services, it has participated in European research projects, it has its own "autonomous system", and it is well known among Italian hackers and geeks.

A community of hackers

One of the key characteristics of ninux is its hacker nature. In the period 2013-2015 (when Italian legal limitations were no longer in place[6] and Snowden's revelations were under the spotlight), ninux almost doubled the number of its nodes and hit the news in many mainstream newspapers and websites. Mesh networks were depicted as a remedy not only for the digital divide, but also for surveillance. Besides a certain degree of journalistic hype, the truth was that around 2010, both the technical and ethical propositions of community networks were extremely advanced. The idea that a mesh network, being technically distributed, could enable the creation of a communication platform with a governance structure inspired innovation and advancements in many directions. Today, the academic community recognises the value of that "avant-garde" period, and community networks have been invited by national and international institutions to document their activities over that time.

3 The Freifunk community is one of the first of its kind in Europe. They meet every year in a get-together called the Wireless Community Weekend. In 2018 it merged with Wireless Battle Mesh, the most relevant European meeting of community networks. See https://wireless-meshup.org/doku.php

4 https://freifunk.net, https://guifi.net, www.awmn.net and https://openwrt.org

5 All the Battle Mesh events are documented at: https://www.battlemesh.org

6 Until 2011 the anti-terrorism "Pisanu" Law, named after the former minister of the interior in Silvio Berlusconi's government, introduced a technical and legal burden on anybody wanting to offer Wi-Fi access to the public in terms of authentication and data retention. The law introduced a legal responsibility for storing privacy-sensitive data. Up until 2012 the "Gasparri" Law, named for the telecommunications minister in the same government, required permission to be granted to operate wireless networks in public places, even using unlicensed frequencies. Today, the Italian system is less severe on Wi-Fi networks, even if issues remain with regards to authentication and data retention. See: https://netcommons.eu/?q=content/community-wireless-networks-intermediary-liability-and-mcfadden-cjeu-case

It is fundamental to understand that without both the technical and social passion of the community network activists, this would not be possible. Hackers made it possible to develop and share the tools that once were only available to them. Today these have been made available to communities with very few technical skills. If a rural community with few technical skills can now use LibreMesh[7] to set up a network, it is thanks to the community network hackers who have worked hard over the last two decades for this to be possible.

Today, the expansion of community networks is remarkable; some of them have reached tens of thousands of nodes, and many new communities have emerged especially in the global South. But what happens when the community network movement starts to lose its appeal to hackers? In the case of ninux, the technical nature of the community has always been a strong driving force. When the community's interest in the emerging technical issues decreased, fewer and fewer people participated in the community.

It may be that the context had a large impact on this evolution. For instance, in the last couple of years, at least five key people, and among the most technically skilled that had participated in the community, simply left Italy as a consequence of the social situation described above. A society that is more unequal and in which it is hard to find economic stability produces isolation and disincentivises participation, and ninux is probably also part of a general decline of Italian community organisations. On the other hand, it is also true what one "ninuxer" said in a meeting in 2017: "Wireless is not cool anymore."

While 10-15 years ago wireless technology was on the rise and attracted the attention of hackers, today, wireless is taken for granted; it is a "commodity", and young hackers are more attracted by other fields (like blockchain, the internet of things, etc.). Similarly, networking, open source and Linux hacking were original and new in the early 2000s, while today students studying information and communications technology (ICT) in universities often acquire those skills while studying. Some people joined ninux as a personal investment in themselves, which later on turned out to be a career in ICTs. It may then be that the specific combination of technical novelty and the status of a "liberation technology" enjoyed by wireless in the early 2000s that made community networks (and ninux) flourish may not be present anymore. It is reasonable to think that ninux, while still being a vivid community (especially on some of the smaller Italian islands), needs to change its principles in order to continue to exist in the years to come.

A parenthesis: Other community networks in Italy

There are several initiatives that may fit the description of a "community network" on the Italian peninsula. Projects like Progetto Neco (Neco Project),[8] GalliaNetwork,[9] Reti Senza Frontiere (literally "Networks Without Borders")[10] and Senza Fili Senza Confini (SFSC, or, literally, "No Wires, No Limits")[11] are small to medium initiatives that may be called "community ISPs". Progetto Neco (Neco stands for "network community") is based in Vietri di Potenza, a town with less than 3,000 residents in the south of Italy. The project was started in 2008 by a group of local hackers with the aim of bridging the digital divide and today has 36 nodes serving roughly 230 families. An association was created, and associates pay a monthly fee to access the network services and the internet. GalliaNetwork is another community ISP, located in the town of Canezza in the north of Italy. Similarly to Neco, it was created in 2011 by a group of residents who had no internet access, before expanding into a network serving several surrounding towns. A group of five to six enthusiasts run the network and offer several services, such as website hosting, a local cloud and internet access. Reti Senza Frontiere is a small association born in 2015 in the countryside outside Rome. It connects a few families to the internet in another digitally divided area.

SFSC stands out from the others for its evolution and the media coverage it has received worldwide. It is another association whose primary purpose is to fight the digital divide in an area north of the city of Turin called Verrua Savoia. From there, it expanded to several small villages isolated from the main city by the mountains. SFSC started as a research experiment led by the Polytechnic University of Turin, one of the most important technical universities in Italy, which had already used a customised wireless device to connect an isolated town. After that first experimental phase, the initiative turned into an organisation, and now serves (according to its president and founder Daniele Trinchero) about 5,000 families in the region for a fraction of the market cost of commercial ADSL service.

The organisation is rooted in the territory and organises courses, skills sharing, and digital literacy

7 https://libremesh.org

8 www.progettoneco.org

9 www.gallianetwork.it

10 retisenzafrontiere.org

11 https://www.senzafilisenzaconfini.org

activities. In 2014 it was featured in *The New York Times*, and later on in Italian newspapers, which gave high visibility to the project. Compared to the other local initiatives, SFSC had the advantage of being born from one of the most important and organised universities in Italy. This offered the necessary technical skills together with network competence and contacts that made it easier to solve the initial challenges to bootstrap the network.

All these experiences tell us that the model of a community network is welcomed in rural areas, in which there is a need for low-cost access to the internet. With their own differences, these networks are growing, or have reached a state in which they could grow more, but are limited by the lack of human resources to make the network scale.

Ninux always tried not to be perceived as an ISP, but as an experimental, hacker network. The reason for this is that ninux was born in an urban area and many people contacted the community hoping to replace their ISP with ninux for free. This utilitarian attitude was discouraged by the community, which clearly stated that while ninux has several gateways to the internet, it was not there just to replace commercial ISPs. Rather, it was a philosophy, a movement that was political, practical and experimental.

Today ninux has expanded into rural areas with poor connectivity. On some islands, its primary purpose is actually utilitarian: to overcome the digital divide. But the original spirit still persists.

Conclusions

The intrinsic innovative value of community networks is their mix of technical and social innovation. Technology (low-cost wireless solutions and open source software to run networks) enabled a new social behaviour, which challenged the status quo in service provision and the monopolies enjoyed by the telecommunications industry. This is true in areas where there was simply no internet access, and community networks showed how this was possible, but also in areas where the big telcos – whose ethical fingerprint is questionable – have a market share. We cannot untangle the technical and the social advances, as the second is enabled by the first, and feeds back into it. Without hackers, there would be no ninux, no Freifunk, no guifi.net, no LibreMesh (just to name a few) and in general, no community networks. If the whole community network movement turns into a "connectivity factory", its original and innovative push will be strongly reduced.

The question that is still open today is how to couple the technical innovation of community networks with the social impact that social enterprises are achieving in other fields (e.g. food cooperatives, to name just one movement that is very active in Italy). A hacker network is, by definition, a moving target, an experimental infrastructure that could be subject to tests, changes and failures. A community ISP, instead, tries to offer a service comparable to the service that a commercial ISP offers. When the ninux community faced the chance of moving to an "in production" network it reacted without much interest. Many people in the community were there to experiment, not to run an ISP. And in fact, running an ISP is a tough job; and most of all, it is a job.

Action steps

The ninux community does not have well-defined decision-making bodies or procedures, and its participants come from heterogeneous backgrounds. Until now, ninux has not had the willingness to try to become a community ISP, even if successful models point in this direction. There are two scenarios we can imagine and we describe them below, with some possible next steps.

In one scenario, the ninux community has no interest in transforming into a community ISP; ninux is then seen as a lab for experimenting with new technologies and ideas, having as outputs innovative distributed infrastructures based on open source software and hardware, and serving as inspiration for new community ISP models. The socio-political motivation is then derived from the mix of these outputs and the open attitude of the ninux community. Ninux would have to update the themes it explores to meet the potential of new technologies that need to be hacked (e.g. the internet of things?), but it may shift away from the goal of being a community network.

In another scenario, ninux takes steps towards becoming an ethical community ISP. The community is increasingly composed of non-technical people whose motivations for participating are derived elsewhere. In this scenario, the ninux goals would shift towards solving the problems reported by the users and the broader local communities. This process requires "technological mediation" skills and the willingness to put aside those practices and attitudes (e.g. techno-elitism) which usually ward other people off from the hackers' domain.

What is not clear is if the scenarios described above are mutually exclusive, or can co-exist to some extent. What we hope is that this discussion takes place in the ninux community, and that the community evolves maintaining its spirit based on socially inspired innovation.

KENYA

CONNECTING THE UNCONNECTED IN KENYA'S URBAN SLUMS

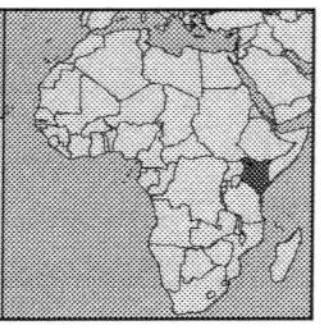

Tunapanda Institute
Josephine Miliza
https://www.tunapanda.org/

Introduction

Kenya, one of Africa's fastest growing information and communications technology (ICT) markets, has a long-term development plan, Vision 2030, that aims to transform Kenya into a knowledge-based economy by utilising ICTs for national development and growth.[1] In the 2017/2018 second quarter statistics report released by the Communications Authority of Kenya, mobile subscriptions recorded a growth of 4.4%, rising from 41 million to 42.8 million. The mobile penetration level reached 94.3%, up from 90.4%.[2] Data internet subscriptions recorded an 8% growth – 33.3 million up from 30.8 million subscriptions.

Yet even with these milestones in mobile connectivity, rural and informal settlements lag behind. Challenges such as the high cost of internet access, a lack of infrastructure, a lack of locally relevant content and a lack of ICT skills are barriers to utilising ICTs effectively for socioeconomic development.

Nairobi is the country's largest and capital city, with an estimated population of 3.5 million,[3] and slums hosting almost three-quarters of this population.[4] Kibera is Nairobi's largest slum, located seven kilometres south of the city with an estimated population of 500,000 to 700,000.[5] Residents of Kibera suffer from poverty, inadequate food and water, rampant diseases and few educational opportunities. A majority of the residents live on under a dollar a day.

Community networks can be defined as a bottom-up approach to meeting community connectivity needs.[6] These needs vary from one community to another; as a result, these networks have different motivations for starting. Kenya, like many countries in Africa, is still new to community networks. This report focuses on the TunapandaNET community network, the only one in the country.

TunapandaNET community network

TunapandaNET is an urban community network operating in Kibera. It is a project of the Tunapanda Institute, a non-profit social enterprise that runs intensive three-month technology, multimedia design, and business training courses in extreme low-income environments in East Africa, such as Kibera and Turkana in deep rural Kenya.[7] These programmes enable young people to become digital professionals, and to gain skills and mindsets to empower other youth in their communities through peer-to-peer learning. The organisation has produced 400 graduates in the past four years.

The TunapandaNET community network was started to help the institute reach more youth in Kibera, since the institute can only accommodate 30 trainees per cohort. In 2015, the institute started developing a gamified e-learning platform called Swag.[8] Swag is an open source software system – for web and Android – that enables individuals and groups to access offline multimedia educational content without needing access to the internet or highly trained teachers. The initial goal was to connect the institute to three partners in Kibera through a wireless mesh network. The partners would serve as hotspots where youth would access the e-learning platform.

Through a video created by the institute, Ubiquiti equipment was donated to help build a pilot mesh network for the project. However, the project implementation team at the institute, consisting of two IT volunteers and three Tunapanda graduates from the previous cohorts, faced challenges such as insufficient knowledge and skills in operating the equipment and few resources online to provide guidance. As a

1 icta.go.ke/pdf/THE%20NATIONAL%20ICT%20MASTERPLAN%202017.pdf

2 www.ca.go.ke/images/downloads/STATISTICS/Sector%20Statistics%20Report%20Q2%20%202017-18.pdf

3 Kenya's population is estimated to have reached 50,827,543 in 2018. worldpopulationreview.com/countries/kenya-population

4 Sana, O., & Okombo, O. (2012). Taking Stock of Socio-economic Challenges in the Nairobi Slums. *Friedrich-Ebert-Stiftung*. https://www.fes-kenya.org/media/publications/2013/Taking%20Stock%20of%20Slums%20Booklet.pdf

5 mirror.unhabitat.org/content.asp?cid=3220&catid=206&typeid=13

6 Rey-Moreno, C. (2017). *Supporting the Creation and Scalability of Affordable Access Solutions: Understanding Community Networks in Africa*. Internet Society. https://www.internetsociety.org/resources/doc/2017/supporting-the-creation-and-scalability-of-affordable-access-solutions-understanding-community-networks-in-africa

7 https://www.tunapanda.org

8 swag.tunapanda.org

result, some of the donated equipment broke. The team was also not aware of any existing projects that they could contact for support for the technical challenges they faced. Finally, the team was able to pilot the network by connecting Tunapanda to one of three partner nodes. However, the e-learning platform did not receive as many users as they had anticipated.

Breakthroughs

Two breakthroughs came in 2016. First, in March 2016, two of the team members were selected to attend a workshop on the internet of things (IoT) and networking for developing countries at the Abdus Salam International Centre for Theoretical Physics.[9] During the training sessions, they presented the mesh network idea to workshop organisers Marco Zennaro and Prof. Ermanno Pietrosemoli, who gave them some pointers on configuring Ubiquiti equipment and donated some routers to the project. The team also met Arjuna Sathiaseelan and Adisorn Lertsinsrubtavee, who also had experience in community networks and offered very valuable advice on how to move forward with the project. Before the workshop, the team was not aware of the term "community networks" and the significance of setting up a community structure that allowed the community to manage its own access needs. Their idea was simply a technological solution to solving the challenges of high bandwidth costs and access to digital educational content. They were introduced to successful community networks such as guifi.net[10] in Catalonia and TakNet[11] in rural Thailand, which they could draw lessons from.

The second breakthrough was meeting Carlos Rey-Moreno from Zenzeleni Networks in South Africa, who was carrying out research on existing community networks in Africa. Following the "Map of the Community Network Initiatives in Africa" report,[12] the first Summit on Community Networks in Africa was held in November 2016, supported by the Internet Society.[13] During the summit, the team met other operators in Africa and shared the challenges they were facing in establishing their network in Kibera. From these discussions, the network operators formed a support group championed by Carlos, Jane Coffin and Michuki Mwangi from the Internet Society to address the issues they were facing.

Going back to the drawing board

These experiences helped TunapandaNET rethink its model. One lesson drawn from successful community networks was that most go for community buy-in first. Using a user-centred design approach, the team started carrying out research on potential network users, a critical step they had failed to take before. This enabled them to understand the unique challenges faced by Kibera residents, and to co-design potential solutions with them. Some of the key challenges were access to computers, lack of ICT skills, and lack of knowledge on how access to connectivity could help them increase their earnings. Most of the youth only accessed the internet for entertainment purposes.

Three areas were identified as areas of great value to the community: education, health and business. Deployment was planned in three phases, with the education pillar being first. Kibera has over 300 educational institutions, most being low-cost, privately run, informal schools.[14] Most of these schools are run by religious institutions and non-governmental organisations serving vulnerable children and youth. They face challenges such as lack of trained teachers, poor infrastructure, and a lack of learning resources due to high costs.

Charles Ochieng, a local community leader and manager at St. Christine Community School, which has over 500 students, said that every year the school loses close to USD 10,000 in unpaid school fees since most parents are unable to pay. Because of this they have difficulties in paying their teachers and rely on the good will of these teachers during tough months. Despite the fact that the school lacks a fully equipped science lab and relies on borrowed equipment, its neighbouring schools rely on it to share its lab with them during the national secondary school examinations. Tumaini School, another informal school with a population of over 300 students, uses downloaded YouTube science experiment videos to teach practical science lessons with only four laptops serving the whole school. The school's principal James Wanyama has expressed his desire for the school to have more computers, but cited the high cost of computers as a barrier. However, despite these challenges, it is

9 https://www.ictp.it

10 https://guifi.net/en/node/38392

11 Lertsinsrubtavee, A., et al. (2015). Understanding Internet Usage and Network Locality in a Rural Community Wireless Mesh Network. *Proceedings of the Asian Internet Engineering Conference 2015*. docplayer.net/24614970-Understanding-internet-usage-and-network-locality-in-a-rural-community-wireless-mesh-network.html

12 Rey-Moreno, C., & Graaf, M. (2016). Map of the Community Network Initiatives in Africa. In L. Belli (Ed.), *Community connectivity: Building the Internet from scratch. Annual report of the UN IGF Dynamic Coalition on Community Connectivity.* https://www.researchgate.net/publication/309291449_Map_of_the_Community_Network_Initiatives_in_Africa

13 Rey-Moreno, C. (2017, 12 January). 1st Summit on Community Networks in Africa Report. *Internet Society.* https://www.internetsociety.org/resources/doc/2017/1st-summit-on-community-networks-in-africa-report

14 openschoolskenya.org

inspiring to see how these schools do their best to deliver learning to a population that seems forgotten by the relevant government institutions.

This phase focused on promoting the use of ICTs for education. This entails providing schools with access to digital educational content through the e-learning platform and digital tools for records management. Teachers would also receive ICT training and be taught how to use the e-learning platform to digitise their lesson plans and deliver learning to students.

In 2017, through a partnership with the Internet Society, the International Centre of Theoretical Physics and Rhinotivity (Denmark),[15] TunapandaNET deployed four nodes serving two schools with over 1,500 students and a youth centre serving 300 local youths. The network provides access to offline educational resources and capacity building for the different network user groups. In 2018/2019, the network will be expanding to 10 additional centres – seven schools, two youth centres and one women's centre. This will be done in partnership with the Internet Society Kenya Chapter[16] through the Internet Society's Beyond the Net Funding Programme.[17] The seven schools have a total of 2,900 students while the community centres will be adding 550 users to the network. It is estimated that the network will have 5,000 users by 2019.

TunapandaNET provides the following services[18] to partner institutions:

- Access to the internet.
- The Swag e-learning platform, which enables users to access learning resources as well as create content. Through the platform, teachers can create digitised lesson plans and curricula. The goal is to provide locally relevant digitised educational resources. The platform also has content that teaches the youth skills in technology, design and business. These micro courses are derived from Tunapanda's three-month curricula.
- Digital skills training for teachers, youth and women, equipping them with relevant skills in education, work and entrepreneurship.
- Technical support for schools and community centres.

15 https://twitter.com/rhinotivity?lang=en

16 https://www.isoc.or.ke

17 https://www.internetsociety.org/beyond-the-net

18 We are offering the services for free for a period of six months, after which institutions will start paying. We are still working out a business/payment model.

The centres will receive a free six-month trial period for all the services, during which the TunapandaNET team will conduct research with partner institutions that will help determine ownership and business models for sustainability.

The network – which uses the 2.4 GHz and 5.8GHz unlicensed spectrum – now has three main layers: core, distribution and access. The core layer is the Wi-Fi network backbone, as well as a data centre, responsible for hosting the Swag e-learning platform, open educational resources and a school management system.

The distribution layer is responsible for connecting the access and core layers. To overcome interference from obstacles such as buildings and trees, there are two base stations set up on high points in Kibera. The hardware used for the base stations are the Ubiquiti LiteBeam AC grid and Ubiquiti LiteBeam AC. The Litebeam AC grid is used for the backhaul connection to Tunapanda Institute. The access layer, which provides the connection to the end-users, has either Ubiquiti LiteBeam AC or Ubiquiti NanoStation M5 and 2.4 GHz D-Link wireless access points. This layer is responsible for distributing internet access and other network services to the end-users.

Beyond access: Building innovative learning and earning communities

> "Building community networks is 20% technical engineering and 80% social engineering."
> –Michuki Mwangi (Internet Society)

With the low literacy levels in the community, most find it difficult to understand the value that the internet would bring to their lives. There is always the "So what?" question. Unless the community members see tangible solutions to their local needs, the network is not sustainable. Going "beyond access" means individuals and communities being able to access better education, health and government services. They should be able to innovate and create local solutions to their unique challenges, such as new markets for their products and services, the preservation of indigenous cultures, and amplifying the voices of the unheard. There is also the challenge of how to incentivise the community to be part of the network as volunteers, or even to take part in the capacity-building programmes. Kibera has many NGOs that give monetary benefits to the community, and this creates an expectation from the community.

There is also some competition among schools and NGOs which can hinder collaboration. The challenge and opportunity ahead lie in building learning and earning communities within the network, and bringing together the different stakeholders to work collaboratively.

The network can be a catalyst in enabling the community to create solutions and boost the local economy. As local schools continue to adopt ICTs, they need maintenance services. The network is working towards equipping local youth with computer hardware, software and network skills to meet the network maintenance needs, and thereby increase their earning potential. Youth in Kibera are very passionate about art and music and the network can be a platform connecting them to local and international markets. The network is also working towards training women in e-entrepreneurship so they can better the business potential of their handmade crafts such as handbags and jewellery.

The gender gap

The gender gap can be addressed from two perspectives: the team and network users. Only two of the seven members of the TunapandaNET team are female. This is a symptom of few women embarking on technology-related careers in Kenya. The main challenge faced by female team members is access to some of the network's partner institutions, due to insecurity and fear of harassment from unemployed male youth. Thus, we female team members have to be accompanied by male team members for all partner visits.

For the network users, very few women and girls are interested in digital skills training programmes or technology-related careers. During the team's interview with form four students at St. Christine high school, only one girl out of 23 girls was interested in a career in science, technology, engineering, and mathematics (STEM). None of the girls knew about technology-related careers or had heard about computer science or coding. Although the school has a computer lab, the school manager said that only the male students used it frequently. When the girls were asked why, they answered that they did not see how learning computers would help their desired careers. I took that opportunity to share with the girls my journey to becoming a network engineer and quite a number took an interest in learning more on technology.

From this experience at St. Christine school and other partner institutions, it was found out that a majority of girls and women in Kibera have very limited knowledge of technology and technology-related careers, and see it as a male domain. Other challenges are high costs of ICT devices such as computers and smartphones. To access the internet, women would have to use local cybercafés, which are mostly used and run by males. Many women have expressed fear of sexual harassment in the cybercafés. Relevance is another issue: women often cannot see how the internet can be used to tackle the challenges they face.

In 2016, Tunapanda started an initiative called Tech Dada (Tech Sisters) to address these challenges through a mentorship programme for girls aged 14 to 20, and offering digital literacy training for women. These programmes are done in partnership with centres connected to the TunapandaNET community network. The initiative has in the past partnered with Plan International Kenya through the Adolescent Girls Initiative Kenya,[19] which provided online safety training for the programme's peer mentors. It has also trained local women champions in Kibera through the Women Voices programme,[20] which used ICTs to amplify women's voices in governance issues in Kibera.

Currently, none of the women from the community is involved in the technical aspects of the network. However, the network is working towards increasing the number of women in the community with technical expertise by encouraging more women to start with the digital literacy training, after which they can proceed to the advanced courses in technology. These courses are to be offered in a local women's centre that the network will be connecting.

Action steps

Community networks are like entrepreneurial startups: it takes resilience, learning from failures, and efficient management of available resources to build one. It is an iterative process, as technology and community needs keep evolving. They also need institutional support to survive.

The following steps would strengthen community networks in Kenya:

- Lower the operating fees: The Communications Authority of Kenya develops the national radio frequency plan and a publicly available National Table of Frequency Allocation.[21] Community networks such as TunapandaNET fall under network facility providers of infrastructure and content providers. The initial licence fees under

19 www.popcouncil.org/research/adolescent-girls-initiative-action-research-program

20 https://womenvoicesictchoices.org

21 www.ca.go.ke/index.php/frequency-spectrum

these categories are between USD 1,000 and USD 2,000 within a county boundary.[22] These fees are difficult for communities such as Kibera to meet. For community networks to have an impact in poor communities in Kenya, these fees need to be lowered considerably.

- Create policy incentives for community networks: In the country's Wireless Broadband Spectrum Policy, drafted in 2017, some of the proposed principles include offering incentives to encourage deployment in rural areas and tax exemptions for initiatives with public interest objectives, such as research and development.[23] Community networks should be given such policy incentives, given their efforts in connecting underserved areas.
- Raise awareness: Finally, given that TunapandaNET is the only community network in Kenya at the moment, it is important to promote more awareness on community networks in the country. This would help TunapandaNET in gaining buy-in from the community and build local partnerships and collaborations to support the network.

22 www.ca.go.ke/images/downloads/TELECOMMUNICATION/LicensingProcedures/New%20Market%20Structure%20Under%20The%20Unified%20Licensing%20Framework%20-February%202017.pdf

23 www.ict.go.ke/wp-content/uploads/2016/03/DRAFT-WIRELESS-BROADBAND-SPECTRUM-POLICY-Final.pdf

KYRGYZSTAN

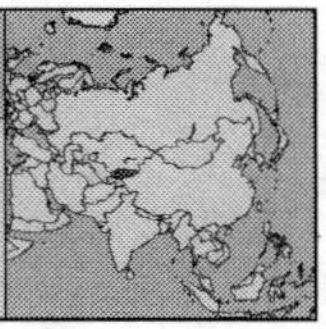

CONNECTING KYRGYZSTAN, VILLAGE BY VILLAGE...

Internet Society Kyrgyzstan Chapter
Isabek Asanbaev and Talant Sultanov
isoc.kg

Introduction

The Internet Society (ISOC) Kyrgyzstan Chapter is setting up a community network in the village of Suusamyr in Chui, the northernmost region in Kyrgyzstan.[1]

Roughly 25% of Kyrgyz citizens are online, with 64% of these living in urban areas.[2] While community networks can help rural citizens access the internet and narrow the digital divide, they are not mentioned at all in the country's information and communications technology (ICT) policies. Radio spectrum is also heavily regulated and one has to apply to register frequencies. The process can take around six months.

The aim of the Suusamyr community network is to create business opportunities in tourism for the community, as well as to offer the community a chance to learn new skills and develop human capital. The internet will also improve the quality and availability of social services, and promote accountability through enabling the closer monitoring of municipal bodies. This report outlines our progress so far, and what we need to do next.

From "barefoot engineers" to building antennas: Getting inspired about community networks

The ISOC Chapter in Kyrgyzstan was established in 2014. ISOC officers from headquarters in the United States then presented the idea of community networks to us, giving the example of the so-called "barefoot engineers" in Nepal who went from village to village and set up community networks. This example got the Chapter members very excited. Later in 2016 the ISOC Kyrgyz Chapter hosted a roundtable on community networks and brought a presenter from India, who spoke about a community network that helped an Indian village enter the e-market with their unique handmade textiles.

One of the participants at the roundtable was a self-employed engineer who built antennas that strengthen the signal of the mobile internet for people in locations with poor coverage. He was very excited about the idea of community networks. The Chapter immediately took him into consideration as an engineer for the community network project. We started applying for ISOC's Beyond the Net funding[3] in the hope of securing a grant.

Applications for Beyond the Net funding are primarily accepted from ISOC Chapters or projects with an ISOC Chapter actively involved. ISOC supports projects that help the community, providing tools and skills and creating visibility for the Chapter.

In search of a village to connect

We started looking for a suitable location. The main criteria was that it should have no internet of any kind, which meant no mobile data coverage, and that internet service providers (ISPs) were not likely to offer services in the community any time soon. We found a promising village near the famous tourist attraction called San Tash – two large piles of stones in the Karkara Valley, which have intrigued archaeologists for centuries. There are several theories about how and why they came into existence.

The history of San Tash goes back more than half a millennium. One explanation of the piles of stones goes like this: When the Mongol warlord Tamerlane (also known as Amir Timur) was going to

1 The ISOC Kyrgyzstan Chapter is involved in a number of other initiatives too. One is the installation of an internet exchange point (IXP) in the city of Osh, located in the Ferghana Valley. This project will help decrease the internet access costs for people in the southern regions of Kyrgyzstan. The initiative also has the potential to impact on access costs in Tajikistan and Uzbekistan, where costs are extremely high. The other project is "Spring of Knowledge", or internet-in-a-box for schools in Kyrgyzstan. It involves the installation of an Orange Pi Zero minicomputer, a Wi-Fi router and a two-terabyte external hard drive in 20 rural schools that have no internet access. This equipment will give high schoolers offline access to Wikipedia, the Khan Academy Lite, as well as many other digital learning materials in local languages. Together with the help of ISOC headquarters we are also organising capacity-building events on IXPs and community networks for regional operators, field professionals and academia.

2 DR Analytica. (2017). *Kyrgyz Republic: The 2016 ICT Sector Overview.* https://analytica.digital.report/wp-content/uploads/2017/07/Kyrgyzstan-The-2016-ICT-Sector-Overview.pdf

3 https://www.internetsociety.org/beyond-the-net

battle, he told each of his warriors to find one stone and throw it onto a pile. They did so, and by the time they finished there was a huge pile the size of a hill. When they returned from the battle, Tamerlane told them to do it again. As a result, there were two hills of stones, both hills of different sizes. That is how he counted his losses.

The piles have been there for centuries. Many people from all over the world come to see them. The locals make business from the tourists, and we hoped that internet access via a community network would help the community attract more tourists, and make searching for accommodation and other services more accessible to the visitors.

Winter passed and fortunately for the villagers a government-owned ISP had begun offering connectivity to the village. For us it meant that we had to once again search for a suitable location. The search was long, as the criteria were hard to match and even harder to verify since we had to physically travel to all the possible locations. We needed a village that was not connected – preferably not even with mobile data – but was still within 50 km from the backbone of any ISP that was willing to cooperate (many do not want the competition of a community network).

We travelled all around the country, which is relatively small but very mountainous. There were several other promising villages. One of them was Sheker, a village in the Talas region in the northwestern part of the country, and home of Kyrgyzstan's most prominent person, the writer Chyngyz Aitmatov. One of his greatest novels is called "The First Teacher", a story about the first educated person who came to a remote village to teach children during the times of Lenin. This year will be the 90th anniversary of Aitmatov's birth, and it will be widely celebrated throughout the Turkic world and the former Soviet Union. It would be great, we felt, to give Sheker another "first teacher" in the form of a community network. However, later we found out that it was impossible to connect the village using our equipment, as it required more than one repeater.

Reading a map of the backbone to find a village

We needed to hasten the search, so we found a wholesale ISP with the second-largest backbone in the country, Elcat,[4] and negotiated with them. They shared a map of their backbone infrastructure with us, which included coordinates of their nodes. This narrowed down our search dramatically and we finally decided on the village of Suusamyr. We visited the village and spoke to the head of the Local Self Government (LSG). He informed us that there was no internet of any kind available in the village. The reason is that the valley – also called Suusamyr – is surrounded by high-altitude mountains, making it hard to connect to the internet. The winter in the valley also lasts for six months instead of three.

The population of Suusamyr is 2,674 people according to the 2009 population census. The length of the valley where it is located is 155 km. The altitude of the bottom of the valley ranges from 2,000 meters to 3,200 meters above sea level. The valley is full of beautiful vistas: mountains, forests, lakes and rivers. It is one of the top tourist destinations in Kyrgyzstan. Some popular tourist activities include hiking, rafting, horseback riding, paragliding, and *kumis*, the seasonal mare's milk detox.

Building a bridge

The first stage of the project involves setting up a wireless link that will connect the village to Elcat's fibre-optic backbone. The wireless link will use AirFiber AF-5X 5.8 GHz radio systems, with 34 dBi antennas installed on 15-metre-high masts. To increase the speed, a MIMO Multiplexer will be installed, allowing us to use four airFiber radios with a single dish antenna.

The second stage is building a fibre-optic network in the village. This will include the installation of several distribution boxes at the crossroads in the village where the fibre-optic highway runs. At first we aim to connect 50 subscribers free of charge. These will include schools and municipal buildings, as well as low-income families.

Expected community benefits and challenges

One proposal that the community made was to increase the presence of local businesses on the web. For example, guesthouses in Suusamyr would be registered on websites such as Booking.com, making them more accessible to modern tourists.

However, overall, most villagers should benefit from the community network, as it is common for people to share resources with fellow villagers, especially given the fact that most villagers are related to each other. For example, one can go to a neighbour's house to use the internet for a short time to search for necessary information or get some social service that is provided by the government online. The community network will increase the availability of a number of social services that are available online and otherwise would require

4 www.elcat.kg/en

travel to the regional centre many kilometres away. These include applying for a passport, registering a vehicle or real estate, registering a marriage or getting a divorce, and for biometrical registration.

The main obstacle that made us lose all our hope was the requirement to register the use of frequencies. We thought that we could use certain frequencies as long as nobody else was using them. When we found out that we needed to register them and that it takes half a year just to go through the application process, we were devastated for about a week. It would be helpful if obtaining frequencies took less time – one month, for example. In comparison, obtaining a data transfer licence takes three to four weeks, and is not that hard.

Because of this we started looking for alternative solutions. It was most feasible to find a company that had registered frequencies and that was willing to cooperate. We discovered Skynet Telecom LLC. Skynet w ill be buying internet bandwidth from Elcat and selling it to the community network at a small margin.

Another challenge, but one we anticipated, was the terrain. There is no direct visibility from the Elcat node to the village. We still need to connect to the Elcat backbone, so that Skynet can buy the bandwidth for the village. We have to install a repeater that has direct visibility to both the Elcat node and the village. We have found a perfect location for the repeater and are currently installing a mast on the hill.

Action steps

We are currently in the process of applying for a licence for data transfer that needs to be obtained from the regulator, the State Agency for Communication. It is also necessary to have the head of the LSG establish a municipal enterprise together with the ISOC Chapter. LSGs were established by law in 2011 in Kyrgyzstan as part of a process of the decentralisation of state power. Municipal enterprises are usually established by LSGs to provide community services such as trash disposal and street lighting among other local-level services.[5]

One of the most important aspects of our initiative is scalability – and we want the community in Suusamyr to be able to help other communities to set up their networks. In this way we feel the project can be replicated from unconnected community to community and village to village, where the community members themselves become the barefoot engineers witnessed in Nepal. During the roundtable on community networks back in 2016, among the participants were a number of heads of LSGs who stated that they have the budget to set up community networks and would love to learn how. We plan to send a concept note on community networks with all the necessary steps described in detail to all of the 453 LSGs in Kyrgyzstan, including the contact information of the engineers in Suusamyr whom we will train to make them able to train others.

5 Post script: We are actually moving away from this concept because of an issue with the State Committee for National Security. Every ISP is obliged by law to buy and install SORM (Russian: Система оперативно-разыскных мероприятий, "System for Operative Investigative Activities"). This is the technical specification for lawful interception interfaces of telecommunications and telephone networks operating in Kyrgyzstan (first implemented in Russia). The current form of the specification enables the targeted surveillance of both telephone and internet communications. It costs between USD 40,000 and USD 100,000. Obviously our little project cannot afford it. We will have to keep Skynet as our alternative solution, including when it comes to the billing of the subscribers.

MALAYSIA

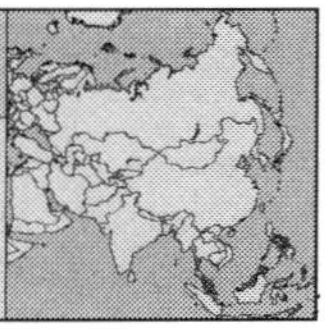

THE GENDER DIMENSION TOWARDS ESTABLISHING COMMUNITY NETWORKS FOR INDIGENOUS PEOPLES IN PENINSULAR MALAYSIA

Persatuan Kesedaran Komuniti Selangor (EMPOWER)
Angela M. Kuga Thas
www.empowermalaysia.org

Introduction

Arguably one of the most marginalised and vulnerable communities in Peninsular Malaysia are the indigenous women (*Orang Asli* women in Malay). While many Malaysians today know about the native land rights issues of these peoples, it took more than 30 years before this level of awareness was achieved. In fact, it is through the internet that the level of awareness accelerated, in particular on the issue of the Baram Dam in Sarawak.[1] It is, however, taking even longer to achieve a greater consciousness about the gender inequality issues faced by Orang Asli women among Orang Asli women themselves, among seasoned activists who have worked on the native land rights of the Orang Asli, and among other human rights activists. This phenomenon is very much symptomatic of how gender equality is still not very well understood even by those who champion human rights or deem themselves progressive.

As EMPOWER we have tried to enhance the analytical and advocacy skills of Orang Asli women since 2013, with the idea that whatever issues they face, they would have to self-mobilise, self-organise and advocate on their own initiative, taking ownership of the problem and engaging directly in strategising and seeking a solution, and eventually solving the problem.

The work of consciousness raising and capacity building is slow and frustrating. This is certainly not something new when it comes to working with communities. We have witnessed the common phenomenon where women fail to acknowledge that gender inequality exists in their community, as "things have always been that way". For example, while getting Orang Asli women to identify and acknowledge the issues of gender inequality that they face has been an uphill task, even getting them to acknowledge that domestic violence or violence against women (VAW) takes place in their communities is challenging. That there are men who prefer to marry and not work and rely on their wives to support them, for the Orang Asli women, is also not a gender inequality issue. They are just matter-of-fact life issues that some women are unfortunate enough to face. In fact, in almost all workshops we have conducted, the Orang Asli women we have engaged with would rather have my colleagues and I believe that VAW and gender-based discrimination do not exist in their communities.

You may be asking at this point, how does this in any way relate to the idea of community networks for these peoples? A lot, in fact, especially in the context of a new government that was voted in on 9 May 2018, and the optimism that there are more allies in government to work with compared to the previous one.[2]

Advocating for community networks

"There is free internet bandwidth in Malaysia" – so I was told, but that was not the full answer I needed. It signalled the beginning of EMPOWER's efforts to explore how best to move ahead with working with the Orang Asli community in Malaysia on establishing community networks with them, with a specific focus on involving the Orang Asli women and girls.

There are only about 148,000 Orang Asli in the whole of Peninsular Malaysia.[3] They are primarily located in the states of Perak, Kelantan, Pahang, Johor, Selangor and Negeri Sembilan. They are also very divided: divided by groups of who knows who, of who can work with who, of who married who and who did not marry who; and they are not necessarily united in a village, not necessarily united through marriage, and certainly not necessarily united

1 See Bardeen, S. (2016, 22 March). Baram Dam Stopped! A Victory for Indigenous Rights. *International Rivers*. https://www.internationalrivers.org/blogs/433-18

2 At the time of writing, the new government has been in power for less than 100 days, but many feel that political will is still weak, as only recently – on 4 August 2018 – another Orang Asli blockade was torn down by a durian plantation corporation in Kelantan. See Mahsinah Abdullah, A. (2018, 16 February). 800 Orang Asli set up blockades against logging, durian plantation near Gua Musang. *New Straits Times*. https://www.nst.com.my/news/nation/2018/02/336207/800-orang-asli-set-blockades-against-logging-durian-plantation-near-gua and Lin, K. J. (2018, 4 August). Orang Asli claim blockade cut down by durian company. *Malaysiakini*. https://www.malaysiakini.com/news/437481

3 2008 data. Expected to be much lower in 2018, estimated at 100,000.

through ethnicity. Officially, there are 18 Orang Asli tribes – Bateq, Jahai, Kensiu, Kintaq, Lanoh, Mendriq, Cheq Wong, Jah Hut, Mah Meri, Semai, Semaq Beri, Temiar, Jakun, Orang Kanaq, Orang Kuala, Orang Seletar, Semelai and Temuan – categorised under three main groups according to their different languages and customs:

- Semang (or Negrito), generally confined to the northern portion of the peninsula
- Senoi, residing in the central region
- Proto-Malay (or Aboriginal Malay), in the southern region.

As many as 76.9% of the Orang Asli are considered poor, with 35.2% deemed "very poor". The literacy rate is only 43% and life expectancy at an average of 53 years old. Some Orang Asli remain on native lands, others have set up homes closer to or within urbanised areas, and some others have relocated to Orang Asli resettlements and poor quality government housing which usually means no opportunity at all for owning land. So, while native land rights issues are very pressing for some, not all face these issues in the same way.

For example, with the advent of the new government of Malaysia as mentioned earlier, after 61 years of rule under the same dominant political parties, two advocacies unfolded. One was the demand that the Jabatan Kemajuan Orang Asli (JAKOA),[4] the government department that is supposed to look after their interests, welfare and development, be completely abolished. The other was to reform JAKOA, to ensure that Orang Asli held key positions in JAKOA and that decision making was more transparent, inclusive and participatory with the community.

The reality is that not all Orang Asli are able to rely on their traditional livelihoods any longer and waged work is necessary. As such, initially there was a prominent clash of positions in civil society as well as among the Orang Asli. However, as consultations start to be organised, with the most recent one held on 21 July 2018, the position appears to be moving towards reforming JAKOA. In line with this, advocating and getting buy-in to the idea of community networks therefore has to be very much needs-based and consultative as well, or it will not be prioritised at all.

To try to unpack the challenges further, I discuss three scenarios, trying to maintain anonymity at the same time to safeguard the privacy of the Orang Asli whom we have consulted.

4 Department for the Advancement of Indigenous Peoples.

Scenario 1

We visited quite a self-contained village. The head of the village is reportedly one of the more educated of the Orang Asli village heads, having been attached to a higher institution of learning before his retirement. We discussed issues faced, and we highlighted that there were three areas that we would persist in working on, and it is only in these areas that we would be able to collaborate with them, as our resources and capacities too were limited. The three areas are: gender awareness raising and advocacy capacity-building training; training women to engage in community schools development; and the establishment of community networks.

It was interesting to see the dynamics as we sat in a circle discussing the issues. There were more women than men, but men appeared to have a more significant weight when they spoke. All were very interested in the training of teachers for community schools, primarily because they saw education as the only way out of poverty for their children, and eventually for themselves.

Prior to arriving at this village, I had heard that the villagers were creating problems for one of the community teacher volunteers, who is perceived as an outsider and who married into the community. She was one of the two Orang Asli women we brought to an educational training workshop on phonics as a teaching methodology and approach. She was not at this consultation, but the second woman, who is seen as one of the women community leaders, was.

The villagers said they wanted a young man to be trained as a community teacher rather than any of the women. The woman community leader whom we had developed a long relationship with, and who has participated in at least five of our workshops, agreed and reiterated this desire. I was a little taken aback, considering how long we had already worked with this community, with at least five of the women having attended two of our trainings. I had also expected that at least one of the two women whom we brought to the educational training workshop on phonics would be put forward as a potential trainee, together with the young man. Instead, the retort was, "Must it be only women?"

The concept of a community network, however, did appeal to one Orang Asli woman at the consultation. She had married an outsider and spoke Mandarin. She already had an entrepreneurial mind-set. She was happily elaborating on how many villagers they had and how the community school could benefit, as well as a number of other villages.

I shared how they could build their own communication tower, and when the village head asked if setting up the communications tower would be legal, and I answered "No", all conversation stopped immediately. Even when I explained that we would need to find a way to work on getting those in authority to make an exception, no one would broach the subject again or asked how this could be done when the village head switched the topic and started informing us of his ideas and potential collaboration with a university for an extended vocational training programme that would include the young people from the village. Discussions subsequently returned to the issue of education, especially English education for the children, and we could not raise the subject of community networks again.

Yet – in something of an irony – the women in the village had been organising group activities among themselves by playing and following tutorial videos from YouTube at the community hall once a week.

Scenario 2

The next village we visited was closer to urban development but it appeared much more run down. Population density seemed higher as houses were packed closer together, and it lacked stable connectivity despite the close proximity to the city. When we broached the subject of community networks here, there was very limited and selective interest from one younger Orang Asli woman. She was more interested in having a computer and a printer and for connectivity to be set up in the village head's home, as they wanted to advocate against the sandstone mining nearby that was affecting the quality of their air and health.

The message we received was that not everyone would be welcoming of a community network, and that it was best to be low-key and small-scale and keep the equipment with the village head.

The wife of the village head attended the consultation, but we could not get a clear reason as to why community members would not be completely agreeable to having a community network, which would benefit many more people rather than just the house of the village head and those closely connected to the village head.

Scenario 3

We also organised a consultation during a training with Orang Asli women who live in villages that are semi-urbanised or urbanised, but we organised this training-cum-consultation outside of their villages. These women were very interested and concurred that there was a need for a community network where they live. They were keen on saving communication costs, developing an income-generating arm, and being able to organise to make demands on the government. This was just after the results of the 14th general election had been announced, so they already knew that there was a change in government. While they persisted in saying that they do not want to have any engagement in politics, these women were more political than the others in Scenarios 1 and 2 in how they wanted updates on how to move ahead with a community network, and other advocacies such as on basic infrastructure issues, health and education.

Looking back at how we conducted the consultations on community networks, I felt that we could get clearer responses and expressed needs from the women if they were outside of their villages, away from not only the men and the gender-power dynamics that come with that, but the prevalent politics in these villages. I felt that even though women were very much present in Scenario 1, only two or three women would speak, and only when prompted. However, bringing women to participate in a consultation outside their villages meant thinking of:

- Who else would need to accompany them? Sometimes their husbands would insist on participating, or the women would insist that their husbands attend as well.
- Providing facilities and caregiving for their children.
- Considering paying an opportunity cost for daily wages lost for those days.
- Ensuring that it only took a day for such a consultation. A two-day event would already be problematic for some, especially if paying an opportunity cost for daily wages lost was not possible.

Action steps

As EMPOWER proceeds with seeking face-to-face consultations with key ministries in the new government of Malaysia, much needs to be done to push for policy and institutional reforms. We have been trying to organise a national Orang Asli women's consultation since the start of 2018. It has been postponed from April to May and in July, yet again.

The resistance comes from the "official structures" of the networks and villages, rather than the women themselves. EMPOWER has spoken with established women leaders of the community, who have expressed support but with little follow-through in real terms. In private, a few Orang Asli women have expressed interest to EMPOWER staff, but appear to not know how to proceed without inviting the ire of those in power around them.

EMPOWER still intends to hold the national Orang Asli women's consultation this year (2018) and at least one more in 2019 so that the Orang Asli women can meet and update each other.

However, the question now arises: Would such a national consultation be considered successful if we only had 15 women, the average number of participants we expect for a workshop with Orang Asli women? To many, this would appear to be a failure, but the sheer effort of trying to bring these Orang Asli women together so that they are better able to express their needs and priorities for advocacies and initiatives that would benefit them, their children and their communities, remains unmeasurable and too often considered insignificant in the whole value chain of development initiatives.

MEXICO

COMMUNITY NETWORKS IN MEXICO: A PATH TOWARDS TECHNOLOGICAL AUTONOMY IN RURAL AND INDIGENOUS COMMUNITIES

REDES A.C., Rhizomatica and Colectivo Ik'ta K'op
Carlos Baca-Feldman, Peter Bloom, Mariano Gómez and Erick Huerta
www.redesac.org.mx, www.rhizomatica.org, www.iktakop.org

Introduction

In 2016, a national survey developed by the Instituto Nacional de Estadística, Geografía e Informática (INEGI)[1] indicated that 47% of citizens in Mexico use a computer, 59.5% are internet users and 73.6% have a cell phone. In addition, the states with the highest index in the digital divide are also those that have the greatest economic challenges: Guerrero, Oaxaca and Chiapas. In these states, where most of the countries' indigenous population is concentrated, broadband access is 17.5% and telephony is 23% on average. Although the data comes from government agencies and may not be entirely accurate, similar observations have been made in studies by non-governmental organisations.[2]

When these statistics are presented as inputs for public policy making, the difficulties are characterised as "market failure" – as if the solution lay in finding ways for large telecommunication companies to develop models that allow them to obtain an economic benefit while meeting the connectivity needs of the poorest populations of the country. This has shown to be unfeasible so far. Another option is for the government itself to seek solutions through its public programmes and budgets, which are usually limited by clientelist dynamics. However, as indicated in the "Toolkit of Best Practices and Policy Recommendations, Module 3: ICT for Indigenous Peoples and Communities" of the International Telecommunication Union (ITU),[3] projects in which the decision making and operation of last-mile networks are made by the community itself are the only ones that have been shown to be appropriate for connectivity in more remote areas.

In this context, there have been several communities that have sought to escape from the logic of the market or the state to solve their connectivity problems. In Mexico, there are many indigenous peoples who maintain organisational, economic and political systems that are not completely anchored in the capitalist mode of production; and their telecommunications projects reflect this way of understanding and being in the world. In this report our intention is to discuss some of the experiences that have been developed in Mexico considering this perspective: a community cellular network in Oaxaca; a community intranet in Abasolo, Chiapas; and Techio Comunitario, a technicians training programme aimed at rural and indigenous communities.

Community networks as a "constant negotiation"

From a legal perspective, these projects are covered by the second article of the Mexican constitution, one that indicates the right of the indigenous peoples to develop their own systems of organisation and indigenous customary law, known as "uses and customs".[4] Their status is also protected by Convention 169 of the International Labour Organization (ILO).[5] In addition to these, Article 16 of the UN Declaration on the Rights of Indigenous Peoples upholds the rights of indigenous peoples to acquire, administer and operate their own means of communication.

As Laval and Dardot[6] point out, in addition to a legal framework, an institutional framework is required that can put into practice the sense of "the commons" in social relations. This occurs from the practice of the communities themselves and the impact on public policies that are generated from the projects that are being developed.

1 INEGI. (2016). *Encuesta Nacional sobre Disponibilidad y Uso de Tecnologías de la Información en los Hogares (ENDUTIH).*

2 An important study on this matter was developed by the Red por los Derechos Digitales (R3D), titled *El estirón de México Conectado: ¿cuánto creció realmente el número de usuarios de Internet en 2015?* and available at: https://r3d.mx/2017/03/12/el-estiron-de-mexico-conectado-cuanto-crecio-realmente-el-numero-de-usuarios-de-internet-en-2015

3 International Telecommunication Union (ITU). (2011). *Toolbox of best practices and policy recommendations. Module 3: ICT for indigenous peoples and communities.* Available (in Spanish) at: connectaschool.org/sites/default/files/pdf_cache/fdof754e2ab3e39da5172bc484cf4247.pdf

4 www.ordenjuridico.gob.mx/Constitucion/articulos/2.pdf

5 www.ilo.org/dyn/normlex/en/f?p=NORMLEXPUB:12100:0::NO::P12100_ILO_CODE:C169

6 Laval, C., & Dardot, P. (2015). *Común. Ensayo sobre la revolución en el siglo XXI*. Barcelona: Gedisa Editores.

The telecommunication projects outlined in this report are community-based, and as such, they contain and reinforce the values and ways of life of the community. They also repurpose and appropriate technology according to each community's economic, political and social organisational forms. In this way, the technology does not determine existing social relations, but rather is transformed to give way to new modes of use, generation of content and particular and novel infrastructure. Jaime Martinez Luna gives us a clear idea of these statements when explaining the concept of *comunalidad* (or communality), the way of life of the indigenous peoples of the Sierra Juarez in Oaxaca, that is based on four pillars: land and territory, collective work or *tequio*, participatory political organisation, and the festival. For him, life in these communities is not completely isolated from the globalised system; on the contrary, it finds itself in a constant negotiation between the local and that which is foreign.[7]

For this reason, following Belli,[8] community networks serve different logics both in their use and in the very way in which they are built. Since each community imprints its way of life through its appropriation of the network, we cannot generalise from these projects, as each one has particularities that make it different from the others explored in this chapter.

The community cellular network in Oaxaca

In 2013, a connectivity project using mobile telephony emerged in Talea de Castro, Oaxaca, from a meeting between indigenous communities, hackers and activists. This experience is based on the relationship that exists between *communality* as a way of life characteristic of Oaxacan indigenous peoples and the possibilities that emanate from the use and transformation of certain technologies. As pointed out by Peter Bloom,[9] the system of community government, the notion of autonomy and the commons are elements that were key in the construction and development of the process. Today there are 14 operating communities with coverage in 63 localities that have, for the first time, built a federated, autonomous cellular telephone network that is managed and operated by and for the communities. In total, 3,500 users are currently served.

7 Martínez Luna, J. (2013). *Textos sobre el camino andado (Vol. I)*. Oaxaca, Mexico: CSEIIO.

8 Belli, L. (Ed.). (2017). *Community networks: the Internet by the people, for the people. Official Outcome of the UN IGF Dynamic Coalition on Community Connectivity.* Rio de Janeiro: FGV Direito Rio.

9 Bloom, P. (2015). *La Telefonía Celular Comunitaria como Alternativa al Sistema Hegemónico de las Comunicaciones en México: Un estudio de caso de las nuevas iniciativas de la Sierra Juárez de Oaxaca.* [Master's thesis not published]. Mexico: UAM-X.

In this process, various organisations have collaborated in the construction of the community cellular model. Rhizomatica[10] is the organisation that created the model and started the operation in the communities. Projects like Osmocom[11] in Germany have been very important for the development of the free software used in the project. The legal framework and systematisation has been created through collaboration with Redes por la Diversidad, Equidad y Sustentabilidad A.C. (REDES A.C.).[12]

In 2016, the operating communities, Rhizomatica and REDES A.C. founded a stand-alone, non-profit entity called Telecomunicaciones Indígenas Comunitarias A.C. (TIC A.C.).[13] This entity brings together the member communities of the network for technical support, peer-to-peer support and maintenance of their networks. TIC A.C. is a fully licensed, social-indigenous operator of cellular services.

For the development of these cellular networks, it was necessary to establish different approaches in the economic, organisational, technological and legal dimensions while drawing on a generic model for the network. For more detailed information on each of them, you can consult the *Community Mobile Telephony Manual.*[14] Some of the key features of the model are:

- Network architecture: It is a hybrid network integrated by three different networks, a first-mile local network managed by the community, consisting of a cellular base station that operates in the 850 MHz band providing cellular coverage and services to standard mobile phones, as well as providing inbound and outbound calling over VoIP; middle-mile IP transport provided by small wireless internet providers; and the fibre backbone and core internet infrastructure operated by large telecoms companies that allow for the interconnection and transportation of data.
- The legal model has two fundamental components: broad self-regulation based on the internal norms of the communities themselves and the association they have created (TIC A.C.); and an indigenous social-use concession, which is comprised of a concession or licence to use spectrum in the 850 MHz band in five states of the country and a single concession to provide telecommunications services throughout the nation.

10 https://www.rhizomatica.org

11 https://osmocom.org

12 https://www.redesac.org.mx

13 https://www.tic-ac.org

14 https://archive.org/details/MANUALTICENGFINAL

- The technological base of the network focuses on two principles: the ease of operation and affordability of infrastructure (approximately USD 5,000). To this end, the project has integrated software from Osmocom in addition to that developed by Rhizomatica such as an administration interface. All code is free and open source software and is accessible through public repositories and documentation.[15] Regarding the infrastructure of the network, the local network consists of a cellular base station which the community owns and manages as communal property.
- The organisational base, as already mentioned, takes as a central element the organisational forms of local indigenous communities. First, the decision to form a new local network is made by the general assembly of each community. Second, the governing structure of the association (TIC A.C.) is composed of a members' assembly and an executive body, which are reinforced by staff working in strategic areas like operations, administration, community relations, innovation and maintenance.
- The economic basis of the project is best understood through the lens of Braudel's writings[16] on the three-level distinction of economies: global, regional/local and subsistence. The local network operates at the subsistence level and is managed by the community itself, the transport network[17] is operated by a local company, and the backbone network is operated by a global company. In this model the community is part of an association that can deal in a global economy given its integration. The cost of the service is approximately USD 2 per user per month, of which USD 1.25 is direct income to the community and USD 0.75 is set aside for the association. This includes unlimited calls and text messages inside the network, with outbound long-distance calling requiring credit top-up.

Although a great advance has been made in the way towards the technological autonomy of these communities, there is still a long way to go and many ways to continue to improve the process. Some examples of these challenges are technological, others are social, such as the participation of women. At present, issues such as the expansion to other states of the country, the development of locally relevant applications, the elimination of legal barriers derived from the current regulation, and the production of content for the network, are being addressed in conjunction with the communities and ally organisations.

Community intranets: The case of the Tzeltal community of Abasolo in Chiapas

Community intranets are projects that indigenous communities have developed in Mexico to cope with the lack of connectivity or limited connectivity. They are characterised by a semi-closed and local system of Wi-Fi links that provides wireless access to locally hosted content that they consider culturally, educationally and socially relevant. Through this type of project, a lack of connectivity or very limited access to the internet becomes an opportunity to define what content is necessary to produce and share locally, and what content from outside the community is relevant. They also provide an opportunity to redefine the characteristics of the network architecture that are the most appropriate, all based on the way of life of each community.

Since 2017, as part of a project supported by the Internet Society's Beyond the Net initiative and awarded to Rhizomatica,[18] a model of community intranets is being developed in four communities in Mexico: Abasolo in Chiapas, Guadalupe Ocotlán in Nayarit and Santa María Tlahuitoltepec and Santa María Yaviche in Oaxaca. This is being done with the support of REDES A.C., the Telecommunications Postgraduate Programme of the Universidad Autónoma Metropolitana Iztapalapa (UAM-I), the Boca de Polen Communicators Network, and AlterMundi, among others.

The project is still in the construction stage in the four communities, and currently the intranets in the communities are disconnected from each other. In the medium term, the general intention is to establish a network of intranets that allows the sharing of content among the communities that participate in the project. The most advanced case to date is that of the Tzeltal community of Abasolo in Chiapas, which is why we take it as a reference in this report.

The precursor to this intranet project arose seven years ago, at the community high school, where a computer science teacher, Luis Ramón Alvarado

15 https://wiki.rhizomatica.org/index.php/Main_Page and https://www.github.com/rhizomatica

16 Braudel, F. (1980). *On History*. Chicago: The University of Chicago Press.

17 The middle-mile or backhaul network.

18 https://www.internetsociety.org/beyond-the-net

Pascasio, developed IntraBach,[19] an educational initiative that allows students to access relevant and quality content to reinforce their studies. This initiative continues to operate today in 15 communities in the region and is constantly nourished by content. Mariano Gómez, a former student of Luis Ramón, and Neyder Domínguez, co-founder of the group, supported the development of IntraBach and later formed a collective called Ik'ta K'op[20] whose purpose is to develop a model of community connectivity for internet access in Abasolo,[21] one which simultaneously provides access to a database of local content resources through the community intranet. This model is known as the Yaj'noptik Intranet, and provides around 1,000 users in Abasolo and five neighbouring communities with internet connectivity and local content.

The elements that make up the model of the Ik'ta K'op collective are:

- The technological aspect works as follows: global internet connectivity is provided through a connection to Telmex, the incumbent telco in Mexico, in the municipality of Oxchuc, 27 km away. Point-to-point links were installed to the community of Abasolo, where it is distributed through 19 nodes, covering 60% of the population. A hotspot server running free software is connected to the local intranet and gives users access to the internet.
- The economic aspect is governed by the principle of collective acquisition of equipment, through what is known in Tzeltal as "Mankomun".[22] The economic model that allows the project to be sustainable is based on providing internet access to users at a relatively affordable cost, with packages from USD 0.50 per hour to USD 10 for a monthly package. However, if a person does not have the economic capacity to pay for the service, they can still use the service in exchange for their collaboration as volunteers in the care and maintenance of the nodes, painting towers, loading equipment, and even as blacksmiths and carpenters that enable the network to continue functioning.
- The organisational base is in the process of being built and includes the project developers and people in the community interested in the project. At the moment it is composed of "caretakers of nodes", people in charge of giving maintenance and having a node in their houses that connects to the network and gives service to the users in that area. Also included are activities such as taking care of the equipment so that it is not stolen, warning the operators about technical issues, selling access coupons, giving service to users and providing a space for the safeguarding of the equipment. Each of these caregivers also has the right to access the internet at no cost as in-kind payment for their services.

The development of this project has allowed the community to make calls to their relatives and acquaintances in other parts of the country or in the United States without the need to walk to the telephone booth or travel approximately 40 minutes to the nearest municipality. Likewise, economic development has been encouraged by allowing the sale of products over the internet, the use of online banking and payment services, and tourism promotion. The promotion of education is another key element in the group's objectives, so the high school has a free connection and the local content server provided through IntraBach[23] is constantly updated.

The challenges currently faced by the Ik'ta K'op collective relate to issues such as the project's relationship with the community, and the creation and integration of local content on the server.

Attention is also being given to the technological, economic and sustainability needs of the project.

The Techio Comunitario training programme

Although this last case is not a community network, we believe it is important to include because the training and capacity building of people from rural areas who wish to develop projects related to telecommunications is crucial to improve access conditions in rural and indigenous communities.

The Techio Comunitario[24] project is composed of three elements that aim to train technicians specialised in telecommunications in rural and indigenous communities: the Diploma for Community Telecommunications and Broadcasting Promoters, four

19 www.intrabach.org

20 www.iktakop.org

21 Before the emergence of this project, the internet was available in a small cybercafé using a voucher system, but the model in which it was distributed was not based on community principles.

22 An analogy allows us to understand this principle. In some festivities, such as the All Souls' Day, a cow is bought for the celebration. This is done by the community sharing the costs, and, subsequently, the animal is sacrificed and the meat is distributed equally. The community selects a day for the sacrifice, and everyone who participates in the work is also fed.

23 At the moment there are two intranets in Abasolo: IntraBach and Yaj'noptik Intranet. IntraBach only serves the high school.

24 https://techiocomunitario.net

technological laboratories, and an online learning platform. In the first diploma course, held between October 2016 and May 2017, a total of 36 people from six states in Mexico participated, all of whom committed to assist ongoing communication initiatives in their communities. In total there were 17 participating organisations and the general coordination was overseen by REDES A.C. and Palabra Radio.

The training process is based on participatory research carried out from 2012 to 2015 with trainers of indigenous and community communicators. The initial intention was to systematise the needs that existed at the training level through a series of interviews and participatory meetings with key actors in these areas. Based on the results of the research, the creation of a capacity-building programme for technological capabilities was envisaged.

The academic structure of the diploma was based on eight modules. First, a common core that covered the topics of community communication and technologies, electricity, electronics and free software. Subsequently, participants had the possibility to choose between one of the three specialties offered: radio broadcasting, wireless internet networks, and community cellular telephony. Finally, an integration module was carried out in which the issues of licensing and regulation and sustainability were addressed. Each of these modules was carried out in different locations that included the states of Oaxaca, Chiapas and Puebla, and the coordination of each was carried out by an organisation belonging to the advisory committee.

The methodology used was based on the educational practices of indigenous peoples regarding the transmission and socialisation of knowledge. The pedagogy was also inspired by the *milpa* (a millennial agrosystem in which the basic ingredients of Mexican cuisine are harvested) that has substantive elements: praxis, daily life experience, creation and sharing. It also drew from general experiences in popular education models and the approaches of free knowledge societies. The methodology of this training process is systematised in the book *Haciendo Milpa.*[25]

The challenges in the construction of this training process have been the breaking of paradigms about teaching and the ways of evaluating learning. Another important challenge was the inclusion of women in the process. Although there was an important effort to get more women to participate in the diploma programme, only one of a total of seven that were accepted to enter the programme finished the course. Despite this, some of the benefits are palpable. These include, for example, the planning of the community intranet through the participation of members of the Ik'ta K'op collective in the diploma, and the installation of community radios by the participants. Finally, the planning of future projects based on the knowledge and relationships built in the training process will be one of the key elements to be developed in the short term.

Conclusions

In Mexico, community networks have been processes that sought to address the digital divide that exists in rural or indigenous communities. That the solutions have come from the communities themselves has ensured the appropriation of technology by communities, and the inclusion of their values and ways of life in the solutions. Using this approach, we can articulate experiences that go beyond the utilitarian vision of the market and the state that have proven to be incapable of serving communities with particularities such as those discussed in each of these projects. In this sense, these projects reaffirm the premise that when the decisions and the operation of technological projects are in the hands of the communities themselves, the projects can be very successful.

Additionally, the three experiences presented contain particularities in which the choice of technologies and ways of proceeding are in constant dialogue with local values and organisational forms. As such – as the first two projects show – technology can become a tool of social transformation, rather than being an end in itself.

Action steps

We cannot generalise about community networks in Mexico from what has been presented here, as space limitations have not allowed us to share all of the experiences, some successful and others not so much.

However, the challenges of community networks in Mexico are many. First, it is very important to encourage the participation of women in technological issues. There is still a very strong belief in the communities that men are the ones who should attend to these issues. Second, in terms of legislation, although a good stretch of the road has been covered, it is still necessary to influence secondary regulations and laws that allow for the application of constitutional laws in regards to the right to

25 https://docs.wixstatic.com/ugd/68af39_802ae1aeee674783bba4cd8dfa102d03.pdf

communication and information. Third, it is necessary to develop even more technologies that allow connectivity at low cost and with simple operating modes for people in general. And fourth, and finally, we believe it is essential to create organisational and economic models that allow these experiences to be sustainable over time and not depend on only one or two people for their realisation.

Although the challenges are many, we believe that the path towards technological autonomy in rural and indigenous communities in Mexico has advanced steadily in recent years. What remains is to continue walking together with the communities and articulating efforts at the local, national and international level.[26]

26 We recommend watching a video on public policies from indigenous peoples that is available in Spanish at: https://fimic.wordpress.com/2017/08/15/videos-sesiones-del-fimic

MYANMAR

THE COMMUNITY NETWORK AS A CONNECTIVITY STRATEGY FOR RURAL AND REMOTE MYANMAR

Alternative Solutions for Rural Communities (ASORCOM)
Michael Suantak
www.asorcom.net

Introduction

Alternative Solutions for Rural Communities (ASORCOM), a community wireless network, was built in the Siyin Valley for 20 villages with around 8,500 people living there. The Siyin Valley is a mountainous region in Chin state in the northwest of Myanmar, with peaks reaching up to over 2,100 metres. The network operated between April 2013 and 2017, before the arrival of mobile services in the area forced the project to evolve to meet new community needs.

The network was set up by the Siyin-Chin Youth Network Organisation (SCYNO)[1] and the Siyin Project.[2] These two organisations were eager to help and develop the communities where many of their members had grown up, and whose needs they were familiar with. They wanted to change the community and build a better, more sustainable society.

The communities involved in the community network were made up of vegetable farmers mostly dependent on shifting cultivation. Their seasonal produce was sold in the nearest town called Kale, almost 100 kilometres away.

This report describes the impact of the network on the communities in the valley.

Technical implementation

When the project first started in 2013, they used the cheapest wireless devices available in town: a TP-Link brand long-range wireless access point. One of the team members, Thomas Khaipi, who studied and lived in Germany, bought two Village Telco[3] devices and tested them as a communication solution for communities in the Siyin Valley. However, they were not useful as a long-range wireless network solution because they only worked well when they were set up close to each other. The project also faced other challenges: for example, the Myanmar government (a military dictatorship) had taken away some devices and warned not to set up private networks without their permission, which had to be passed by the union ministry from Naypyidaw. At that time internet censorship was taking place and building private networks was not encouraged.

Later on, the team got the opportunity to buy and install Ubiquiti[4] devices. They changed the network topology and redesigned the network, using 5 GHz for the backhaul connection and for the main connection between the villages, while 2.4 GHz devices were used for the connection inside the villages. The old TP-Link devices were also used for wireless access point connectivity inside the buildings. They were gradually replaced with Ubiquiti devices which supported mesh firmware. For some areas, such as school compounds, Commotion[5] mesh firmware with Ubiquiti devices was used.

Community engagement and impact

The community members supported the initial pilot project from the beginning. Many times they had discussed and shared alternative ways to solve their communication difficulties, their poverty, lack of skills and knowledge, and educational challenges. Many joined the team and shared their physical energy to cut down tall trees in the jungle and bring them to the villages to use as masts for the Wi-Fi antennas. Some shared their houses for accommodation while we set up the networks, and prepared food for a series of community meetings in the different villages. Some shops donated tea, coffee, cold drinks and cake for lunches when there were activities. Other shops discounted the cost of petrol and diesel used for the project.

As part of the project, we sold some used computers and mobile phones with credit and instalment plans to the communities. We organised some basic computer and mobile training workshops on how to use the internet and social media. When they had their own devices they could play and learn much more effectively. They built their

1 SCYNO was formed by Siyin youth with the aim of doing social work for communities in Myanmar.

2 The Siyin Project was also formed by Siyin youth to raise funds outside of Myanmar to support SCYNO and other activities.

3 https://villagetelco.org

4 https://www.ubnt.com

5 https://commotionwireless.net

confidence in using information and communications technology (ICT) devices and software tools. They could communicate with their family members and their relatives who live in other places in Myanmar and abroad.

As their main income related to cultivation, they learned techniques and methods for agriculture and food production by doing online research. They also learned how to check the weather forecast. They could communicate with buyers from cities and negotiate the vegetable prices before sending their produce to the town markets. School teachers shared what they learned about the internet with their students. Students also had the opportunity to use the internet, and to read, see and learn many new things.

As the women's population was bigger than the men's in the Siyin Valley, most of the hard physical work was done by the men, but the tasks of organising and communicating with other communities in the other villages were mostly done by women. Financial and administrative work, such as bookkeeping and playing the role of treasurer and secretary for meetings and discussions, was in the hands of women. As the shop owners were women, donation and discounting decisions were made by them as well.

Given that the project planned to collect a small amount of money as a contribution from users to pay for the shared internet bill and other expenditures, women in the community led the collection process and the management of the income every month. Some young women and men even met partners online, and ended up getting married. Some women learned textile design and techniques for making clothes, and made them for themselves and sometimes sold them to other community members.

As the project had its own local server, it could download and host Khan Academy[6] and World Possible[7] content for educational purposes. Entertainment videos were also hosted on our local server, so that network users did not need to buy DVDs and use the internet for entertainment anymore. Instead, they could enjoy their selected entertainment "channel" on their smartphone or laptop from their home.

One unique story from our project involves a teacher who was a young woman and who worked at a high school in Thuklai village in the Siyin Valley. She had come from a city called Monyua, which is more than 800 kilometers away. This was the first time she had travelled to Chin state, and she felt alone and scared. She said that she used to cry every day in the beginning, but because of the ASORCOM wireless internet she could now connect with her family and friends who were far away. She also said that she was now very happy in Thuklai village. Her friends, who are also teachers in the other villages in other parts of Chin state, wanted to transfer to Thuklai because of the easy access to the internet and the support from the local community.

Lessons learned

Freedom is not the same as getting things for free. Sometimes people do not value services that they do not have to pay for. To empower people to have responsible behaviour when using the community networks and maintaining them in a sustainable way, we asked for contributions, not only of money, but also their time, whether through physical labour or other forms of volunteer engagement.

Sometimes users would fight amongst themselves. Some users downloaded videos and games that affected connectivity for everyone. Sometimes people wanted to charge their neighbours for connecting to their router. We have had to offer counselling to resolve these disputes. We have also had to install software to monitor and control the system to limit heavy downloads and access to "adult" content, and to restrict some users during school hours.

Some reported that community users only spent time using social media, which they felt was not improving their lives in significant ways. In order to change this and encourage users to read more widely, we set up a system where users first had to complete a quiz with general knowledge questions before being granted a week's internet access. If they did not know the answers they would have to go do some research in the local library folder on the network.

Sometimes parents complained that teachers used social media during school hours instead of focusing on school lessons and teaching. The team met with the communities and teachers and agreed on limiting the use of social media and internet access during school hours, which, as mentioned, was configured on the system.

There was no national electrical grid in the project area, so the project had to depend on hydropower[8] and solar power. During the rainy seasons, hydropower stations were washed away by flash floods, and solar power was made ineffective by heavy cloud cover. The network did have a

6 https://www.khanacademy.org

7 https://worldpossible.org

8 Private hydropower stations were built by individual households in Chin state to get electricity.

backup generator, but the diesel needed to run it led to extra costs for the communities. This meant that the use of the network was limited in the rainy seasons.

Changing needs

The ASORCOM community Wi-Fi network was active until the end of 2017. In mid-2017 the area received mobile telephone coverage by the MPT GSM network. Later Ooredoo 3G and Telenor 3G also offered their services to the area. Because of this, the network needed to develop new plans, which included upgrading the network so that it could be used for local community media including audio podcasts, video streaming and IPTV. A data server was set up to assist with local administration.

At present, the team is working on supporting community radio stations in Myanmar. It is setting up a community FM radio station in the area and has been engaged in conducting surveys, meeting with communities, offering radio journalism training and workshops, and drafting constitutions and bylaws for community radio stations which will contribute towards reforming the national broadcasting law. The team has also been forming a development committee for the communities in the valley and building a music recording studio.

Action steps: The need to create a community wireless cooperative

In the absence of strong business cases for traditional network expansion, community wireless mesh networks have been successfully implemented in other industrialised and developing nations to address major barriers to internet adoption and universal connectivity. But deployments of community-owned networks often require outside support, including capital grants or technical assistance. Rather than relying on external mechanisms for assistance, a cooperative seeding organisation formed in-country could continuously provide the necessary support for multiple community networks, and act as an intermediary between the communities and high-level stakeholders such as wholesale internet service providers (ISPs), equipment vendors and the government.

In order to create such a cooperative, ASORCOM has proposed the formation of the Community Wireless Cooperative (CWC) with the aim of promoting a sustainable, low-cost approach to expanding internet connectivity to rural, remote and underserved areas of Myanmar.

At the moment there is only one community network in Myanmar and ASORCOM encourages and supports setting up many others. When there are many autonomous community networks in the country, they will effectively form a new digitally linked constituency of alternative network operators that provide ICT services to their communities and catalyse new social and economic development opportunities.

The new "information constituencies" will also be in a position to represent the interests of Myanmar's digitally underserved communities and contribute to policy recommendations for the design and implementation of Myanmar's ICT infrastructure and proposed universal service fund.[9]

9 To build community networks in Myanmar, financial support from such a fund is necessary.

NAMIBIA

THE SEED OF A COMMUNITY NETWORK

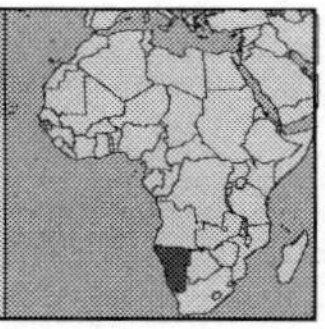

Internet Society (ISOC)
Quentillian Louw in conversation with Jane Coffin and Colin Muller
https://www.isoc.org

Introduction

How is a community network born? In many cases, through someone who experiences the limitations of living in a remote area, and at the same time realises the potential that the community could experience if only it could have a better connection to the net. That is, a community network needs a champion.

This report is the story of such a champion: someone who is trying to bring change, even if it is a small change, to his community. He is doing this recognising that small change can lead to bigger change and to complete transformations of the lives of the community.

This is the story of Quentillian Louw and his project that is the seed of a community network in Namibia. It is based on an interview conducted with Quentillian by the Internet Society's Jane Coffin and an ISOC intern, Colin Muller.

How it began

Getting kids through school...

In 2015, while Quentillian was finishing his studies at university, he met a professor, Nic Bidwell,[1] who taught him about wireless networks that were being set up across the globe. He read about a project in Zambia called Macha Works,[2] a community network that started with a very small link to the internet. The impact that that link had on the community is what convinced him to take on a research project on the theme. After the research was finished in 2016, he wondered how he could turn theory into reality, to assist his community.

That is how the project started, in a village called Groot Aub located about 60 kilometres south of Windhoek, where 6,000 people live. According to Quentillian:

> In this village you have a lot of elderly people looking after their many, many grandchildren that are going to school. Where are they going to get the money to buy internet data to do the school work? Because they are depending on their grandparents and the little their parents send them. And you have some of the parents here as well, but they are unemployed and do not have incomes. They are trying to get their children through school, to at least go to high school, because here we only have to go to grade 10. So they are trying to get them to go to grade 11 and 12 and finish, and at least try and get a job and let's say "rescue" the rest of the family. But if you do not have the means, and there is no one helping you, then what you are trying to do to help your child is difficult.

There are two mobile networks accessible from town, a 3G network from Telecom Namibia and a 4G network from MTC. But access speeds are not fast: 2 Mbps or 2.5 Mbps in 4G. According to Quentillian, data access can cost from one Namibian "dollar something" per megabyte, to about two dollars per megabyte:

> But for us that is a bit heavy. It all comes down to affordability. People do not feel included if they do not have the money or the means to access a network because they know that other people access it. The price for data is expensive, especially in Groot Aub.

Other than the mobile networks, there was no other way to access the internet in the village. While the schools ran a "one laptop per child" project, they were closed after hours when the kids actually had time to use the computers. Quentillian also noticed that some of the school kids had got hold of one school's Wi-Fi password and were squatting around the school yard at night or in the afternoons connecting through their mobile devices. He saw the possibilities: he wanted to help grade 10 learners pass their exams by providing them with the necessary online resources, because there is no library in the town:

> Many of the kids leave the school at 12 or 1 p.m. and they are not allowed back in to go use the

1 Author of the thematic report "Community networks: Stories and power" in this edition of GISWatch.

2 See the Zambia country report in this edition of GISWatch.

> labs and the internet connection – because there is no teacher willing to go sit there with them and monitor them. There is the one laptop per child project, so they have those laptops and internet connection, but what I see most here is the teachers in the school houses using the connection after hours, rather than the children being allowed to access the internet on the school premises and make use of the laptops.

From the start, Quentillian was searching for ways to get the students online. He couldn't provide internet access, so he provided a service to download files for them upon their request:

> It feels really bad when you walk in the street and you see the person you were in class with; they are a bum now, they did not make it. And you wonder what was the difference between you and them. And I saw that my mother was a teacher and I was lucky to have that. She could afford to get me the necessary resources I needed and what I needed to pass. And you can see that that might be the reason why the other person did not make it, because they did not have that. So it is kind of like a mission, trying to provide those who don't have, to provide them the chance to get out.

...and helping startups

A second objective, besides helping students, was to try and get small local entrepreneurs information about businesses that they wanted to start up. For example, if somebody wants to start a car repair business, they might need information on specific engines, which Quentillian tries to provide to them by searching online.

Quentillian is also working with a very active community member who has a lot of fresh ideas to contribute to the project. However, his work can get lonely sometimes. The project team is made up of Quentillian, Nic Bidwell as an advisor, and sometimes there are others who are willing to help sustain it:

> They just simply help out when they can, then they move on when they have the opportunity to make money or get a job or anything like that. So my position is basically just, how do I put it? The only one; yeah, a start-up.

Building for the future

There is a big demand

Currently Quentillian is searching for ways to share his internet access more broadly, though the connection he has does not have good bandwidth right now. He also feels limited, sometimes finding it difficult to get community members everything they ask for.

Because of this he wants to turn this project into a mesh network, which implies getting people involved and installing devices at their houses. "This will make the network robust," he explains. "When one device goes down, you have the other one moving around it."

Quentillian thinks the network has to be reliable, and to be available when people need it. It also needs to be convenient, covering as large an area as possible:

> There is a big demand. I have noticed that among the youth, the first thing – or these days, the basic necessity to them – is a smartphone. They want to be able to go online and use Facebook and Instagram and all of that. It is funny that for most of them, the next step after the mobile phone is a laptop or a desktop computer. But what I have seen with most of the kids surrounding me is that they are able to access the internet from their houses. They actually use their mobile phones.

He thinks inclusion is what will drive the adoption and growth of the network. If more people benefit from it, more people will want to keep it up and running, and more people will be interested in keeping it growing.

Funding and sustaining a free network

The main challenge at the moment is funding – all of the devices he has are ones he has saved up for. He also taught himself how to install and manage them. He feels this is just the beginning of a shared network, and hopes to find enough funding to be able to install point-to-point connections in key areas of the village:

> So you have point-to-point connections from the different locations, and at each location that is where you can now provide mesh from the main node that you have for that location, then just mesh in that entire area. But then once the main node fails, the mesh for that area is disconnected from the server and the rest of the locations.

Quentillian would like to start a formal organisation to manage the network:

> If funding was available to pay for a connection, for let's say a certain amount of time into the future, and I knew that there was no need to pay that, and I knew that there were backup devices available, I wouldn't mind giving my

> time, on weekends, to keep the network up and all of that, because that is what I want to do now, and I am already giving my own money from my side, my own internet connection, and I am sharing it. If an organisation were to start it would have to be non-profit, because I don't see me saying that providers are charging and it is too expensive for people to afford it and then yet I come still charging. It is not like the money is coming to me, it could still come back to the organisation. But the whole point is having an organisation to help self-sustain the network. So let's say it is non-profit, so you just have people contribute; I don't know in what way, but that is still something to work out.

He wants to provide the connection freely, and at the same time think of ways for it to generate income:

> Once it has all the devices and it is a fully running network and you have people in the community working on it, and you have educated them on how to maintain the network and possibly grow the network, you also have to find some way of funding all of that without having people pay for access to the internet. I still have to look at other examples of how people did it. How community networks sustain the network without having to get extra funding just to do that.
>
> I want to look at other models, at how people sustain this over a long period of time without charging. I know that that is not a good thing in terms of sustainability, but I am sure there is some way. Someone has done it before, and they have managed to keep the network up without people having to pay for it. I want to learn what other community networks are doing.
>
> We are also talking about starting an NGO to just make sure that funds don't get misused and that no one person has power over funds or to make all the decisions, because let's say I get to make all the decisions, and I think I am making the right decision but it is not the right one in the eyes of others, and it is not benefiting others. You don't want that. You want it in a way to be a mixed group of people from different parts of the village. For the mesh network to work we need to have devices in certain houses in each of the different locations in the village. So you need to have one person from that location that you can use to reach people in that area interested in participating in the project, and all of that.

Getting the community involved

Another challenge is how to make the community get involved:

> So in terms of social challenges, try to get people involved who are like-minded; you should always remember that. Get people in who want to do the same thing. Don't get people in who you believe have good contacts, because that doesn't work at the end of the day.

Quentillian is exploring several ways to involve community members:

> The way I would put it, is that it is to your own benefit. If I feel like a person cannot see that it is for their own benefit, then I cannot work with that person, because they are not trying to benefit the entire community now, they are just looking out for themselves. Because if someone comes to tell you, "I have something here, it is not going to directly benefit you immediately, but it is for the greater good, it is here to benefit everyone around you, and in so doing also benefit you in a way," it is in that sense that you are supposed to grab that with both hands and go with it. Those are the kind of people I look for, because if I go to you and explain it to you one time, and you feel like you are not part of it, then we just go on and explain it to the next person until you find like-minded individuals who actually want to do something for the community and develop it. That is why you need to find that one person in the location that is like-minded and willing to do it for the people around him. When you have that person, he is now more able to tell you which people in that area are more willing to also do that.

Show the benefit and grow it from there

Quentillian was given advice to go out into the community, informing them and teaching them how to use Wi-Fi and convince them to support the cause, but he is wondering if this is the right approach. The community has been disillusioned by many failed projects that did not bring them real benefits, he explained, adding:

> In Groot Aub it is not going to work like that. What you have to do is put it up, set the network up, have it running, get the people to tell you what content they want. Then show them how to use it. And show them that it is actually benefiting them. Tell them that if they use their mobile devices they can access these resources

> and can even go in and request more if they need something else. Because what happens is that arguments will calm down. If you approach them in the beginning they will want to know how they are going to benefit monetarily, what money they can get out of it. If you are in charge, what are you getting for doing it. That's how it is, sort of a mentality problem. That is the mentality here. So in my opinion, I believe you set it up, you have it running, and then you show the benefit of it and then grow it from there.

He plans to start a train-the-trainer project where community members can train other community members:

> Once you step away, you give it over to the community completely, and everyone is trained and they can continue training each other. You can always be there to help in emergencies or when it is really needed, but the plan is for it to be owned and run by the community itself.

The will to fight

Quentillian has been attending regional meetings, such as the Third Summit on Community Networks in Africa[3] in the Eastern Cape in September, to meet others that are setting up similar projects. He concludes:

> It makes it feel a bit small compared to the others that I have seen. But meeting other people doing the same thing gives you sort of like something to look forward to. It kind of tells you that there is help out there. If you need help there are people you can ask. So it helps you. It makes you feel like you are small, but it kind of gives you the will to fight. Like you want to become like those networks.

3 https://www.internetsociety.org/events/summit-community-networks-africa/2018

NEPAL

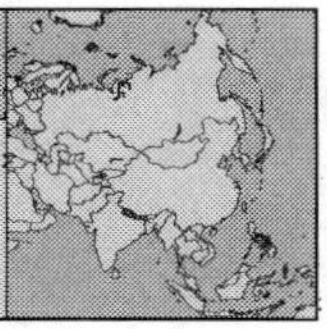

CONNECTING HIMALAYAN VILLAGES IN NEPAL TO THE OUTSIDE WORLD: THE TALE OF THE NEPAL WIRELESS NETWORKING PROJECT

National Innovation Centre
Mahabir Pun
https://www.nicnepal.org and nepalwireless.net

Introduction

This report describes the experience of the Nepal Wireless Networking Project (NWNP), a grassroots initiative in remote Himalayan communities in Nepal. Launched in 2001 in Nangi, the project has been extended to more than 200 villages over 17 years. It is currently the only community network in Nepal – although others are emerging and looking to draw on NWNP's experience.

The initial objective of NWNP was to address the dire need for communication services in the remote mountain villages. At that time, in 2001, there was not a single village that had modern telecommunications and transportation facilities. Instead, villagers had to walk several hours – or even days – to exchange messages, or just to make a phone call. Most people in the villages were not aware of the potential offered by either computers or the internet.

The project might have been just a dream, if social activist and team leader of NWNP Mahabir Pun had not written an email to a BBC group forum[1] out of curiosity, asking for help from volunteers to connect a Himalayan village. BBC subsequently published an article on his work.[2] To his surprise, he received an overwhelming response from volunteers around the world. The idea of NWNP started to materialise after that.

With the help of volunteers using the then-emerging Wi-Fi technology, the villagers in Nangi informally started NWNP in 2001. After doing various phases of trial and error for one and a half years using quite basic indoor Wi-Fi devices and homebuilt antennas, they managed to get the connection to work in the village. As the network grew, the project started to receive media attention, and funding from individuals started coming. As a result, a simple project that was started to connect a village in a mountainous region ended up offering education and telemedicine, and an opportunity to generate income through local e-commerce platforms, to communities in remote regions of Nepal.

Policy, economic and political background

When NWNP was launched, there was great political instability in Nepal due to a Maoist insurgency. As a result, Nepal's government policy for the development of wireless networks was very strict. At that time, the import and use of any kind of wireless networking equipment in Nepal was illegal. Because of this, the equipment was brought with the help of international volunteers and university students. After the restoration of democracy in 2006, NWNP lobbied with the association of internet service providers (ISPs)[3] and demanded that the 2.4 GHz and 5.8 GHz frequencies on the ISM band be de-licensed. As a result of this the government de-licensed those frequencies in September 2006.

A second regulatory obstacle was that to become an ISP in Nepal, it was necessary to pay a huge licence fee. NWNP lobbied the regulatory body, the Nepal Telecommunication Authority (NTA),[4] to reduce the licence fee. As a result the NTA issued a new law that made the licensing procedure simple. It also reduced the fee to just 100 Nepalese rupees (around USD 1) a year for those who want to work as rural ISPs. These regulatory changes helped community networks to easily get licences in Nepal, and also to legally provide internet access and IT services in the rural areas.

An innovative experience

NWNP certainly has had an innovative and interesting beginning to share with wireless communities around the world. Back in 2001, while villagers in Nangi were looking for different ideas to bring the internet to the remote mountain communities using Wi-Fi technology, most of the engineers and experts in the communications field had told them that it was not possible to make a long-range wireless link using normal Wi-Fi routers. Their main concern was the lack of funding to buy high-end equipment

1 BBC. (2004, 25 May). Wi-fi lifeline for Nepal's farmers. *BBC News*. news.bbc.co.uk/2/hi/technology/3744075.stm

2 BBC. (2001, 22 October). Village in the clouds embraces computers. *BBC News*. https://news.bbc.co.uk/2/hi/science/nature/1606580.stm

3 ispan.net.np

4 https://nta.gov.np

and the distance of over 40 km that the radio signal had to cover to connect Nangi to the nearest city, Pokhara.

Despite negative feedback from the experts, the NWNP team members decided to continue conducting field experiments using basic equipment, such as a 2.4 GHz indoor wireless router with 60 MW transmitting power, normal switches, and solar power sources. These devices were donated by individuals from abroad. At the beginning, NWNP had to even use home-built dish antennas because these were not available on the market.

After conducting several field experiments over a period of one and a half years, in May 2002, the NWNP team managed to connect the village to the city of Pokhara using a dial-up internet service. The technical experts, who were sceptical of the project, were amazed because – technically – it was not possible to make a 40-km link using indoor routers and home-built antennas, especially when the area is surrounded by high mountains.

NWNP has now acquired powerful wireless equipment for connectivity, and has rolled out to more than 200 villages in 15 districts of Nepal. Solar power systems have been installed at the repeater stations to make the networks robust and reliable. The project is gradually adding various applications and services for the benefit of the villagers.

Technical configuration

NWNP was started from scratch, and was built gradually over a time span of about 17 years – and it is still growing. The project has set up base stations in three cities, Kathmandu, Pokhara and Gorakha, to which the villages are connected. The base stations have routers and servers that are linked to the internet through a leased optical fibre line. From the base stations, a series of repeater stations have been built on the mountain tops to connect different villages. The access points on the mountain tops work as relay stations that distribute the internet to end-users in the villages. All the villages are connected to the access points through point-to-point or point-to-multipoint wireless links.

High-speed backhaul radios operate on a dedicated core local-area network (LAN) that reaches from the base stations to different districts through the relay stations. The longest point-to-point link NWNP has made is 59 km from one mountain top to another. The distance between an access point and end-users ranges from 2 km to 10 km. The villages that are connected to the network are divided into different sub-nets to manage the network smoothly. Routers have been installed at each of the relay stations. These provide DHCP[5] services to end-users and serve as interfaces between the backbone, wireless LAN (WLAN), and local distribution LANs.

For the point-to-point backhaul links, unlicensed 5.8 GHz wireless devices made by different companies have been used. However, for the last-mile connectivity and for the hotspots, 2.4 GHz wireless devices are being used. These devices using unlicensed bands are now easily available on the market. This is affordable equipment that can be used to bring broadband services to rural areas. However, the main constraint of using 2.4 GHz and 5.8 GHz frequencies is that it requires a clear line of sight to connect the rural areas, whereas the villages in mountainous regions are scattered across a diverse topography. Furthermore, if there is a big tree, small hill or mound or harsh weather along the path of the radio waves, the signals get blocked or the signal strength decreases drastically. Because of these hindrances, NWNP is facing some problems in its efforts to connect the most remote villages using wireless equipment in the 2.4 and 5.8 GHz bands.

To identify alternatives, NWNP conducted a pilot project in 2016. The objective was to test the potentials of TV white space using 460 MHz to 478 MHz, and VHF technology using 192.5 MHz to 202.5 MHz. The pilot project connected eight villages using these technologies. NWNP had been issued with a licence to use those bands; however, getting the licences was not easy.

The pilot project is still running and it is working quite well. We found that TV white space signals and VHF bands can travel much greater distances and overcome obstacles, such as buildings and vegetation.

It would have been good to use devices using the 900 MHz band for the last-mile connectivity in Nepal; however, the government has not made it available for community networks. For last-mile connectivity in community networks it is recommended to use the 900 MHz band if it is open for public use.

Community involvement in NWNP

From the very beginning NWNP has been working in remote and rural areas. These areas of Nepal are sparsely populated, and the majority of the inhabitants are subsistence farmers. The poorest and least developed part of the country is in the mountainous region. Most of the young people (between 20 and

5 https://en.wikipedia.org/wiki/Dynamic_Host_Configuration_Protocol

30 years of age) leave the villages in search of jobs in the cities or abroad. Because of this, mostly women, children, the youth and old people are living in the villages.

At the beginning, the villagers in Nangi were not even aware of the internet. Therefore, it was a challenge to involve community members in building NWNP. However, it did not take long for people to understand the benefits of having access to an internet connection, especially once they learned to communicate using VoIP, text chat, emails and bulletin boards.

There are many community stakeholders involved in NWNP. These include local schools, healthcare clinics, businesses, local government and individuals. The users of the community networks include farmers, teachers, students, health workers, development workers and local government officials. Users are mostly younger people such as students. This is because the students at the schools get to learn about computers and the internet first hand.

Project management

For the smooth operation of the community networks, network management committees made up of community members have been formed to manage the different services provided by NWNP. The local management committees include members from mothers' and women's groups, social clubs, school management committees, and communication centres.

The key role of NWNP is to provide technical support to build the network and connect facilities such as rural schools, healthcare clinics, community communication centres and local government offices. The servers and routers at the base stations are maintained by the NWNP system administrators. It also provides training for the capacity building of local technicians so that they can build, run and maintain the network smoothly.

Each local management committee appoints a technical person to troubleshoot and fix the technical problems and to provide support for the users in the village. In case the local problem cannot be solved by local technicians, NWNP sends technical team members to help fix the problem.

Because of the above structure, NWNP does not provide services directly to the end-users; instead, the services are provided to end-users through the network management committees of each village. The management committee is responsible for maintaining and operating the network. The project, however, has to generate income for maintenance and operation.

One of the key factors contributing to sustainability is that the rural communities should be involved from the start. If possible, it is important to then transfer the ownership of the network to communities. To ensure financial sustainability, the management committees charge reasonable connectivity fees of about USD 15 to USD 30 per month depending on the bandwidth used. This monthly connectivity fee is a bit expensive for the rural population but has been coming down every year. The monthly fee, paid by the community centres, individual users, local businesses, rural schools and rural clinics, is used to pay for the internet bandwidth cost and to incentivise the technical support team.

Conclusions

Most rural and remote communities in developing countries face similar key challenges, such as poor infrastructure, lack of resources, and a lack of skilled person power. Understanding the local context, and securing the involvement of local and international actors and local communities to complement financial resources, are important.

Governments should also access universal service funds to support community networks and make them sustainable. However, this has not happened yet in Nepal.

Mobile phone services have now penetrated around 90% of the population of Nepal. At the same time, Nepal has a federal system of government now and many of the newly elected state and local governments are trying to introduce ICT-related services in the villages. Because of this, NWNP has shifted its priority from communication service to public services, such as e-education, e-health, e-agriculture, e-commerce and digital literacy programmes. NWNP is now working with eight local governments in developing user-friendly web-based applications and mobile apps which will help to make the lives of the rural population easier.

Action steps

The following key lessons have been learned through the NWNP project:

- Use affordable equipment in conjunction with the 2.4 GHz and 5.8 GHz unlicensed bands to build community networks in sparsely populated areas. These bands are unlicensed in most of the countries in the world.
- Involve local stakeholders, such as local businesses, local community organisations, local governments, and individual community members. They must all be engaged for the creation

of a smooth operation. Likewise, the network should be owned by them.

- Train local technicians for the technical sustainability of the network, so that the community can expand the network when necessary, and troubleshoot the problems in time.
- Produce local content in local languages and integrate digital literacy programmes into the network roll-out plan to help the rural population appropriate the internet.
- Acquire subsidies from the government if possible for community networks. This is helpful at least for a few years at the beginning. It will be difficult for rural and remote communities to build and operate community networks using their own resources.
- Lobby the government to use the universal service fund to support community networks.

NIGERIA

THE REHABILITATION OF A RURAL COMMUNITY NETWORK IN WAR-TORN NORTHERN NIGERIA

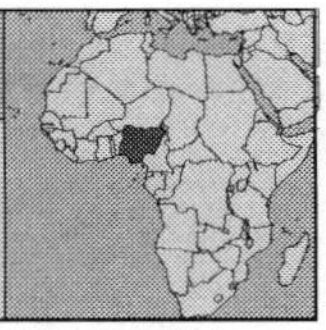

Fantsuam Foundation
John Dada
www.fantsuam.org

Introduction

Nigeria boasts high ambitions for broadband penetration. For example, the Nigerian Communications Commission (NCC), whose mission is to promote universal broadband access,[1] has listed this as the first point in its eight-point agenda.[2] With such commitments coming from the highest information and communications technology (ICT) policy bodies in Nigeria, there seems to be an enabling environment to fast-track the provision of affordable access supported by skills and tools to enable people to solve their connectivity challenges.

However, to what extent is this the case?

At the moment, Nigeria has achieved 21% broadband penetration, most of which is urban based and relies on 3G and 4G mobile technology. The country hopes to increase its broadband penetration to 30% by 2018[3] and to "increase rural access coverage by 40% by 2020."[4] This modest target has to contend with the absence of appropriate first-mile infrastructure.

The Fantsuam Community Wireless Network, which was set up in 2008 and operated until 2011, was established to meet the needs of the subsistence farming communities and peri-urban slums of Kafanchan in the Jema'a Local Government of Kaduna State. These communities are typical in that they lack electricity and road access, and do not have educational and health institutions or many small-scale businesses.

The experience of the Fantsuam Community Wireless Network will be of particular interest to African countries that have experienced or are in the grip of sectarian violence, with its disproportionate negative impact on women, girls, older persons and people with disabilities.

Policy context

Nigeria operates a complex system of regulation for its communication infrastructure, with different regulatory functions located in different agencies within the Ministry of Communications. The ministry's mandate is to "facilitate universal, ubiquitous and cost effective access to communications infrastructure throughout the country"[5] through its departments and agencies: the National Information Technology Development Agency (NITDA),[6] Galaxy Backbone,[7] Nigeria Communications Satellite (NIGCOMSAT),[8] Nigerian Communications Commission (NCC)[9] and Nigerian Postal Service (NIPOST).[10] The main government regulators are the NCC and NITDA.

The NITDA regulates, monitors, evaluates and verifies progress on all national IT policy implementations. Although one of its mandates is "to serve as a clearing house for all IT procurement and services in the public sector,"[11] this has been interpreted to mean that both public and private network equipment deployed in Nigeria must get the approval of the NITDA. This may be because the agency is focused on the local high-demand hardware and software market,[12] not on the low-cost, open source resources typically used in community networks.

The NCC and NITDA were responsible for setting up 867 networks intended to serve marginalised and underserved communities. Fantsuam Foundation's efforts to engage these regulators in dialogue on how to make these networks viable and self-sustaining have yet to yield any positive results.

The nearest projects that may look like community networks in Nigeria are actually networks

1 https://www.ncc.gov.ng/about-ncc/who-we-are
2 https://www.ncc.gov.ng/documents/728-8-point-agenda-milestones/file
3 https://www.ncc.gov.ng/stakeholder/media-public/speeches-presentations#2016
4 Federal Ministry of Communications. (2017). *Nigeria ICT Roadmap 2017-2020*. www.commtech.gov.ng/Doc/Nigeria_ICT_Roadmap_2017-2020.pdf

5 www.commtech.gov.ng/the-ministry/mct-mandates.html
6 https://nitda.gov.ng
7 www.galaxybackbone.com.ng
8 www.nigcomsat.gov.ng
9 https://www.ncc.gov.ng
10 www.nipost.gov.ng
11 https://nitda.gov.ng/mandate
12 Amanze-Nwachuku, C. (2016, 12 September). Nigeria: Coscharis Tech Gets NITDA Approval for Production of Cosmos Computers. *This Day*. https://allafrica.com/stories/201609120558.html

set up by the government, the private sector and philanthropic organisations for disadvantaged, underserved and marginalised communities.

The titles of these projects may give the impression that they function as community networks, but this is not the case:

- Community Resource Centre
- Community Learning Centre
- Community Communication Centre
- Community ICT Centre
- Public Access Venue
- Rural Information Technology Centre.

The top-down mode of establishing these projects, almost entirely funded externally, does not make them qualify to be recognised as community networks. The communal ownership of infrastructure is one of the identifying and sustainability-enabling features of community networks. The absence of this critical element may explain the non-viability of most of these projects.

For example, the NCC, which funded several of these projects, observed that "many schools and communities where the NCC had set up resource and school knowledge centres, have not put them to adequate use and this is affecting internet penetration in rural areas."[13]

The size and political complexity of Nigeria creates a challenge for accessing information for marginalised communities. The licensed private sector operators do not have the incentive to provide services to under- and unserved rural communities, especially because they are able to make substantial profits from their urban operations.

The Ministry of Communications has developed an ICT Roadmap[14] which is intended to create two million jobs by 2020. Achieving this laudable target can be facilitated if community networks are given the enabling environment for their establishment and operations.

There is currently no regulation on community networks in Nigeria. Frequencies are allocated from the office of the regulator and are sometimes allocated by auction, and are also managed from the regulator's office.

One of the yet-to-be explored communication technologies that can be available and accessible to marginalised communities in Nigeria is television white space (TVWS). However, this resource is yet to be made publicly available. Recent research by APC[15] also showed that the process for obtaining the regulator's approval for access to this resource can be quite complicated and too expensive for interested community networks. In this regard, Fantsuam Foundation, in partnership with the Centre for Information Technology and Development (CITAD),[16] another member of APC, have lobbied the regulator, and have also enlisted other civil society organisations working in the IT field in Nigeria. Fantsuam's application to the regulator for permission to do a pilot using TVWS was turned down.

The Fantsuam Community Wireless Network

The cycle of sectarian violence that started in 2011 and continued into 2018 in our host communities has rendered thousands homeless, and led to an increase in marginalised, vulnerable and impoverished families and individuals. The war also led to the deaths of thousands of young men, leaving behind older persons, traumatised adolescent girls and young women. However, the cycle of violence has also destroyed what little communications infrastructure was available in these communities. The hub of the Fantsuam Community Wireless Network was located in the Fantsuam offices in the peri-urban slum of Bayanloco. From there, radios were set up within a 10 km radius of the hub.

In this context, Fantsuam Foundation provides a suite of integrated services as part of its mission of poverty elimination in northern Nigeria. Our key activities lie in sustainable livelihoods, health, education and social protection, with gender, volunteering and ICTs as cross-cutting concerns.

Our constituency is the rural poor in northern Nigeria,[17] who are among the poorest in the world.[18] Literacy levels in the region are also among the lowest in Nigeria. With little disposable income, and just emerging from the period of prolonged sectarian violence, the provision of basic necessities such as food, clothing and shelter remains the over-riding

13 Nigerian NewsDirect. (2017, 29 September). Schools, Communities not using ICT Centres – NCC. *Nigerian NewsDirect*. nigeriannewsdirect.com/schools-communities-not-using-ict-centres-ncc

14 Federal Ministry of Communications. (2017). Op. cit.

15 Association for Progressive Communications. (2012). *Spectrum for development: Nigeria*. https://www.apc.org/sites/default/files/countries/factsheet%20nigeria_eng.pdf; Jensen, M. (2013). The role of TV white spaces and dynamic spectrum in helping to improve internet access in Africa and other developing regions. In E. Pietrosemoli & M. Zennaro, *TV White Spaces: A pragmatic approach*. Abdus Salam International Centre for Theoretical Physics. wireless.ictp.it/tvws/book/8.pdf

16 https://www.citad.org

17 Amzat, A. (2017, 24 July). Despite decades of funding, literacy level in the northern states remains low. *The Guardian*. https://guardian.ng/news/despite-decades-of-funding-literacy-level-in-the-northern-states-remains-low

18 Eweniyi, O. (2017, 10 November). Nigeria Is Set To Become The Poverty Capital Of The World By 2018. *Konbini*. https://www.konbini.com/ng/lifestyle/nigeria-poverty-capital-world-2018

priority. While our activities are always determined by the host communities' priorities, we have been exploring ICTs as an enabler of these priorities.

Making communications accessible and affordable to these populations provides a lifeline for rural communities. Among other things, it allows them to stay in contact with distant relatives who send remittances. The low disposable incomes of the survivors, coupled with the sparse remaining communications infrastructure, dictate that information that rebuilds the local economy ought to be prioritised.

Fantsuam's experience with provision of microfinance services has also shown that women and girls, while being the most traumatised from the crisis, remain the most resilient and reliable with regards to the use of their loans. Therefore, the same approach was taken in our efforts to set up the community network. As far as available records show, the Fantsuam Community Wireless Network was the only rural internet service provider (ISP) in Nigeria.

Physical infrastructure

In 2008, when the network was set up, it provided intranet and internet access to local partners in the community.[19] The community network was formed by community-based organisations such as educational institutions, faith-based institutions, health services, small enterprises and individuals. We connected to the internet using satellite dishes and paying a subscription to Broadband Global Area Network (BGAN), a global satellite network,[20] but this was ultimately unsustainable due to the high costs.

Since 2008, the location, size and operations of the network have changed in response to the unstable political situation and socioeconomic challenges of the rural host communities. Now the only surviving equipment is located in the network operation centre at the Fantsuam premises. There are, however, no clients such as rural farming communities and small businesses currently connected to it. Its user sites have been lost to vandalism and the violence.

From 2008 to 2011, when the Fantsuam Community Wireless Network was fully operational, it served a major hospital, two private clinics and three educational institutions, and was also frequently used by the security services. The network served 23 individual homes within the 10 km radius of its operations

The power infrastructure consisted of a hybrid system of a deep-cycle battery bank and 2 KW solar panels. The system charged from three different sources: the grid when electricity was available, a diesel generator and an array of solar panels. The network operation centre ran solely from solar energy. The solar power system had been designed to provide 12V and 24V DC output in order to fit the input voltage of all low-power servers and workstations that ran the centre infrastructure and training classrooms. The centre provided training in basic computer literacy, computer maintenance and network configurations, among others. The network operating centre was designed to host a battery bank of approximately 70 batteries and 24 south-facing solar panels on its 20-degree roof.

In order to reach the participating communities, a 45-metre-tall mast, equipped with earthing and lightning protection, together with a mandatory signal light, was erected at the network operating centre.

The wireless backbone was built with smartBridges airClient multiband point-to-multipoint outdoor wireless links which included integrated multiband sectoral antennae that could operate both in 2.4 GHz and 5.1/5.8 Ghz frequencies. The airClient equipment used the IEEE 802.11e standard to support traffic prioritisation and bandwidth management per client.

Now we continue to provide training in basic computer literacy, and internet and computer maintenance at the centre. The centre is now also government accredited to offer online university entrance examinations. As mentioned, the equipment at the centre survived the sectarian violence, including the solar panels. However, we now have to rely on mobile phone connectivity, which is expensive.

Conclusions

There is an increasing international interest in the promotion of community networks, and Fantsuam Foundation's participation in the various interest groups may provide an avenue for reaching out to the Nigerian regulators to engage in a dialogue on the topic. It is also important to engage the regulators on how marginalised communities can get access to TVWS. Our international partners include APC,[21] the Net Rights Africa Coalition,[22] the Dynamic Coalition on Community Connectivity (DC3),[23]

19 Balancing Act. (n/d). Nigeria's Fantsuam pioneers wireless service dubbed Zittnet. https://www.balancingact-africa.com/news/telecoms-en/3562/nigerias-fantsuam-pioneers-wireless-service-dubbed-zittnet

20 www.bgansatellite.com

21 https://www.apc.org

22 https://rightscon2018.sched.com/event/EHks/africas-digital-rights-secretariat-net-rights-africa-coalition-and-its-prospects

23 https://www.comconnectivity.org

ISOC[24] and the ISOC Community Networks Special Interest Group (CNSIG).[25]

The specific issues of post-war reconstruction, rehabilitation and reconstruction have to be addressed within a framework that ensures that the injured and traumatised female populations of the affected communities are supported so that their voices are heard in the unfolding development agenda of their communities. This has led us to explore the use of TVWS for which Fantsuam Foundation has made efforts to get the regulators' approval. As mentioned, Fantsuam had applied for permission to undertake a pilot of the TVWS to serve its internally displaced population, but this has not been approved yet.

The survival of the Fantsuam Community Wireless Network – even if it is just the centre – throughout the period of crisis represents a hope for a better future and a reminder of what has been lost and needs to be replaced in order to restore normalcy to this rural economy. The deployment of ICTs for use in education, health and agriculture are the current priorities. The process is being made as inclusive as possible by providing a range of ancillary services to complement the communication mandate of the network. Basic digital literacy programmes have now been established for older persons above the age of 70, while information and support services, including reproductive health both for rape victims and teenage mothers, are being provided.

Action steps

The absence of a coherent policy and strategy for achieving the stated objectives of the regulators[26] may benefit from the diplomatic engagement at ISOC and other international organisations. As a keen participant in many of these international forums, the Nigerian regulator may be persuaded to support local initiatives such as those by Fantsuam Foundation, and liberalise access to TVWS.

While the Fantsuam Community Wireless Network is actively engaged in the social and community development services of reconstruction, rehabilitation and reconciliation, Fantsuam Foundation is lobbying the regulators for improved access to affordable communication infrastructure for its partner communities. The absence of clear policy guidelines on the processes of setting up community networks means that permission has to be obtained for every major community network activity.

Local initiatives such as the Fantsuam Community Wireless Network are already focused on marginalised communities. In our case we have a special interest in facilitating access to ICTs and supporting the development of internally displaced women and girls. But an enabling environment is needed to help similar local initiatives to come on stream.

24 https://www.internetsociety.org

25 cnsig.info

26 See https://www.ncc.gov.ng/about-ncc/who-we-are and www.commtech.gov.ng/Doc/Nigeria_ICT_Roadmap_2017-2020.pdf

PAKISTAN

WIRELESS FOR COMMUNITIES (W4C) PAKISTAN

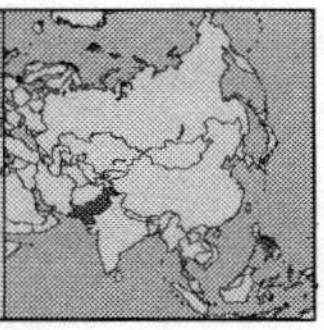

Wireless for Communities Pakistan
Naveed Haq
https://wforc.pk

Introduction

One of the key obstacles to improving internet penetration in rural and remote areas is last-mile connectivity. The lack of commercial viability, as well as huge network roll-out costs, worry operators who are reluctant to make the necessary investments. Although mobile broadband has emerged as an excellent alternative to wired last-mile connections, its usage is heavily weighed to benefit urban and semi-urban areas, rather the rural lands.

For example, in Pakistan, even with over 74.21% mobile penetration, only 28.14% of the population uses broadband internet.[1] In villages – where more than 55% of the population lives – the Pakistan Telecommunication Authority told us that internet usage is likely to be less than 8%.

Community networks are proven to be an excellent way to address the gap in last-mile internet connectivity for underserved communities.

Wireless for Communities (W4C) Pakistan is part of a larger community network programme managed by the Internet Society (ISOC) in Asia-Pacific, together with its partner organisations in India, Pakistan and Nepal. The programme involves deploying line-of-sight wireless technology and low-cost Wi-Fi equipment to create community-owned and operated wireless networks.

Established in December 2015, our pilot community network is situated in "Chak-5 Faiz". "Chak" is a term used for a scattered community that includes multiple small villages. Chak-5 Faiz is located 25 km from the city of Multan.

The network

Unlike the traditional, "top-down" commercial approach, in a community network environment, deployment starts from the end-user or the "last mile". However, your network does require reliable backhaul connectivity (either wired or wireless) in order to carry packets to and from the global internet. Typically, a telephone company or internet service provider (ISP) offers backhaul bandwidth.

To keep our operational costs at a minimum and to ensure robust backhaul, we preferred to partner with an existing local ISP to build our community network. This approach would not only save network running costs, but also minimise respective regulatory and/or policy processes including registration, licensing and compliance.

COMSATS Internet Services (CIS),[2] established in 1996 as a pioneer ISP in Pakistan, is our local partner, providing a dedicated 10 Mbps bandwidth for our community network, as well as on-ground technical support through its network engineers.

Our community network is centred around a 55-metre-high tri-pole tower holding sector antennas which gives a two-kilometre area of coverage in a scattered community comprising several small villages and local schools. The tower acts as base station, linking clients through both the 2.4 GHz and 5.8 GHz frequency bands – these are categorised as "free or unlicensed spectrum" in Pakistan, meaning that they can be used by anyone without having a licence or paying a fee. Following a point-to-multipoint (P2MP) topology, the base station broadcasts to several receiving antennas using time-division multiplexing for maximising bidirectional traffic flow over longer distances.

The base station set-up also includes network management devices to distribute the total allocated bandwidth (10 Mbps) to connected clients. These management devices also control the overall network access through MAC addresses and usernames/passwords. The access control is implemented to ensure optimum usage of this community network and prevent an open public Wi-Fi environment.

Connecting the community

Following the deployment of core network infrastructure, our next goal was to connect the community. In order to try to maximise the 10 Mbps bandwidth, we conducted a community survey to gauge the wireless signal strengths as well as the needs of the community.

1 https://www.pta.gov.pk//en/telecom-indicators

2 www.comsats.net.pk

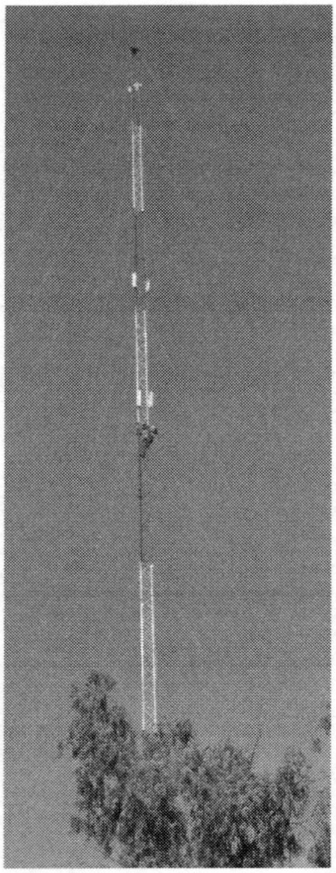

Network base station tower; with attached wireless antennas (right).

Our survey identified a local government school for girls located within the network range. The school had a non-functional internet connection, running over a telephone line that had been disconnected for months. The school computer lab was equipped with computers – but because there was no internet nor teaching staff trained in computers, most of the machines were sheltered in beautiful protective covers that had rarely been removed. The situation was a bit ironic and sad: despite having computers, the school girls were not able to use the internet.

Since this is a government school, we first had to secure permission from the local district education department – this was given instantly. While we were installing the equipment, including a wireless antenna, the students were heard whispering with joy: "These people have come to connect us with the internet." Yes, the internet is a luxury and still a dream to many! According to a recent International Monetary Fund (IMF) report, Pakistan is among the top eight countries where the majority of the population cannot access or afford the internet.[3]

The connectivity for the school was established in a few hours and the whole computer lab was provided with internet access – our first client for our community network. After this the school started running computer and internet classes once a week.

We also found that a local polytechnic institute offering courses in agriculture technology needed connectivity. Agriculture constitutes the largest sector of Pakistan's economy. The majority of the population, directly or indirectly, is dependent on this sector. It contributes about 24% to the country's gross domestic product (GDP), accounts for half of the employed labour force, and is the largest source of foreign exchange earnings.[4]

The primary purpose of this institute is to organise diploma courses on agriculture sciences and promote research on agriculture matters. Here, again, we came to know that a computer lab exists, but that due to limited internet connectivity both students and staff were not able to perform basic internet functions such as email. The students had no choice but to rely on places outside their institute to complete assignments, either at their homes or at internet cafés in the nearby city.

3 International Monetary Fund. (2018). Chapter 2: Digital Government. *Fiscal Monitor, April 2018: Capitalizing on Good Times*. https://www.imf.org/en/Publications/FM/Issues/2018/04/06/fiscal-monitor-april-2018

4 www.pbs.gov.pk/content/agriculture-statistics

Project team installing network antennas at the agricultural institute.

Our community network project was warmly welcomed by the institute. They later told us that the first thing they did when they got connected was to open a Facebook account[5] and start publishing about their activities.

The third and last site connected through our community network was a nearby village of 20 to 25 households. Permission to install roof antennas and other equipment at both the school and the polytechnic institute was easy; however, we were a bit sceptical about how easy it would be in the case of the village. Our first challenge was to map the most appropriate house to sync with our base station, and then, the perhaps more difficult challenge of getting acceptance and permission to install an antenna – it may, for example, look strange to villagers.

However, all it took to secure their permission was to say that our community network would "bring internet to your village."

We got to know that many households in the village have relatives working in the United Arab Emirates (UAE), and the internet for them holds a great value in that it allows them to make free calls using Skype. They were facing limited connectivity: they had no access to wired internet, and the available mobile internet did not work because of weak signals. As a result they had to travel approximately two kilometres to use an internet signal. Although using the internet for Skype calls sounds like a bare minimum usage of the real potential of the internet, for the villagers the ability to talk every day for free with a family member is of huge value.

To avoid any misuse of the network through connecting through unknown devices, we donated a number of pre-programmed tablets to the village.

By the time we had completed our antenna installation and connectivity testing, the news of us "bringing internet" to communities had spread to several nearby villages. A number of villagers appealed for similar internet access, and said they would be willing to pay a monthly internet access fee. Unfortunately, these villages fall outside our network diameter, and we had to give them our regrets. Their requests showed that there is a strong need to find alternative and sustainable solutions to address the connectivity gap that exists between urban and rural areas – and that a community network is one solution.

Owning the network

The most important element to achieve the desired success from a community network is the ownership of the community network by the connected community. It is very important to ensure that those who are connected value the network and online services – in our case, supplementary educational content – provided over it. Community ownership of the network is also critical to ensuring the sustainability of the network, which in many cases can be large, complex, costly, and potentially fragile.

In the case of our community network, where operational costs are at a bare minimum (due to the in-kind contribution of backhaul bandwidth as well as technical support from our partner ISP), the prospect of the network being sustainable is greater. However, equally important is that users get to understand that the internet can be used for more than just browsing, sending emails and making free Skype calls. The sustainability of a community network is also about getting the community to understand the "value add" that the internet can offer.

Community training

For the first six months after providing our three clients with internet, we just let them use it in any way they wanted to. A team of engineers from our partner ISP would make monthly visits to resolve any

5 https://web.facebook.com/pg/Govt-Pak-German-Polytechnic-Institute-For-AgricultureChak-5-Faiz-Multan-936555679712809/posts/?ref=page_internal

Team installing antenna for connecting the village.

technical issues, and also to get regular input/feedback from the community. Yes, there were instances when wireless router cables were accidentally unplugged, resulting in a panic. These regular visits included basic troubleshooting exercises for users so that they could begin to resolve minor issues themselves.

Now, more than two years since deployment, the network has never had a major breakdown – our antennas and equipment operate under the supervision of the community, and users in the connected village have bought an uninterruptible power supply (UPS) unit to keep the internet running during power load shedding hours.

To develop the capacity of community members further, as well as to demonstrate the added value of the internet, we provided training to the school girls, students at the polytechnic institute and community members in the connected village. This training was designed using the "train the trainer" methodology – we train community members and build the capacity of a local trainer, who then can train others in the community, gradually reducing dependency on us.

The training of the school girls, who were in the 10-14 years age group, was about how to apply the internet in their learning processes. We had some interesting discussions with the girls and, yes, it is true that girls (especially in rural areas) lack ownership, access and control over the use of technology, in addition to cultural barriers that they face. Some of the girls have computers in their homes, but the males in their families do not allow them to use them. They also do not have permission to use mobile phones.

Because their exposure to the internet was very limited, it was imperative to include training material that offers a basic understanding of how the internet works and what it can offer.

The training provided to the polytechnic institute was about using the internet in agriculture research and studies. We also provided training to staff on advanced-level computer usage that helped them run a community training course for local residents using their computer lab.

The training provided to the village community was on retrieving useful information about crops, weather conditions and farming practices that can improve their harvest – wheat and cotton are two key crops in the area.

At this point, we have completed two different training sessions with the groups. We have produced 10 trainers who are ready to deliver their first training sessions in September and October.

Training sessions at the school, at the polytechnic institute and in the village.

School girls taking an online class.

Online supplementary education

Last year, after analysing bandwidth usage patterns, we saw an opportunity for running a small online project using unused bandwidth. Our desire was to do something that not only improves the use of our community network, but also brings value to the community.

The quality of education in Pakistan's rural areas is not up to the mark in comparison to urban and semi-urban areas. The 2015 "Education for All" review report[6] by UNESCO and Pakistan's Ministry of Education lists the poor quality of education as one of the key challenges in Pakistan. The report further points to the poor quality of teaching due to an acute shortage of well-trained and motivated teachers.

Starting phase three of our community network project, we decided to run an interesting online test project offering supplementary education for the school girls. This project would provide extra academic help to school students to help them achieve a better understanding of their study material. Our objective was not to disturb their regular classes, but rather to design a different learning experience that complements their existing courseware.

To do this, we needed school teachers who preferably had experience in running an online class environment. The other requirement was to get approval from the district education department, since the permission we had was to connect the school to the internet only. This time we were also well received by the government department, and approval was given in a couple of days.

FIGURE 1.

Online classes baseline vs end-line scores

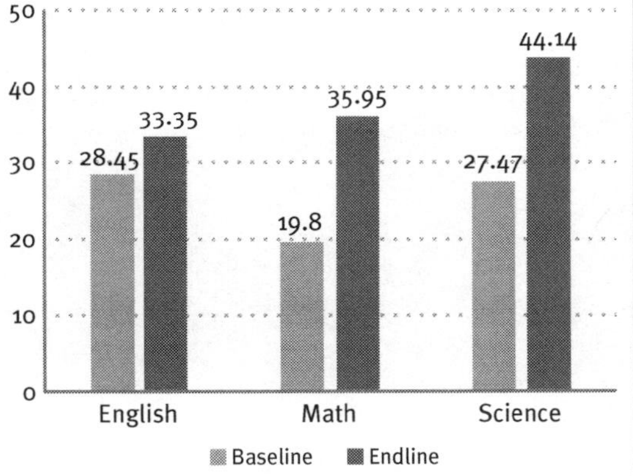

We ended up partnering with TeleTaleem,[7] a local commercial organisation experienced in tele-education. They delivered real-time, online lectures on English, mathematics and science for three months to grade six girls. The classes were completed a month before their final examination.

To record the impact, a baseline assessment was carried out before the course started and an end-line assessment was performed at the course end. The overall impact assessment results (see the graph in Figure 1) are really encouraging, with a substantial improvement in mathematics and science.

This truly was a great experience for us in capitalising on the potential of the community network, and moving beyond just access. We have plans to run similar classes this year too, as well as offering online classes to the polytechnic institute in the near future.

6 Ministry of Education, Training and Standards in Higher Education. (2015). *Education for All 2015 National Review Report: Pakistan.* unesdoc.unesco.org/images/0022/002297/229718E.pdf

7 www.teletaleem.com

Conclusion

According to the UN Internet Governance Forum Dynamic Coalition on Community Connectivity:[8]

> Community networks [...] are a way to develop future business by creating "digitally savvy" communities, hungry for more local content and additional services. These often are not super high-tech networks. They serve a local community-driven purpose to connect within and to connect from the village or community "out". They might be local open-source 2G solutions, or Wi-Fi mesh solutions using license-free spectrum. The aim is to build capacity for both the demand and the supply of digital tools.[9]

In countries like Pakistan, where around 70% of the population is still not online, community-owned networks are not widely seen as a way to bridge the digital divide and achieve internet access for all. While the country's Universal Service Fund Company[10] has embarked on projects to improve broadband connectivity in rural areas, the digital divide remains. It is not uncommon for villagers to have to walk two kilometres to get online, like the villagers in Chak-5 Faiz had to before we arrived.

Action steps

To promote community networks in Pakistan, the following steps are necessary:

- Streamline or eliminate related regulatory requirements, especially those that are not applicable to small, community-based networks.
- Expand universal service and other public funding opportunities and include community networks as eligible for funding from the universal service fund.
- Introduce approaches to provide spectrum access and innovative licensing for community network operators.
- Encourage community initiatives to build networks aimed at reducing the digital divide in Pakistan.

8 https://www.comconnectivity.org

9 Brown, K. (2017). Preface: Putting People at the Heart of the Internet. In L. Belli (Ed.), *Community Networks: The Internet by the People, for the People. Official Outcome of the UN IGF Dynamic Coalition on Community Connectivity.* https://bibliotecadigital.fgv.br/dspace/bitstream/handle/10438/19401/Community%20networks%20-%20the%20Internet%20by%20the%20people%2C%20for%20the%20people.pdf

10 https://usf.org.pk

PERU

A TELECENTRE IN THE JUNGLE

Communication Department, Pontificia Universidad Católica del Perú
Juan Bossio
departamento.pucp.edu.pe/comunicaciones

Introduction

This report is not going to be on community networks. In Peru community networks are rare, and are not resilient or sustainable. This is because of a conjunction of reasons.

Firstly, while laws do not prohibit community networks, regulations and bureaucracy make it almost impossible to create a community network able to openly provide telecom services to citizens: legal procedures need to be done in Lima, most of the technical requirements such as tests or security procedures have been established with big telecoms in mind, legally authorised professionals do not live in rural areas, and so on. What has existed until now are local communication systems able to deliver communication services to an organisation's network, as in the case of the Huaral Valley Agrarian Information System. This is a Wi-Fi network connecting more than 10 agrarian organisation offices in the Huaral Valley, giving some of the settlements their first and for some time their only telecommunication service, consisting of a telecentre and a phone terminal.[1] However, this is not a community network in the strict sense, since the system cannot openly provide services to other organisations, institutions or the public at large.[2]

The second reason is financial: the equipment is expensive and far from a community-based organisation's financial capabilities.

Thirdly, telecom services – especially mobile – have grown very fast in recent years, reaching most rural and deprived settlements by now, even though the availability of the service does not mean that it is also accessible and affordable, which would be necessary in order for it to be useful for social development.

Recent studies by DIRSI,[3] a Latin American research network on the social impacts of information and communications technologies (ICTs), show that on average, the *quality* of internet usage is diminishing while usage is growing. Use of the internet for things other than entertainment or communication is on average less common now than years ago (i.e. a lower percentage of internet users are accessing educational services, e-government services and so on), and even less common in demographics with fewer resources or less education, which are typically groups which access the internet only through a mobile phone.[4]

Privately owned public internet access points called "cabinas públicas" – a Peruvian model of cybercafés – have been the main internet access points in Peru over the past 20 years, since their popularisation among students and young professionals in the 1990s, up until the moment when mobiles become the main way to access the internet in Peru in 2016. Telecentre projects and strategies seem to be old fashioned now, both in Peru and Latin America more generally. Different factors have contributed to this, one of the most important being the extension of mobile access.

During the 1990s and the beginning of the century, universal access policies promoted rural access through subsides. Nowadays, public policies in Peru and most Latin American countries favour market solutions for increasing access to ICT services, but access does not seem to be the only issue to look at.

1 See the case study on the Huaral Valley Agrarian Information System at https://www.apc.org/sites/default/files/APCProPoorKit_CommunityModule_CaseStudy_HuaralValley_EN.pdf; Belo, R. (2004, 15 December). Wi-fi web reaches farmers in Peru. *BBC*. news.bbc.co.uk/2/hi/technology/4071645.stm; Bossio, J. (2018). *From an Information System and Telecenter Network Project to a Community Network: Building Financial Sustainability through Social Sustainability.* https://www.researchgate.net/publication/238777579_From_an_Information_System_and_Telecenter_Network_Project_to_a_Community_Network_Building_Financial_Sustainability_through_Social_Sustainability

2 Nowadays, communications services (telecentre and telephone) are not provided by Huaral's system because the Huaral Valley is now covered by private services.

3 https://dirsi.net/web/web/en/about-dirsi/history

4 See the reports "Panorama del acceso y uso de Internet exclusivamente desde terminales móviles: El caso de Argentina, Colombia, Guatemala, Paraguay y Perú" and "Panorama general de acceso y uso de Tecnologías de Información y Comunicación: El caso de Argentina, Colombia, Guatemala, Paraguay y Perú" available at https://olatics.net/2018/02/06/descargas

To summarise: community networks are not common in Peru. Here access to communication services in general and the internet in particular has been growing across the country and for all social sectors, filling gaps in terms of access and availability, firstly through public services (public phones and cybercafés) and now through private services (mobile). Public policies are now moving from universal access to promotion of private access. However, a decrease in the diversity of use and the usage of the internet for social development purposes has been observed. This report presents the case of a telecentre network, and shows how it takes more than access to make a difference. It is hoped that this will offer insights into community networks, and enhance the purposeful use of ICTs in the community context.

A telecentre network in the Amazon

Rural and poor areas have less access to ICT services in Peru, and the Amazon is the geographical zone with the worst access to communication services in the country. As a result, rural Amazon settlements are an important focus when it comes to those with fewer opportunities to get connected. *Red de Telecentros de la Amazonía* (RTA, Amazon Telecentres Network)[5] was a project promoted by CEDRO,[6] a Peruvian NGO, which included 39 centres catering for community access to ICTs. The telecentres were first set up by CEDRO and are now owned and maintained by district-level local governments. Most of them are in really small and rural localities (14 have less than one thousand inhabitants) and had been the first way for the communities to access the internet.

These purpose-oriented telecentres train users and provide information useful for development such as market and product information for farmers. The network has also organised hackathons and teleconferences.[7] A recent evaluation shows that 33% of the users have had contact with a government officer through the internet, 32% accessed information for their businesses, 16% promoted their products, and 87% used online education opportunities. Nowadays, telecentres from the RTA are participating in a new project by CEDRO on small-scale financial services. They collaborate on the organisation of training workshops on savings, budgeting, debt management and the selection of financial service companies.

It is interesting to mention here that the project has drawn the attention of a Peruvian communication provider that would like to offer services to the remote areas included in the project. Yachay,[8] an internet service provider (ISP), is going to offer private service in 70 localities using part of the project's installations and equipment. Yachay is part of the RCP group (RCP is an NGO which was the first ISP in Peru).

Telecentres in the Monzón Valley

This report is based on the experiences of RTA telecentres in the Monzón Valley, which is the territory of a district with the same name. The Monzón District comprises 1,521.4 square kilometres, situated in the high jungle (mountainous jungle), and is part of the Humalies Province and Huánuco Department. It has an estimated population of 28,605 people, chronic malnutrition affects 44.3% of kids under five, and around 80% of adults work in agriculture.[9] This valley was one of the bigger producers of coca plants, which are used to make cocaine, until the beginning of this decade. Thanks to its complicated geography and lack of transport infrastructure, it was ruled for around 25 years by Shining Path and drug traffickers. Since 2011 it has been mostly brought under the control of the state; far from doing this through police action alone, the government found that the best strategy to sustain control was a social and economic one: crop substitution. Peasants are now cultivating other crops besides coca, such as cacao and coffee.

María Teresa Delgado is the coordinator of a CEDRO project on digital inclusion being implemented in the zone, which includes the Monzón Valley, with six telecentres, four in settlements with less than one thousand inhabitants. She started to work with telecentres in 2012 and says that one of most important outcomes of the RTA was to become sustainable by getting the support of local governments and the participation of the community through the "Allies Committee". The Allies Committees (ACs) are community-based organisations (CBOs) promoted by the RTA project in each rural settlement that has a telecentre. ACs are independent from both CEDRO and local governments, the latter of which operate from the central town of each district and not the rural settlements where the telecentres are located. ACs look after the day-to-day administration of the quality and provision of services. They help telecentre operators when it comes to organising training and other activities,

5 https://telecentrosamazonicos.com

6 https://www.cedro.org.pe

7 Videos on these and other activities are available at https://www.youtube.com/user/TelecentrosAmazonia

8 https://yachay.pe

9 https://www.inei.gob.pe

organise fundraising activities, and help spend these funds on small expenses such as computer accessories or peripherals. Delgado said that the local government is in charge of the area of social and economic development at the telecentres.

Richard Clemente is now CEDRO's digital promoter on finance for the Monzón Valley telecentres, but has been working for the project for some years. According to him, the most important effect of the project is that trainees always want to learn more. He also mentions some cases in which one would find a direct impact on the living standards of users, such as the case of a young hairdresser and stylist who states that she learned much about her business though the internet.

Sachavaca

Sachavaca is a small rural settlement in the Monzón Valley. It has 300 inhabitants. Its telecentre was launched in January 2014.[10] Katherine Alvarado, the telecentre operator, has been working at the telecentre for just a few months. She was trained by the former operator. She describes her duties as training new users and helping other users to find information. The telecentre offers some training programmes with certificates for those students that complete them.

Kids and teenagers visit the telecentre to look for information related to their studies; peasant farmers visit the telecentre to find out about cocoa cultivation and trade – diseases, prices, places to sell and so on; school teachers and other state professionals go there for digital training; housewives go to help with their children's homework; and small entrepreneurs look up information on a range of topics, from cooking to hairdressing to handicrafts. People from other settlements close to Sachavaca also visit the telecentre.

Katherine has attended local CBOs meetings and visited the school and other places to talk about the telecentre and its services. Finally, she coordinates with the AC and meets with them on a regular basis.

As mentioned, the AC is the main link to community – it is critical for sustainability in several ways, and ensures that the telecentre is community-driven. Milton Sánchez Gopia is the coordinator of the AC of Sachavaca. He was mayor[11] of Sachavaca when the installation of the telecentre started. He was part of the first groups of local people trained in the telecentre and was invited to join the AC together with four other members of the community. Milton explained to us that the AC coordinates with the telecentre operator (Katherine) and oversees the quality of her work. They organise activities held in the telecentre, such as Mother's Day activities in May, buy small stuff needed for day-to-day services, ask the district local government for other things they need, and hold fundraising activities.

But these are just the "formal" things they do. Through talking to Milton, one realises that a critical function for members of the AC is to act as champions of the telecentre – promoting its benefits and "spreading the word" about the telecentre from a community member's perspective. Members of the AC share contact details with local CBOs because most of them are also part of those organisations. These include the local mothers' club, soup kitchen ("comedor popular"), or the producers' organisation. They promote the use of the telecentre by the school because their children go there; they appreciate that school teachers and other public servants enhance their capacities using the telecentre, because those professionals work with and for them.

They get to know about impact on productivity or trade from farmers themselves, because they are their friends, relatives or neighbours. By using the telecentre, organising fundraising or promotional activities, and sharing information or anecdotes, they encourage the rest of the community to use the telecentre too.

Milton says that the telecentre contributes to the social, cultural and economic development of his locality. He says that some young people trained at the telecentre are now working outside of the community, thanks to the certificate that they received in the training. There are women-driven small enterprises, whether involving *bijouterie* or cooking, that have been created after getting information through the telecentre, and several peasant farmers have a accessed information on pest control, cropping and trade issues. He mentioned the case of Teófilo Cierto, a fish farmer and banana grower who is offering and selling his products on the internet. Last but not least, Milton outlines the videoconferences held between different telecentres who are part of the RTA. During these events, peasant farmers from different localities in similar regions share their views and experiences growing and selling their products so that they may learn from "others like them".

10 https://www.facebook.com/TelecentroIDSachavaca

11 In rural areas there is a kind of local government in small settlements ("centros poblados") that serves as local representation at the district-level government which would be many kilometres away. However, this local government does not have any significant budget or capacity to take important decisions.

Reload telecentres strategies

As this case clearly shows, telecentres may be a way to enhance purposeful and useful usage of the internet, making access a real tool for local development. Sachavaca had no mobile service when the telecentre started. While it now does, the kinds of things the population uses the internet for would not have happened by itself: young people and some adults would be using social networks, but we feel that hardly anybody would be using the internet for development purposes. The growth in mobile access in countries with big social inequalities such as Peru is closing a divide, but opening another: the constructive use of that access for bettering the socioeconomic well-being of communities. Markets and technologies will not close this alone.

Because of this, and because community networks are difficult if not impossible to sustain at this point in time in Peru, it is important that policies promote community-driven telecentres to encourage the community appropriation of technology for socioeconomic betterment.

To end this report, it would be fair to restate its opening thoughts. This report was not about community networks in Peru, but it was about a case that outlines a series of lessons that would serve community network projects in other countries: the need to promote the useful application of ICTs, that is, the importance of getting community members to champion or promote purposeful usage of the internet. It also points to the value of engaging local public offices in local access projects, among others.

PHILIPPINES

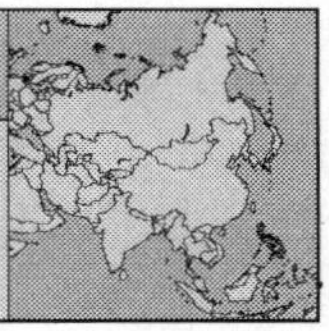

CONNECTING COMMUNITIES THROUGH MOBILE NETWORKS: THE VBTS-COCOMONETS PROJECT

University of the Philippines – Diliman and University of Washington
Mary Claire Barela, Josephine Dionisio, Kurtis Heimerl, Manuel Victor Sapitula and Cedric Angelo Festin

Introduction

Despite the rapid expansion of mobile coverage throughout the world, roughly 10% of the world's population lives beyond the reach of a cell tower.[2] This is particularly problematic in the Philippines, with just under 70% mobile phone penetration.[3] Because these isolated and relatively poor communities are not considered commercially viable within the current business and technology model of commercial operators, bridging this "last mile" connectivity gap requires innovative technological solutions.

To combat this, our team, in partnership with a local telco and local communities, has developed and deployed GSM community cellular networks (CCNs) in the rural Philippines. The CCNs deliver basic mobile telephony at a fraction of the capital and operational expenses of traditional cellular networks by including local agents in the operation of the network.[4] To prove the importance of cellular access, we are also evaluating the impact of cellular connectivity in our partner communities, specifically across gender and social networks, through the use of a longitudinal randomised control trial and participatory qualitative research.

Our CCNs are deployed in the rural and isolated *barangays*[5] of Aurora Province, Philippines. Each barangay has a cell site that provides cellular coverage to the entire community. The system is powered via solar and is connected to a VSAT backhaul to route calls and SMS to the broader public telephone network. The network, dubbed "VBTS Konekt Barangay", has been in operation for over seven months and with more than 1,200 subscribers. To date, four out of the seven pilot sites have been launched (Brgy. Umiray in September 2017, Brgy. Dikapinisan in October 2017, Brgy. Dibut in February 2018, and Brgy. Diotorin in May 2018). All seven pilot sites are expected to be operational by the end of 2018.

Policy, economic and political background

Our biggest challenge is the lack of an operational framework that supports last-mile service delivery in the context of community networks. The Public Telecommunications Policy Act of the Philippines,[6] which governs the development and delivery of public telecommunications in the country, states in its policy declaration that "expansion of the telecommunications network shall give priority to improving and extending basic services to areas not yet served." Since our network uses regulated GSM frequencies to operate, another large challenge is the lack of available spectrum that can be assigned to CCNs and allow them to operate legally. In the Philippines, all channels on GSM mobile cellular bands (900/1800/2100 MHz) have been allocated to incumbent mobile network operators. On top of this, licences cover the entire country with no "use-it-or-lose-it" provisions, forbidding local actors. As such, even in areas where these telcos have no presence, small-scale cellular networks cannot just operate without coordinating with the regulatory agency and the assigned frequency owner. The creation of policies to support rural connectivity (such as a universal access fund or spectrum sharing) would help in extending basic communication services in the country and is actually part of the

1 The Village Base Station Project - Connecting Communities through Mobile Networks (VBTS-CoCoMoNets) is a collaborative research project between the University of the Philippines and the University of California, Berkeley, with linkages to the University of Washington, University of California, Davis, and Aurora State College of Technology. It is being carried out through a project grant from the Philippines Commission on Higher Education, through the Philippines-California Advanced Research Institutes initiative.

2 GSMA Intelligence. (2015). *Rural coverage: strategies for sustainability.* https://www.gsmaintelligence.com/research/?file=53525bcdac7cd801eccef740e001fd92&download

3 https://www.statista.com/statistics/570389/philippines-mobile-phone-user-penetration/

4 Heimerl, K., Hasan, S., Ali, K., Brewer, E., & Parikh, T. (2013). Local, Sustainable, Small-Scale Cellular Networks. Paper presented at the International Conference on Information and Communication Technologies and Development (ICTD), Cape Town, South Africa, 7-10 December. https://kurti.sh/pubs/vbts_ictd_13.pdf

5 The *barangay* is the smallest political unit in the Philippines. It is abbreviated as Brgy.

6 Philippine Republic Act No. 7925. (1995). An Act to Promote and Govern the Development of the Philippine Telecommunications and the Delivery of Public Telecommunications Services.

mandates of the law. The Philippine Department of Information and Communications Technology and the country's regulatory agency, the National Telecommunications Commission (NTC), have yet to formulate a new policy on "refarming" and redistributing the spectrum that will foster better delivery of communication services in the country[7] or on the creation of a universal service fund.[8]

The current framework presents myriad barriers for small, community players to participate. To be able to operate telecommunications services, an organisation will have to first secure a congressional franchise and a certificate of public convenience and necessity (CPCN) which will prove their financial, technical and legal capability to offer the proposed services.[9] On top of this, carriers (including small local operators) also have a number of other licences they must acquire for operation, including environmental clearance certificates, height clearances, and the local mayor's permit. All in all, there are as many as 25 permits needed for each cell site.[10] The sheer number of required permits is very difficult for a small entity to complete.

In the past, access initiatives in the Philippines have concentrated on providing internet access to remote and rural parts of the country. These networks typically employ point-to-point, long-range Wi-Fi links to connect institutions, such as schools, to the nearest internet point of presence. One example is the Digital Provide initiative,[11] where secondary schools in Batanes were connected to the internet service provider using long-range Wi-Fi links. These initiatives were largely "top-down" in nature and focused on basic IP connectivity rather than community ownership. Over time, the expansion of mobile internet access in suburban and provincial areas has rendered some of these networks inactive.[12]

Engaging partners and community participation

Our first major innovation is our public-private partnership for sharing cellular spectrum with a large mobile operator. Given the absence of regulatory support and spectrum access for community cellular networks in the Philippines, we found it necessary to find a partner that shares the project's vision and that would allow the community network to operate under their frequency licence. We found that partner in Globe Telecom,[13] a major telecommunications company in the Philippines. Since our sites have a smaller subscriber base than what it would consider viable, our community network deployments are placed under its corporate social responsibility programme.

Our engagement on the ground is founded on the sustained collaboration between several institutions: higher education institutions such as the University of the Philippines (UP) and Aurora State College of Technology (ASCOT), the telecommunication company (Globe), local government units at municipal and barangay levels, and operational cooperatives in the area. Through top-level agreements, Globe consented to the use of 2G frequency under their franchise. In addition, Globe allowed us to use their SIMs and cloud services which provide the interconnect from the VBTS network to other phone networks. Globe is also assisting the project to comply with the required NTC permits, such as radio station licences and pricing approvals.

Partnership with local government units at the municipal level is crucial because of their administrative authority over the project sites. In rolling out the infrastructural requirements of the CCNs, municipal mayors facilitated the legal appropriation of lands where the towers were allowed to be built, and helped expedite the issuance of various permits and clearances for construction of the village base stations within their areas of jurisdiction. The municipal governments helped mobilise local labour that assisted university-based engineers and also helped identify potential partner cooperatives based on track record in project management. In due course, the municipal governments allocated funds in their annual budget for the maintenance of the village base station towers and for the mobilisation of community-based civilian security forces to protect the towers and CCN-related equipment from possible theft and vandalism.

Cooperatives with municipal-wide operations are one of many types of organisations at the

7 Marcelo, P. (2018, 4 March). DICT drafting policy on frequency reallocation. *BusinessWorld*. bworldonline.com/dict-drafting-policy-frequency-reallocation

8 National Economic and Development Authority. (2017). Philippine Development Plan 2017-2022. pdp.neda.gov.ph/wp-content/uploads/2017/01/PDP-2017-2022-07-20-2017.pdf

9 King-Dominguez, R., & Acebedo, R. (2013). Philippines. In L. Garzaniti & N. Good (Eds.), *Telecoms and Media: An overview of regulation in 44 jurisdictions worldwide*. London: Law Business Research Ltd. www.syciplaw.com/Documents/T2013%20Philippines.pdf

10 Camus, M. (2018, 20 January). Gov't pushes new cell tower scheme for level field, better service. *Philippine Daily Inquirer*. business.inquirer.net/244506/govt-pushes-new-cell-tower-scheme-level-field-better-service

11 Ramos, M. C. Jr. (2008). Addressing the digital divide in Philippine education. Paper presented at the 1st Workshop on Wireless Broadband Access for Communities and Rural Developing Regions, Karlstad, Sweden, 11-12 December.

12 Gonzales, C. (2018, 5 May). Batanes now LTE-connected. *Rappler*. https://www.rappler.com/technology/news/201837-batanes-lte-4g-smart-connection

13 www.globe.com.ph

ground level, alongside tribal councils for the indigenous peoples' communities, and associations of fisher folk and farmers. Municipal-level cooperatives were chosen as acceptable project partners, especially for telecommunications, because of their established juridical personality which made them eligible to establish business transactions with Globe, and because of their familiarity with administrative requirements in managing income-generating activities. They were enjoined to handle the day-to-day operations, business management, and maintenance of the community network, while systematically building the capabilities of community-level associations so that these associations may eventually exercise greater control and accountability over the CCNs.

The cooperatives are in charge of the SMS-based electronic load (e-load) distribution from Globe to the local retailers. The cooperative orders the e-load from Globe on a monthly basis. After the cooperative's mobile number has been topped up, the cooperative then distributes it to authorised local resellers in the communities. The cooperative receives a wholesale discount from Globe, which in part is also passed on to the community retailers. Finally, the gross revenue from all charged calls and texts on the network are split based on an agreed revenue-sharing scheme between Globe and the communities. In the revenue-sharing scheme, the cooperative gets 80% while the remaining 20% goes to Globe. Earnings are used by the cooperative to pay the honoraria of community personnel and as savings for operating expenses beyond the subsidised first year.

The majority of our retailers are women (eight out of ten) and are already running their own *sari-sari* or general merchandise stores. Aside from the retailer's discount, the retailer also charges an additional convenience fee per transaction to the subscriber, which is a common practice in the Philippines. On top of e-load sales, some retailers grabbed the opportunity to sell mobile phones and mobile phone accessories in their area.

During onboarding, local stakeholders are first oriented on the goals of the project, with emphasis on community ownership and public service over profit. Prior to launch, social science researchers facilitated social enterprise training sessions with cooperatives that had no prior experience in conducting business-related activities pertaining to the selling of SIM cards and cellphone load. VBTS engineers conducted trainings with community maintenance personnel (Level 1) and with ASCOT engineers (Level 2). The scope of Level 1 revolves around daily maintenance and basic troubleshooting of the CCN tower and equipment, which will be performed by community site operators. The scope of Level 2 consists of tasks which require intermediate technical knowledge or tools to complete. Technical personnel will coordinate with higher levels for incidents or issues that they cannot resolve at their level. It was later clarified during a technical breakdown that municipal engineers will need to be closely involved in troubleshooting hardware-related concerns, which will aid in the efficiency of the CCN system.

On the day of the network launch, we hold a small programme wherein we introduce the community network, its capabilities and limitations. The launch events are well attended by community members, and become useful venues to introduce our on-ground partners (retailers, maintenance) and address questions and concerns from the community. Post-launch, we also open an SMS-based support hotline. This SMS-based service is free and open to all subscribers in the community. Aside from being a channel for community members to send network-related inquiries, this is also being used to send questions, suggestions or other feedback to the project team.

Usage trends

Currently, we have more than 1,500 subscribers, equivalent to more than 81% of the total eligible population (15 years old and up) across all operational sites. About 40% of the subscribers top up monthly, spending USD 1.20 per month. Monthly average revenue per user is around USD 0.60. We have seen strong usage and adoption in sites that are tremendously isolated and where the VBTS network is their only means to communicate out of the barangay. In areas where an alternative service is less challenging to access, subscribers are not dependent on the community network, as they can get an incidental signal from other networks by walking several kilometres.

In terms of network traffic, we have observed that subscribers take more inbound calls than they make outbound calls – the number of inbound call minutes is six times greater than the number of outbound call minutes. This is indicative of a "call-me" behaviour where subscribers in the community let their outside contacts call them instead of making a call themselves, taking advantage of the fact that receiving inbound calls is free. SMS traffic, on the other hand, has the same volume in both directions. Over time, we have observed a general trend of outbound calls and bidirectional SMS traffic reaching a steady state, but the level of incoming calls continuing to increase. Subscribers are also budget-sensitive and, as a result, there is a request for value-for-money promos, such as the "unlimited" call and text bundles offered by conventional networks.

Technology platform

The second innovative part of this project is the technology platform. We have chosen to leverage the CommunityCellularManager[14] (CCM) stack, a novel IP-based cellular core. CCM allows for multiple individuals to run separate community networks under one technical domain and is split into two technical portions: the client and the cloud. In our case, Globe manages the CCM cloud installation and provided us an account for our networks. We manage the client installations, porting the client to a variety of compatible hardware platforms (Nuran Litecell, Endaga CCN1, Fairwaves UmSITE). We developed additional software-based client features that are necessary for field operations and to assist with the implementation of our research and evaluation needs, including call and text promo support. The CCM cloud handles the routing, interconnect, and phone numbers for our network (and other Globe community-style installations) – generally anything having to do with the integration into the Globe network. This is the first large-scale CCM deployment in the world. Our team led the integration work with Globe and assisted it in the installation and operation of both the cloud and client inside of its network.

Social impact

Our mixed-methods approach to social impact evaluation of the installations, using a randomised control trial and participatory qualitative research, is another novelty of this network. In the past decade, roughly four billion individuals in developing countries have started using mobile phones for the first time. While qualitative evidence of the impacts of network access is abundant and clear, outside of a few well-cited papers[15] about phones and agricultural market inefficiencies, there is a lack of rigorous (e.g. randomised control trial) empirical evidence[16] on how mobile phones affect the social and economic outcomes of individuals and households.[17] This is in part due to the difficulty in measuring the causal effects of mobile phone adoption, since the market-driven spread of mobile phones is not typically exogenous to economic outcomes. Together with a team of economists and social scientists, we are working to develop a rigorous body of evidence of the social and economic impacts that the mobile phone network brings to communities in the rural Philippines.

Though much of the evaluation has yet to be completed (with the endline survey expected six months after the installation of the last site), we have noted some immediate effects. The most obvious impact is in the enabling of communications in the communities. Prior to the arrival of the network, locals would need to travel several hours by boat or rough roads before they could reach an area with a cellular signal. Now they enjoy a more accessible and convenient way to get in touch with family and contacts outside their barangay. The community cell service has also helped the barangay council report faster to the municipal government and vice versa.

But more than communications, the CCNs also enabled social capital formation. The presence of the CCN has the potential to contribute to a more vibrant local economy as alternative sources of income are spurred through the selling of mobile phones and mobile phone accessories, and the provision of repair and maintenance services. Also, the community cellular service has the potential to connect previously disparate communities into an established network of value chains.

Our trainings are not only limited to transferring technical knowledge, but are conducted on different levels and on different dimensions that involve imparting the message of community benefit and ownership. Aside from training the partner cooperatives and on-the-ground personnel, we also extended our training to the local government units and to nearby higher education institutions. More than being respondents to the technical difficulties of the installations, the advancement of their skills and knowledge is crucial for their future research and interventions that would attend to the needs of the community. The collaboration of knowledge and skills from the institutions brings new ideas and appropriate applications that matter to the local community, which is the goal of this inclusive effort.

Our experiences in the implementation of the project render more visible the social infrastructure

14 https://github.com/facebookincubator/CommunityCellularManager

15 Jensen, R. (2007). The Digital Provide: Information (Technology), Market Performance, and Welfare in the South Indian Fisheries Sector. *The Quarterly Journal of Economics, 122*(3), 879-924; Aker, J. C. (2010). Information from markets near and far: Mobile phones and agricultural markets in Niger. *American Economic Journal: Applied Economics, 2*(3), 46-59.

16 The is largely due to the immense level of coordination between academic researchers, government regulators, and commercial operators that is required to conduct a careful randomised roll-out.

17 Aker, J., & Mbiti, I. (2010). Mobile Phones and Economic Development in Africa. *Journal of Economic Perspectives, 24*(3), 207-232.

requirements for replicative and sustainable CCNs. We have encountered challenges in setting up the business model and structure, as well as creating a feasible trade and distribution network, as these sites are far away from traditional financial institutions like banks and remittance centres. Our partner telco also has internal vendor certifications, trade requirements and processes that our on-the-ground partners need to satisfy. Because of this there is a need to acknowledge the distinct nature and "business model" of CCNs which should be viewed differently from how traditional and large telecommunications organisations normally operate.

Conclusion

We believe our experiences contribute a unique perspective on community networks in three ways: 1) by involving a novel public-private spectrum licensing partnership between the UP, Globe Telecom and the local communities; 2) using the CommunityCellularManager cellular stack to provide 2G basic cellular service including voice, SMS and, eventually, data; and 3) a mixed-methods approach to social impact evaluation of the installations using randomised controlled trial and participatory qualitative research.

These factors have allowed us to sustainably provide cellular coverage to one of the most remote areas of the Philippines, and an area that incumbent telecoms are unable to profitably serve. It has also allowed us to help empower the communities themselves to own and operate their own telecom equipment. We plan to continue to scale the CCN solution in the Philippines and hope to continue to have deployment experiences to share.

Action steps

Our experiences in rolling out CCNs showed that more needs to be done for small network operators to thrive in the country. As it currently stands, it would still be difficult for small communities to set up their own CCN without intervention, such as from our project, or with a public-private partnership with a mobile operator. We would therefore propose the following action steps:

Formal and legal institutionalisation of CCNs as a mode of community-based social entrepreneurship telco service delivery

The current legal framework has no category where community networks could fit in. As a result, small operators are forced to adapt to the model used for telcos and other large organisations. There is a need to acknowledge the distinct nature and "business model" of CCNs which is a different model altogether from traditional and large telecommunication organisations. We suggest providing community networks with a legal character that is bound by regulatory parameters, entitlements, and sets of applicable national and local government standards for their facilitation and growth.

Designation of an exclusive frequency licence for CCNs

Our biggest challenge in starting up our CCN deployments is the lack of a dedicated spectrum licence that can be used for last-mile service delivery. Other countries have taken the initial steps of opening up some spectrum so that small networks can operate legally. For example, the Netherlands has set aside a portion of the 1800 MHz band for licence-exempt mobile communications.[18] Mexico has set aside 2 x 5 MHz of spectrum in the 800 MHz band for "social" use.[19] While our networks are able to operate through our partnership with Globe, we believe that it would empower Philippine community networking in the long run if a swath of frequencies were available for development, research and social efforts. With the growth of LTE[20] and its support for over 40 different bands, it should be possible to provide access to CCNs.

Support for scaling up community networks

The number of permits and licences that need to be completed is already a large barrier that discourages new and small community operators from venturing out and starting their own networks. To foster more community networks, such barriers must be simplified and streamlined. It would also help if telecom equipment were more accessible in the country. Currently we import most of the telecom equipment from foreign suppliers and manufacturers. It would help support sustainability if parts could be easily sourced from a local distributor. Lastly, building a community network still entails some capital costs which may be out of reach for some community organisations. Access to seed funding or capital financing would help encourage the grassroots deployment of CCNs in other unserved communities.

18 Netherlands Ministry of Economic Affairs. (2016). Radio Spectrum Policy Memorandum. https://www.government.nl/documents/reports/2017/03/07/radio-spectrum-policy-memorandum-2016

19 Song, S. (2015, 17 April). How To Let GSM Serve The People That Other Networks Can't Reach. *Many Possibilities.* https://manypossibilities.net/2015/04/how-to-let-gsm-serve-the-people-that-other-networks-cant-reach

20 Long Term Evolution (LTE) is a standard for broadband wireless communication for mobile devices.

PORTUGAL

HOW TO BUILD A WIRELESS NETWORK AGAINST ALL ODDS: THE CASE OF WIRELESSPT

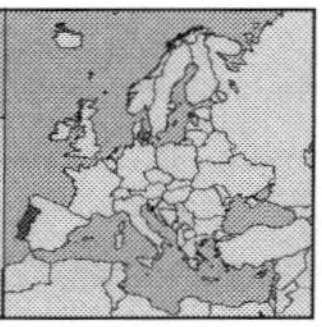

WirelessPT
Miguel Vieira
https://wirelesspt.net

Introduction

Over the years developing WirelessPT, I found that the biggest difficulties people mention are how to start a wireless network, and uncertainty over how people will embrace the concept and, ultimately, adhere to it. Everyone at some point has specific difficulties, but we all share these deterrents.

To give you an idea about the adversity I faced, this was the scenario: I was 5,500 km away from the town of Moitas Venda where the project would be deployed. Moitas Venda is a civil parish in central Portugal, 6.70 km^2 in size and with a population of 866 people. The dominant employer is a local tannery, but the parish also manufactures tarpaulins and textiles, and produces marble used in construction. Despite this industrial dynamism, the parish has not seen a lot of investment, and has little infrastructure to attract new industries.

Due to its geography,[1] the town has always been known for difficulties in implementing projects using radio frequencies. I had only Pedro Maximiano, a long-time friend and project partner, to help me, as I faced a disgruntled community with low tech skills, displeased about wireless services following a previously failed wireless project in which some had invested, and old blocked and damaged hardware from that project that I could reuse. No one wanted to or could spend money on wireless networking anymore. On top of it, I had a prejudice towards wireless networking, and knew very little about it.

I was told by someone in the town that since I was not living there I was clueless about how things needed to be done.

The country context was also hardly amenable to setting up a user-developed and managed community network that, in the case of WirelessPT, drew on the potential of open source software to achieve its aims. Policies in favour of community networks had never existed. The idea of sharing resources in a community was always looked down on with prejudice or at least seen as something that could not make money and was therefore unimportant. Any potential political champions one could find would always want public credit and visibility for their personal brand in exchange for their support, sometimes demanding control and trying to dictate how the project would work.

The obstacles could be summed up as: no funding, archaic politics, isolation, personal egos, selfishness and no support. And without help, for the end-user the investment would be costly.

At least – and although with some boundaries – the 2.4 and 5 Ghz spectrum was free and available for use by citizens.

What exactly is WirelessPT?

WirelessPT.net is an open source mesh network project created, developed and registered by me. In a mesh network, all routers and/or routing devices that are added to it will automatically communicate between themselves in a similar fashion to how the internet works, in this way creating an identical communication system like the internet but without cables or wires or the need for human maintenance. This kind of network concept allows us to create a network and expand it every time we activate another router nearby.

A key advantage is that it does not cost much to develop and implement. At the same time, the network cannot be completely shut down since no one controls all routers or access points and there is no central control point. If one access point goes offline, the network automatically reorganises itself without the dead access point.

The router or routing device used is exactly the same type of router that we can buy in any regular computer store. However, we use a very specific type of software and firmware developed by me which I have named "mvwrt". This software makes the network "self form" and "self heal" in case of any connectivity rupture – any router added to it will automatically detect the rest of the nearby mesh network, and add itself to it while simultaneously expanding it.

The more routers the network has, the wider its coverage will be. Some of these routers can be or will be connected to the internet and in this way all WirelessPT.net users will also have internet access.

1 It is situated between several small mountains and is elevated compared to other towns.

This type of escalation has been successfully implemented in other places such as Moita, near Marinha Grande, and Videla, which established a 3.5 km link to Moitas Venda with up to 6 Mbits at 2.4 Ghz.

The project is designed to be deployed in a similar fashion to a franchise system where every new local network will follow the documentation on the project wiki and will have its administrator as the first "node owner". A node owner is any person that at his or her own expense sets up a wireless access point and either joins or initiates the project in a new location and who will abide by a user and participant-specific peer agreement.

At the time of writing, only two open source, community-managed wireless projects had been successfully implemented in Portugal. Only WirelessPT is still actively maintained and regulated by a comprehensive user agreement. The other community network has always been closed to a group of developers who administrate it like a private internet service provider (ISP) and have never made it attractive to the end-user.

Making the project relevant to the community

In 2009 I realised that although one or two people could execute the project, I needed to make it open and attractive to people who were not tech-savvy and for whom internet access would be for watching YouTube, updating social media accounts, and using email and voice services (VoIP). So I decided to build an online community using the moitasvenda.net URL to promote the town. This became a portal with a forum, a chatroom where users could communicate, and an encyclopaedia where information about the town's history, culture, traditions, memories, its past and its people were published.

Then a small section for the wireless network was added.

During the next two years a plan was drawn up. I was going to develop an independent and non-commercial wireless network to provide free-of-charge, open and democratic access to the highways of information technology among communities. I wanted to allow those without access, due to poor telecommunication infrastructure or limited financial resources, to gain access to the internet through a cheap, affordable method of telecommunication.

All this would be built by the ordinary citizen, using everyday materials, low-cost equipment owned by the users, and any available resources.

I decided to develop an open source software solution for the network, and also build an online training platform specifically for the wireless community. The training platform includes everything needed to learn, educate and train anyone who becomes interested in expanding the network in its franchise-type system to other towns. Ultimately the platform would be used to unify all community wireless networks in Portugal.

Getting your feet wet, even if you can't swim

My first trip to Moitas Venda to start the initial deployment was the hardest. I had only three weeks to fix and deploy old broken hardware that was left abandoned by the previous community wireless project, and I had no skills or knowledge on how to manage it. We also had to talk to a number of disgruntled wireless users and investors who had been involved in the previous project – it had failed to ensure sustainability and basic functionality in the town – and we had to convince them to try something different.

In January 2011 – with the assistance of Nuno Carvalho and the extremely crucial help from Pedro Maximiano who always helped set up access points and promoted the project over the years by talking to people in person, by email, using social media and on forums – the network was launched. The technical specs for the network were the following: five broadband fixed-line nodes with access to the internet served nine wireless routers flashed with a DD-WRT operating system, which then distributed the connectivity using WDS Wi-Fi links in the 2.4GHz band via 9-15 dBi omnidirectional antennas.

Thanks to Pedro, all the participant members who had invested in the previous wireless project and bought hardware to set up their nodes decided to be part of WirelessPT. It was for them a way to have internet access for their families at the cost of the hardware investment they had already made and was already installed at their homes. Three of the nodes we set up were owned by local businessmen and women who wanted to extend their home internet connection to their stores, or vice versa.

Given the initial low resources available, the network was open to a limited number of people: only to the node owners, their family and friends and whoever contributed in any way to the project. The initial average bandwidth was 5 to 10 Mbits, allowing the community to browse sites like YouTube. Later, an accidental connection of 3 km was established to a laptop using a wireless USB pen which achieved 1 Mbit.

I left the country in the same month as the launch, and during the next two years the network performance was audited.

Building the network community node by node

Phase one was a success and gave me valuable information for future development. For example, as soon as the user count increased to an average of 30 users, or a gateway would go offline, occasional technical problems would happen which revealed the DD-WRT and WDS were limited. Manual administration to fix problems became a burden, hard or impossible to do when remote access was lost.

Simultaneously, a few services that had been implemented, such as IRC chat, a discussion forum, a mail server, a network-attached storage system, a community wiki, VoIP, torrent and even a second-life server, were removed since the network and users were not ready for some of them.

By the end of 2012 a new trip was scheduled to Moitas Venda, and based on lessons learned over the preceding two years, I decided to develop the training platform parallel to the town's previous digital community, but now just for the wireless network and with the objective of sharing my knowledge to provide information and training on how to deploy a wireless network node, whether for a complex solution or do-it-yourself for non-tech-savvy people.

Plans for 2013 had challenges. New firmware and routing protocols were needed in order to eliminate the need for human administration given the lack of expertise in the community. Setting up a node should be possible by non-technical people. Hardware had to be replaced, meaning higher costs, which, given the financial possibilities of the community, and non-existent external financial resources, would be a problem.

Eventually I found the perfect cheap hardware, but at a "cost". I was able to find people selling their home wireless routers on internet community sales sites, many times for a fraction of their store price in either Portugal or where I was. Without any funds to start, the investment in hardware had to be done by me in hopes that people would be interested in what I was developing for them. This led me to scavenge the city where I was based for years to meet strangers selling the hardware I needed.

The next step took into account network performance studies, observations and usage tests done during the first two years in order to develop new firmware specifically and purposefully designed for the environmental characteristics of the community. Based on OpenWrt[2] and using Batman-adv[3] as the routing protocol, a beta version named "mvwrt" was developed.

After three weeks of hardware upgrades and replacements at my own expense, in January 2013 I created a do-it-yourself wireless node kit which facilitated non-techies to plug and play it anywhere.

With the new online wireless portal and Pedro's community communication skills, the network got more people interested in participating as node owners as well as users. The nodes were now 300N wireless routers, their count increased to 14, bandwidth went up to 20 Mbits – there was better throughput and even a few monetary donations for hardware from members.

Carefully planned technical administration boosted the network resources and allowed the number of users to increase. With the cooperation of the community, the wireless spectrum usage of all types of wireless routers in the town was organised in order to maximise the effectiveness of possible bandwidth availability provided by the network while minimising radio frequency "overlapping pollution".

Free to the community

It was now time for another remote audit until the next three-week visit and upgrade in January 2014. Based on a year of study and observation with regard to the network performance upgrades previously deployed, this resulted in a new and stable version of mvwrt firmware, which was now 100% plug-and-play and self-managed without the need for manual administration, which was crucial to solve all the local network administration problems previously found.

The node count increased again, as well as the participation of four women who owned their local business and more than one wireless node which they shared with the community.

Other key people in the network who provided valuable donations either in hardware or money and help were mostly women. For example, my mother proved to be an extremely valuable asset in the town, who – without any technical skills, but following a few simple sets of instructions and stickers – was and still is able to help keep crucial parts of the network running as per remote access needs.

2014 was the year that access to the network became freely open to anyone in the community, with the number of users ranging between 50 and 200 at a time.

Plug and play

The next two visits planned for 2016 were going to be crucial due to the characteristics of the town, the network, its participants, as well as my role as the developer.

Two thirds of the node owners live in Moitas Venda. Half of these people own two nodes. These they use to share their internet service with their second owned property, either a business or residence, using a mesh topology as the travel path to overcome distance and obstacles, and, as a result, saving money since they had no need for a second ISP account.

2 https://en.wikipedia.org/wiki/OpenWrt
3 https://en.wikipedia.org/wiki/B.A.T.M.A.N.

A third of node owners live outside of Portugal, using the network when they return for vacation to the town but keeping their node working during the year. The last group of node owners includes me. I own five nodes, one gateway that serves the community, and several IPCams to monitor the gateway location, my property and weather conditions.

This last third is the tech-savvy group that still works on the network when present in the town and which now also includes Pedro and Ruben Vieira, another network member. Given that the project is located 5,500 km away from me, without any possibility of me fixing any hardware problems that the network encounters, new enhancements were required to the physical structure of the network in order to ensure continuous operation without the need for human intervention. These included the use of self-sustained power supply units in case of technical malfunctions in the electrical grid, as well as rewiring equipment to the electrical grid to prevent downtime due to equipment being accidentally unplugged or damaged because of a poor quality plug or faulty power outlet.[4]

As 2016 arrived, and still operating in the 2.4 GHz frequency, new and improved firmware was deployed. The active node count increased to 22 nodes, new and different antennas were implemented, and bandwidth went up to 30 usable Mbits. Later, with the work of Ruben, the project expanded and created a planned 3.5 km link to the town of Videla.

At the same time, and given the growth of the project, I was invited to present mesh technology at the University of Minho and participated at the international Battlemesh event held at FEUP, the faculty of engineering at the University of Porto, presenting WirelessPT and promoting community networks built by the ordinary citizen.

Since I left the country in May 2016, the network has been running completely on its own: it is self-sufficient, self-managed, self-healing and self-maintained. Contact with the community has been mostly online on social media and using the digital platforms mentioned before. Now the deployment of any wireless node kit can be done by anyone just by plugging it to an electrical outlet.

In 2016 WirelessPT became a registered trademark and after seven years of operation it still performs as planned and all due to one ingredient: "If there is a will, there is a way!"

4 Hiding power and network cables from humans proved to be very effective in preventing something being unplugged by mistake. In some places, the only way to shut down an access point is by shutting down electricity in the whole house or specific rooms. If a router needs rebooting, the node owner is told to shut down and turn off the electricity at the main electrical supply source. The node owner will not have to look for any plugs anywhere.

Considerations to take into account

Over the years and despite competitive, closed projects and new technologies, open and self-managed community networks still have a place in our society, and great potential in rural regions, even without progressive or supportive country policies.

The key to success when developing an independent community network is not to overload the community with all the bells and whistles, but simply to listen to their needs and how those needs can be met with new, open and shared technologies and resources at a very low cost. Expert developers are not the most important ingredient, and in fact they can even scare the community and lead it to reject the project due to fear and a lack of understanding of the project's technical aspects.

What is imperative is to have trustworthy, key people in the project who are immersed in the community, and are able to engage the population in common community interests, even if these are not about technology, but about the use of technology as a path to achieve a common goal. This will ensure the project's sustainability.

Action steps

In order to succeed, community networks need to develop ways to:

- Be open, transparent and enjoy it.
- Be revolutionary and dare to take a chance.
- Educate and engage communities about resource-based shared economies.
- Envision new ways to achieve a better sustainable society at all levels.
- Be active in causes that are important to the community.
- Learn from other network implementations and do not be afraid to re-invent the wheel.
- Look for individuals who are passionate about relevant interests of the community.
- Work with local businesses for mutual gains or in partnership in order to have their participation in the project.
- Have members engaged in local politics, either as constituents or even as candidates running for office who, if elected, will have a stronger voice that can make a difference to help support these types of community projects.

Perhaps the most important thing that needs to be urgently done is to have the European Union develop regulations that work for community networks.

SEYCHELLES

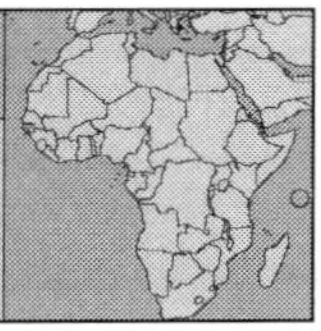

WHY ARE THERE NO COMMUNITY NETWORKS IN THE SEYCHELLES?

Janick Bru[1]

Introduction

There are no community networks in the Seychelles, an archipelago whose population is found essentially on four granitic, mountainous and densely forested islands. The telephone and internet services for fixed points or mobile phones offered by private telecom companies and internet service providers (ISPs) reach virtually all corners of these islands.[2] There are currently 33,000 internet subscriptions in the country, nearly all of them broadband, for around 97,000 inhabitants[3] living in about 25,000 households. Statistics show that 58% of the population are regular or intermittent users of the internet.[4] This means that there is still a significant group of people who do not have access to and may not use the internet, although many may own and use mobile phones. It is believed that this segment is composed of i) the elderly, ii) those with limited education and/or social problems, and iii) those who cannot afford to pay for the internet.

The rate of development in the Seychelles is rapid, and reliance on information and communications technologies (ICTs) is also increasing rapidly. Could community networks be helpful for the 40% or so of the population with no ready access to the internet and the wealth of information available there? This report looks at the situation in order to understand why there are no community networks in the country.

Legislative and regulatory context

According to the Seychelles Broadcasting and Telecommunication Act of 2000, no one is allowed to "provide a broadcasting service, or a telecommunication service, except under, and in accordance with, a licence granted under the Licences Act."[5] The licence specifies:

- The radio frequency or frequencies allocated to the licensee.
- The description of the antenna and transmitter to be used.
- The geographical area in which a mobile transmitter, where applicable, may be used.
- The location of the antenna and fixed transmitter.
- The obligation, if any, to share the frequency allocated with any other person.[6]

These regulations were somewhat softened by the Policy on the Use of Fixed Broadband Wireless Access (FBWA) Frequency Bands (FBWA Policy) of 2004, which set out requirements for operating in the 2.4 GHz (2400-2483.5 MHz) and 5.7 GHz (5725-5825 MHz) frequency bands. The Department of Information Communications Technology (DICT) says in the introductory paragraph of the FBWA Policy that it was "created to allow other users to operate low powered[7] short range devices, in the two frequency bands on a licence-exempt basis, whilst protecting the high powered FBWA operation of the two Internet Service Providers (ISPs) in these bands."[8] It adds that the purpose is so that "all ISPs [...] can contribute towards the deployment of wireless access nationwide." It includes the "general public" under its definition of "other users". These provisions are nonetheless subject to Part 2, which lists several points emphasising that any such service must "be operated according to the operational requirements set by" the DICT.

In practice, a representative of the Seychelles Licensing Authority (SLA) said that anyone who wanted to offer a broadcasting or internet-related service needed to prepare and submit a project to the DICT, which would determine the acceptability of such a project. If the project is approved, then the DICT informs the SLA as to the kind of licence that

1 The author liaised with the National Institute for Science, Technology and Innovation (NISTI) and the Seychelles Community Training Institute (SECTI) for information about their projects.

2 Although the quality and speed of connection can vary according to location.

3 Zoom. (2018). Seychelles Broadcasting Corporation. This TV programme informed viewers that for 2017, population estimates include 17,500 foreign workers.

4 www.ict.gov.sc/ReportsStatistics/Reports.aspx

5 www.ict.gov.sc/documents/BTA2000.pdf

6 Ibid.

7 According to information received, this expression refers to equipment emitting "low frequency" signals under 900 MHz.

8 www.ict.gov.sc/documents/FBWA%20Policy.pdf

should be issued. Details of the approval process are not in the public domain.

Collaborative efforts in the Seychelles in the area of ICTs

There are some community access initiatives in the Seychelles that seek to use the potential of ICTs to improve the lives of ordinary citizens. A few of these could, as a next step, consider setting up a community network to further strengthen the community's engagement with, and appropriation of, technology.

There are a number of local community initiatives in the country but they all operate in social areas, where the only formal requirement is to register as an association. To complete registration, the group needs a membership of at least two individuals who agree to work together towards a common goal. Such groups include the Cancer Concern Association (for cancer victims/survivors and their relatives), Night Pastors (for people on the street), and Friends of Prison (to contribute to the rehabilitation of prisoners), to mention a few among the numerous ones that exist.

For example, while researching this report, I was introduced to a recent initiative that aims at providing farmers with access to market information and helping them source buyers. The idea stemmed from a conversation that Manfred Laporte, one of the initiators, had with two friends who are farmers and who were finding it difficult to sell, or find places to sell, their products. Laporte, who owns a UK-based IT company, later met a representative of the Seychelles Agricultural Agency (SAA) and together they approached the Seychelles National Institute for Science, Technology and Innovation (NISTI) for support. They felt that building a knowledge-based economy, as promoted by the Seychelles government, required the existence of platforms for knowledge sharing.[9] The team is currently working on a platform that will link farmers and producers with buyers of agricultural produce while maintaining records of the availability of various items in real time. They are receiving the active support of the DICT which, according to Laporte, has essentially agreed to provide funds for the subscription needed for hosting and operating the platform. This subscription will cost approximately SCR 300,000 (about USD 22,300) annually.[10] As I suggest below, this is a crucial commitment from the government to a grassroots-initiated ICT project, because cost is a key barrier for any ICT initiative to be developed in the Seychelles.

The project is still being developed and Laporte hopes that the first phase will be operational within the third quarter of 2018. Farmers and others should be able to use their mobile phones to access the platform and work offline in order to keep costs low. Laporte states that a number of farmers have agreed to be part of the pilot. "This is now a community effort with members from SAA, NISTI and the farming community," he says. It remains to be seen what steps will be taken to continue building momentum within this community in order to attract and, more importantly, maintain a high level of participation.

The Seychelles Community Training Institute (SECTI) is an NGO which has developed a social action project to provide basic skills and knowledge to specific groups of citizens so that they become self-dependent "instead of continuing to depend on the state social funds."[11] One of the specific objectives is to "strengthen or encourage the use of ICT" by developing ICT skills that can be used for various purposes including small projects at the community level.[12] The SECTI's ICT projects target unskilled young people, those with social problems (including drug addiction)[13] and the older population. Regarding the last group, the organisation feels that ICTs are alien to many individuals aged 40 and above, and that this causes them considerable fear and anxiety. The SECTI's overall goals may be somewhat broad, but its objectives for e-literacy and skills development are specific: they wish to offer training to introduce the use of ICTs to members of the community who would otherwise not have the possibility to acquire such knowledge and skills.

Marie-Nella Azemia, the director of the SECTI, argues for the importance of unconnected communities getting online:

> They need to have access to information and for this you need to use all relevant tools, the internet, mobiles and all apps, to keep abreast. In today's world, everyone, without exception, needs to know what is going on. You need to

9 Seychelles Nation. (2018, 20 April). Farmers Digital Engagement Platform: An understanding. www.nation.sc/article.html?id=258535

10 SCR – Seychelles rupees. Approximate exchange rate: 1 USD = SCR 13.40 in May 2018.

11 SECTI Think Big Project.

12 Ibid.

13 An article from May 2018 gives the results of a recent study showing that the population of heroin users in Seychelles aged 15 years and above is around 4,800, which is equal to 5.6% of the population. Laurence, D. (2018, 5 May). Alarming drug results for Seychelles: survey shows 6 pct of population has used heroin. *Seychelles News Agency*. www.seychellesnewsagency.com/articles/9090/Alarming+drug+results+for+Seychelles+survey+-shows++pct+of+population+has+used+heroin

network and to have contact with other people, organisations and communities which are in different localities. If you are not connected, if you do not understand what is going on, if you cannot talk and engage with others then you are lost. [...] All people need to be empowered so that they take their rightful place in society. This means being able to engage, being able to understand, being able to participate and to contribute – for the development of yourself and of the community.[14]

While there is no experience of community networks in Seychelles, both Azemia and Laporte believe that such networks could be useful for their projects. According to Laporte:

> Community networks can bring stakeholders closer in order to achieve common goals, for example as for a health community forum or, in this case, a Seychellois farmers' community for buying and selling agricultural products, and to contribute to the development of agriculture. It is a convenient way of bringing together all sides to create viable economic progress.[15]

Laporte also highlights the fact that while there may be well-developed infrastructure for ICTs in the country, there is actually very little relevant local information online, and until this is rectified, people in the Seychelles cannot fully benefit from associating at a local level.

Other difficulties highlighted are the slow pace of bureaucratic processes for implementing such ICT projects and the perception that these processes are somewhat opaque. At this point, though, it is felt that the main hurdle for connecting and for accessing information in the Seychelles is the cost.[16]

Key challenges faced in setting up community networks

Overall, the factors that are seen as the main deterrents to the establishment of community networks in Seychelles are:

- High prices for anything ICT related – equipment, services, maintenance
- Regulations regarding equipment and operations that are not always clear and may sometimes seem contradictory.[17]

Despite the wide coverage offered by commercial ISPs, it is evident that areas of need regarding access to the internet still exist, unrelated to geography and location.

The hurdles to overcome, including the cost of services and the cost of equipment as stipulated by existing regulations, might seem too high for a group of average citizens to choose to venture into establishing community networks, and perceived benefits from such networks may not seem worthwhile. The current lack of relevant local content available online is certainly no encouragement – even though developing this content could be a result of setting up a community network.

Benjamin Sonon, who operates a computer services company, believes that it is difficult for individuals to cooperate in ventures such as community networks for various reasons, the main one being the general socioeconomic climate which does not lend itself to cooperative effort, especially when resources, which are scarce for the average person, have to be invested.[18] Daniel Socrate, a Seychellois telecom professional,[19] considers that the absence of community networks in Seychelles represents a lack of perceived need rather than anything else. He feels that while there are technically competent people who could deliver "traditional" cooperative solutions, there is no interest in linking up as communities via such solutions because, in his view, people see neither the value nor the potential of such traditional cooperative efforts.

Another drawback mentioned by some is the fact that the main service providers in the country, fiercely commercial entities, have considerable control over the facilities that exist, as they are joint owners of the fibre optic cables that link the Seychelles to the rest of the world.[20] One individual referred to a major ISP as a "monopoly from colonial times".

Socrate's perception, on the other hand, is that "people are linking up as communities via the use of internet through social media platforms such as

14 Meeting with Marie-Nella Azemia and Rose Mary Dogley of SECTI, May 2018.

15 Meetings and email exchanges with Manfred Laporte, May 2018.

16 Prices range from SCR 49 for 150 MB for mobile internet to post-paid capped packages which can cost SCR 5,748 for 100 GB. There are anecdotes about lower-income earners who spend up to SCR 3,000 (about USD 230) a month on communication costs (phone calls, texts, data) despite the fact that their salary may only be around SCR 8,000 (USD 615).

17 For example, one of the individuals with whom this was discussed had difficulty reconciling the idea of "licence exempt" operations with the fact that these have to be submitted for approval and must be granted authorisation to operate.

18 Meeting with Benjamin Sonon, May 2018.

19 Telephone and email exchanges with Daniel Socrate, a telecom professional currently working for a Seychellois ISP.

20 Office of the Vice President. (2010, 11 October). Seychelles to be connected to the world through the Optic Fibre Submarine Cable System. www.gov.sc/PressRoom/DisplayPressRelease.aspx?PRLID=78

Facebook and WhatsApp" and that since "more than 60% of the population have access to the internet via their smartphones, the various groups are linking up through the most effective and modern channels."

Would the existence of community networks benefit communities in the Seychelles? One can argue that they would. They could help to foster greater community spirit and increase collaboration within communities if people were to see the actual benefits – such as lower access costs, the "upskilling" of the community, revenue generation, and an increase in social capital – that could be derived from them. What is necessary is for a pilot project to be set up in the Seychelles to show proof of concept – but for that to happen, it needs government support, and buy-in from the communities themselves.

Action steps

Steps to be taken to promote the idea of community networks in the Seychelles include:

- Raising awareness of community networks and their benefits in the media.
- Fundraising for a pilot project to show proof of concept.
- Approaching policy makers, service providers, and communities themselves to participate in a multistakeholder discussion on the topic.
- Lobbying for a change in the application of regulations that impact negatively on the cost of equipment and services in the Seychelles.

SOUTH AFRICA

CHALLENGING INEQUALITY IN POST-APARTHEID SOUTH AFRICA: A BOTTOM-UP, COMMUNITY-LED BUSINESS MODEL FOR CONNECTIVITY

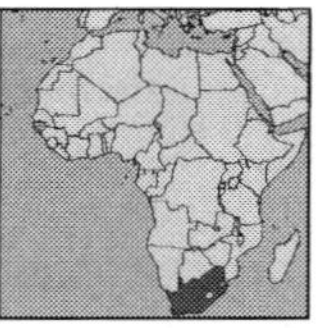

Zenzeleni
Sol Luca de Tena and Carlos Rey-Moreno
zenzeleni.net

Introduction

South Africa is a beacon of industrial capacity and strong political development in Africa. Despite this, the country has still not managed to deal with the devastating socioeconomic legacy of apartheid, and while politically emancipated it has been unable to escape the dubious honour of being one of the most unequal societies in the world.[1] Around 55.5% of the population continues to live in poverty,[2] with this number reaching 80%[3] in the rural areas. Wealth and poverty are largely racialised, with a majority poor black African population often unable to afford daily essentials or access resources, decent education or basic services.

Poverty and inequality in South Africa are also expressed through the digital divide. This divide is a reflection of the historic and current socioeconomic reality, but it also threatens future development: people and communities who lack connectivity, digital opportunities and associated skills and resources will be further marginalised over time unless interventions are made to change this.

Community networks may present an avenue to challenge socioeconomic disparities by offering rural and impoverished communities a chance to own, manage and sell their own affordable, reliable communication services. Zenzeleni – a community network that is based in one of South Africa's poorest provinces – is demonstrating the developmental power of community networks by enabling the creation of local businesses and allowing communities to participate in the development of their own communities. Centred around the specific needs and values of their respective communities, community networks such as Zenzeleni are powerful tools that can offer communities much needed connectivity, infrastructure, skills and revenue. They are a mechanism through which localities and people who are structurally marginalised from the greater economy can participate in the economy as equals, and on their own terms. This last fact – that community networks allow communities to organise themselves according to their own priorities and needs, and distribute their benefits in the same manner – means that community networks offer an appealing alternative to top-down development and public policy.

This report discusses the context, approach and challenges faced in setting up Zenzeleni in an attempt to inform and encourage others to learn from and expand on its experience.

Policy, regulatory and telecommunications context

South Africa's telecommunications regulatory framework includes a number of concessions which can be used for the purposes of greater social inclusion. These exemptions are significant in that they create some flexibility for community networks to establish themselves with fewer bureaucratic burdens and costs, and enable them to play a role in the telecommunications sector. Perhaps the most important is the formal licence exemption for operating certain types of telecommunications infrastructure and services. Zenzeleni has made use of this exemption.[4]

South Africa also offers a licensing exemption with regards to Wi-Fi and, recently, television white space (TVWS) spectrum, also allowing for higher radiated power than in many other countries. In the event that a community network runs out of channels in its licence-exempt spectrum, the national regulator, the Independent Communications Authority of South Africa (ICASA), also allows a very flexible and low-cost fee structure in other bands that could be used in backhaul links.

While crucial, this flexibility is nevertheless not sufficient to allow communities to establish

1. Sulla, V., & Zikhali, P. (2018). *Overcoming Poverty and Inequality in South Africa: An Assessment of Drivers, Constraints and Opportunities.* World Bank. documents.worldbank.org/curated/en/530481521735906534/Overcoming-Poverty-and-Inequality-in-South-Africa-An-Assessment-of-Drivers-Constraints-and-Opportunities
2. Using the upper-bound poverty line (UBPL). Statistics South Africa. (2017, 22 August). Poverty on the rise in South Africa. www.statssa.gov.za/?p=10334
3. Nassen Smith, M. (2017, 16 October). "Confronting Inequality": A report on IFAA Conference. *Institute for African Alternatives.* https://ifaaza.org/2017/10/16/confronting-inequality-a-report-on-ifaa-conference
4. South Africa does not have specific legislation for community networks.

community networks. Several other obstacles pose a barrier. For instance, communities in rural areas are unlikely to know the details of national telecommunications policies, or even if they do, they are unlikely to understand the complex language of public policy and regulations, or the technical processes needed to access opportunities. Such communities may also not have the access to resources to allow them to travel to regulatory offices to complete application forms, or to contract the legal services needed to ensure they are compliant with various regulations, or the technical services needed to establish the infrastructure to launch their community network. A complex set of knowledge and skills are required to understand and exploit these opportunities.

For its part, Zenzeleni has had a wide range of collaborators who stepped in to assist in overcoming these obstacles. They include organisations such as the University of the Western Cape (UWC)[5] that helps build evidence, Ellipsis Regulatory Solutions[6] which offers expertise to navigate the regulation landscape, the Association for Progressive Communications (APC)[7] and the Internet Society (ISOC)[8] who provide support through their networks and resources, and other community networks around the world with whom to share experiences, among others. These collaborations have raised the profile of Zenzeleni, accessing assistance for it at national and international levels. Their joint advocacy has been critical in getting community networks recognised as mechanisms for positive change and connectivity in South Africa, and it is partly through their advocacy that in May 2018 the South African government's Department of Telecommunications and Postal Services (DTPS) formally indicated its intention to collaborate with and support Zenzeleni during its annual budget speech in parliament.[9]

Zenzeleni: Generating capital and opportunities through connectivity

Zenzeleni currently provides affordable, reliable connectivity within several communities in the rural Eastern Cape province of South Africa, specifically in Mankosi, Mcwasa, Nomadolo and Zithulele. Like many other areas in the Eastern Cape and other parts of South Africa, these communities were deliberately and systematically underdeveloped by the racist and oppressive colonial and, later, apartheid regimes. They are characterised by extremely high unemployment, deep levels of poverty, high rates of out-migration by people of economically active ages, and a lack of economic infrastructure beyond the presence of general stores (which stock very basic products) and some limited tourist accommodation. The result is that today, more than 20 years after the fall of apartheid, unemployment in the area is around 98%.[10] Most residents live on around USD 1 per day, relying on government old-age and childcare grants or on remittances from relatives working in urban areas.

Telecommunications constitutes a major portion of the monthly expenses of these village residents – around 25% of monthly costs, according to a five-year study by UWC.[11] This is partly a consequence of the need for telecommunications in a context where families are so dispersed around the country, and also because the telecommunications tariffs for the area, charged by formal service providers, are some of the most expensive in the country.

To illustrate: residents typically buy the smallest prepaid voucher, ZAR 5 (South African rands)[12] at a time, which equates to around five megabits (MB)[13] of data, or six minutes of call time. Better rates can be accessed by paying for moderate monthly contracts – such as ZAR 199 for five gigabits (GB) or ZAR 499 for 20 GB[14] – but both are unaffordable to local residents. The costs are exacerbated by the fact that vouchers in such areas retail through local distributors at a 40% margin (hence a ZAR 5 voucher is actually sold for ZAR 7). Residents also have to pay an additional ZAR 5 to recharge their phones – possibly at a neighbour's or local shop – as many do not have electricity at home. Yet once they have purchased vouchers and charged phones, residents then struggle with unreliable network coverage.[15]

It is estimated that 15 communities within the broader area around Zenzeleni collectively spend over ZAR 20 million (over USD 1.5 million) on

5 https://www.uwc.ac.za

6 https://www.ellipsis.co.za

7 https://www.apc.org

8 https://www.internetsociety.org

9 Ndabeni-Abrahams, S. (2018, 17 May). Budget Vote 32 Address by Ms. Stella Ndabeni-Abrahams, Deputy Minister of Telecommunications and Postal Services. *Africa Newswire*. https://africanewswire.za.com/deputy-minister-stella-ndabeni-abrahams-telecommunications-and-postal-services-dept-budget-vote-2018-19

10 https://wazimap.co.za/profiles/municipality-EC155-nyandeni

11 Rey-Moreno, C. (2015). PhD dissertation, Rey Juan Carlos University.

12 The currency conversion at time of writing was ZAR 1 = USD 0.08, so ZAR 5 is USD 0.40.

13 Costs from field research and compared with Stolp, J. (2017, 16 May). Feeling ripped off about airtime? You probably are. *Fin24*. https://www.fin24.com/Tech/Opinion/feeling-ripped-off-about-airtime-you-probably-are-20170516

14 www.shop.mtn.co.za/crs/browse/productFilter.jsp?categoryName=SIMOnly

15 Rey-Moreno, C. et al. (2016). An in-depth study of the ICT ecosystem in a South African rural community: unveiling expenditure and communication patterns. *Information Technology for Development, 22*, Supplement 1, 101-120. https://www.tandfonline.com/doi/full/10.1080/02681102.2016.1155145

telecommunications annually.[16] Much of this capital is not retained locally: it flows back to large telecommunications companies in urban centres, leaving little capital or other value within the impoverished communities.

Zenzeleni, however, creates local internet service provider (ISP) businesses that allow the communities to retain their spending internally, and hence enables them with affordable access to digital resources and a local source of income.

Building up a community network

Zenzeleni ("Do it yourself" in isiXhosa) was born in 2013 through a friendship between a UWC doctoral student doing action research in the area and a local community activist. It led to a partnership between UWC and the local tribal authority of the Mankosi community. The project was co-developed and co-created over a six-year period, during which local buy-in and support were nurtured and developed. The project has evolved by placing the community at its centre, and by responding to the community's self-defined needs and opportunities.

At first Zenzeleni was simply a local wireless intranet providing free voice services between analogue phones connected by solar-powered routers. It later included an external connection to the internet via a 3G modem to enable these phones to make calls to national numbers. During the process, the community recognised the opportunity for the solar station powering the routers to also charge phones, and they began to offer mobile phone charging services from these routers at a cheaper rate than the local retail shops (ZAR 3 rather than ZAR 5). It proved a valuable opportunity for the community to generate income.

These interventions resulted, among other things, in a training programme where 12 young local people were guided to use the internet to apply (successfully) for national tertiary education grants. It was a massive achievement in an area where completing secondary education is rare.

Until this point all of these interventions were conducted under the auspices and funding of UWC research. Parallel techno-economic and social studies were undertaken to gauge the barriers that the community experienced in accessing, using and benefiting from telecommunications.

In 2014 Zenzeleni established a local cooperative with full ICASA licence exemptions to operate and offer communication services. However, efforts to access towers – both public and private – and to collaborate with existing telecommunications networks to access fibre were unsuccessful.

Yet other successes were achieved: in 2016 Zenzeleni was recognised internationally and received an ISOC Beyond the Net grant;[17] in 2017 it was a finalist in the Mozilla Equal Rating Innovation Challenge;[18] and also in 2017 it was awarded the South African national award for Best Innovation with Social Impact.[19] Funds from these awards allowed Zenzeleni to create its own wireless backbone; at first its internet connection was to the National Research and Education Network via the University of Walter Sisulu (60 km away from Mankosi as the crow flies).

At the end of 2017, Zenzeleni secured its first private sector client (or anchor tenant), the local branch of a large corporate – and through a like-minded company, Zenzeleni was able to secure uncontended access to a fibre connection. Apart from servicing its four communities, Zenzeleni now offers internet to seven local businesses and three schools.

Currently, Zenzeleni's solar-powered, wireless network can carry up to 200 megabits per second (Mbps). While there are still limited connection points (or hotspots) to the network, the average monthly traffic has been 1.5 terabytes (TB), with over 5,000 different devices connected to the network. Since devices are often shared by several people in a household, the actual number of users is higher. Recently Zenzeleni has started selling uncapped data vouchers which are active for a month. This voucher model is still being tested. Connectivity for local businesses and schools occurs at a monthly fee.

Governance structure: A tool for integration and development

As Zenzeleni has grown, it has evolved two parallel functions: an umbrella non-profit company (Zenzeleni), and the local community-owned and operated ISP (the Mankosi cooperative).

The members of the cooperative are elders – men and women – from different community villages. Among other things, they decide who hosts mobile charging stations and hotspots, as well as who sells the vouchers. They have met monthly since 2013 to understand and shape Zenzeleni, and to use it to contribute to the development of their communities and families. In local Xhosa culture the elders

16 Based on extrapolating the numbers in the conclusion of Rey-Moreno, C. et al. (2016) to the population of the communities in Nyandeni (www.wazimap.co.za/profiles/municipality-EC155-nyandeni) and King Sabata Dalindyebo (www.wazimap.co.za/profiles/municipality-EC157-king-sabata-dalindyebo).

17 https://www.internetsociety.org/beyond-the-net

18 https://challenge.equalrating.com

19 University of the Western Cape. (n/d). Zenzeleni Project Wins Social Innovation Award For Connecting Rural Community. https://www.uwc.ac.za/News/Pages/Zenzeleni-Project-Wins-Social-Innovation-Award.aspx

are responsible for caring for their community's well-being, and this social vision is as essential to Zenzeleni's sustainability as the economic and technical aspects. The cooperative's constitution clearly states the intention for the business to develop the area and grow opportunities for their children. In the words of one of the cooperative's directors, who spent his life working in the mines: "I didn't know that you could get a little bit of money without having to go outside of the community to work hard. That is the future I want for my children".

The non-profit company acts as a support mechanism for the cooperative and as a bridge between the local community and the established telecommunications sector. It has four directors drawn from a range of relevant fields and experiences. Their role is to guide the cooperative through a myriad of issues including acquiring licence exemptions, brokering deals, accessing the relevant technology, building resources, gathering evidence, advocacy, managing the network, accessing funding, and creating partnerships, among other things. These tasks are all necessary for Zenzeleni to operate in the telecommunications sector; yet they are totally removed from the daily lives of the community. In rural South Africa, where there are often no services, even the simplest tasks take days of travel and frustrations with systems which do not cater for communities. The intention is that one day the cooperative will have the skills, resources and access to online services to undertake these tasks autonomously. For now the non-profit company is required as a support structure.

At this stage the cooperative generates enough income to pay for its own bandwidth, replace infrastructure and grow its network by adding more access points. As it expands, so will the revenue available to the cooperative. The non-profit company has until recently run on a volunteer basis, with intermittent support from grants. Zenzeleni's ecosystem (a non-profit company and different cooperatives providing internet access) will reach sustainability when several cooperatives serving different communities within regions contribute a fee towards maintaining a network that is treated as a common-pool resource and managed by the non-profit company.

This year Zenzeleni will test its ability to scale towards sustainability by expanding to a new area and seeding a new cooperative. This is partly funded by the South African Department of Science and Technology and the Technology Innovation Agency, and will allow Zenzeleni to undertake a more systematic – and resourced – incubation process. Key to this is setting up systems and transferring skills for cooperatives to take a further step toward autonomy. Here there is a strong emphasis on the inclusion of women in taking the lead in the new cooperative and also further recognising their role within the Zenzeleni ecosystem. Another key focus area for Zenzeleni is defining its common-pool resource (or commons) approach[20] to managing its network infrastructure.

Essentially, the commons is an approach through which the network assets are jointly owned by all who utilise them. New users can connect to the existing network, and the costs of maintaining and upgrading it are shared. This is significant, as the start-up and fixed costs of a network are very high, whereas those of increasing bandwidth or expanding a section of the network are proportionally much lower. This allows users to avoid duplicating infrastructure and services. It also allows them to access economies of scale as a large body made up of smaller individual entities. Since fixed costs are shared among all users, as more users join the network, costs become increasingly lower for all.

Partnerships with various players in the telecommunications sector are important to increase the efficiency and benefits of the commons approach. There has been overwhelming interest in Zenzeleni from other communities in rural and urban townships[21] as well as from local businesses and institutions. The integration of different players, with communities at its centre, is an opportunity for positive social, racial and economic integration in the South African context. However, one of Zenzeleni's challenges is to offer a system that promotes benefits for all, but also accounts for socioeconomic differences in the country so as not to further propagate these.

A model for co-creation

Zenzeleni's community-centred model holds significant promise to address the digital divide. In these rural and impoverished areas, traditional technology interventions fail. For instance, donations of computers to schools in the area have seen the technology remain unused, as people do not have the skills to use them, or due to a lack of electricity or internet; or they are stolen soon after being donated, since such technology is a valuable resource in an impoverished area. The failure of such interventions, however, is often due to poor or non-existent consultation with the so-called beneficiaries. Zenzeleni has been different. It is a model of slow co-creation with and through the

20 Zenzeleni bases its common-pool resource management on the model developed by guifi.net, which services more than 34,000 active nodes. https://guifi.net

21 South African term for informal settlements found outside cities where people often live in tin shacks and where there are few services.

community, and its network has been cared for and used.

Integration will require time and recognition of different ways of working. Zenzeleni offers hope that there is an alternative to isolation and marginalisation. Nowadays it is common to see people huddled around Zenzeleni hotspots: Bongani,[22] a boy watching soccer and analysing each players' moves intently to "up his skills"; young Lundi's face suspicious at first, then lighting up when shown how to access maths learning support videos; Sipho, a keen local rapper, searching the online music world and finding something of himself in others. Sharing content that reflects the identity of the amaXhosa is a next step... each one a step toward a more connected and integrated South Africa.

Conclusion

Zenzeleni's case shows how economically marginalised communities that otherwise depend on government grants and remittances can, through a community network, both access and own high value services in South Africa. Internet connectivity offers access to resources that are otherwise not available in these communities, and the new ISP business provides poor people with a new income stream.

Through ownership, transparency and collaboration, Zenzeleni also shows that technology can be readily accepted and integrated into rural life. It has the potential to offer equal resources to all people of South Africa, at least online. However, the effective uptake of such technology depends foremost on input and acceptance from the community. It is important to move slowly, to build trust, acceptance and knowledge, and to seed and incubate the ISP businesses.

Collaboration has also been important to overcome the multiple legal, technical, financial and social barriers experienced by communities. An approach responsive to challenges, research and a multidisciplinary team have been critical to the success of Zenzeleni.

All of the above has not happened without challenging the status quo in telecommunications and norms around development and transformation. But then, things cannot change and stay the same.

This year Zenzeleni will test its model by scaling to new areas. We are working hard for this vision, and following in the words of another local cooperative director: *Sifuna uZenzeleni anwenwe nje ngomlilo* – "We want Zenzeleni to spread like fire."

22 Although these examples are real, the children's names have been changed to protect their privacy.

Action steps

The following key steps are necessary to strengthen Zenzeleni and community networks generally in South Africa:

- *Implementation and growth:* Zenzeleni needs to answer the pull from communities that want their own community network. This will require some seed funding and a bigger support team (seed funding, partnerships and volunteers are welcome).
- *The importance of local relevance and international collaboration for community networks:* Core to the sustainability of the community network is maintaining its relevance to the local people and context. This requires ongoing reflection and work. Likewise, participating in community networks regionally in Africa and globally offers us a chance to develop a robust system of peer-to-peer support and sharing of experiences to sustain our community network model. This requires ongoing participation in and nurturing of international people networks.
- *Collaboration with the South African government in its policy response to community access:* It is important for community networks in South Africa to engage with the government in its policy response to local-level access. This will help create more innovative and responsive policies that enable the potential that community networks offer for South Africa and the greater region.
- *Enabling communities to access infrastructure, spectrum and regulation:* A key regulatory change relates to infrastructure sharing and cutting red tape. Sharing public and private telecommunications towers, services and spectrum – and simplifying the bureaucracy to access them – will increase the potential for communities to deploy their own community networks, and amount to significant resource and cost savings for communities and the sector.
- *Engage and support the community:* Finally, it is worth restating a simple fact about community networks: they are for the community. Many of the past community access programmes such as telecentres or e-schools have failed due to a lack of consultation with their beneficiaries, poor participation of the communities in the initiative, and a lack of proper technical and other support. Programmes need to allow communities to determine their priorities and enable them to own and operate their community assets – they should not just be passive customers paying for access to the internet. Long-term capacity building and mentorship support are essential.

TANZANIA

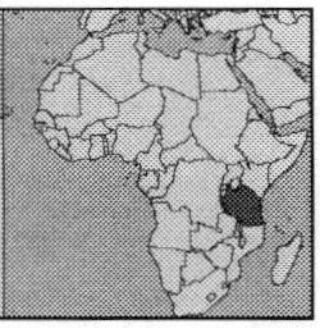

PILOTING THE USE OF TV WHITE SPACE FOR COMMUNITY NETWORKS IN RURAL TANZANIA

The University of Dodoma, College of Informatics and Virtual Education; Internet Society
Jabhera Matogoro in conversation with Jane Coffin and Colin Muller
https://www.udom.ac.tz; https://www.isoc.org

Introduction

The Kondoa Community Network (KCN) is the first community network to pilot the use of television white space (TVWS) in a rural area to address the issue of the internet gap in Tanzania. Jane Coffin and Colin Muller from the Internet Society spoke to Jabhera Matogoro, assistant lecturer at the College of Informatics and Virtual Education of the University of Dodoma, who set up the network. The overarching objectives of this project are two-fold: firstly, to pilot the use of TVWS as an alternative solution to deliver wireless broadband in rural Tanzania; and secondly, to encourage a bottom-up approach to connect the unconnected in Tanzania through the community network model.

Colin: Could you tell us how long the Kondoa Community Network (KCN) has been operating and how long it has been an idea of yours and everyone else working on it?

Matogoro: Actually we started working on TV white space (TVWS) concepts in 2014, and then in 2017 we undertook experimental spectrum measurement to understand how the ultra high frequency (UHF) spectrum band is being utilised in Tanzania, especially after analogue to digital migration. In early 2018, a team of two members from the University of Dodoma visited Kondoa District to engage the community in setting up a community network to connect the unconnected in the district. The team conducted a sensitisation and awareness workshop with community members around Kondoa so that they could own the project and later be able to sustain it in the future. So, in short, studies on TVWS started in 2014 but the official operation of KCN started in May 2018. The official operation in this context means when internet access was made available to KCN. In a measurement study[1] that we conducted it was found that there is a huge potential for using the UHF spectrum band, especially in rural areas where no transmission is currently available. The installed wireless link achieved an internet speed of 4.53 Mbps and 4.83 Mbps for download and upload respectively.

Colin: So it started as an idea in 2014, and then did you give it a name and formalise it later on? Could you say when you applied for the two-year authorisation letter[2] for experimentation?

Matogoro: Yes, it started as an idea in 2014 when I was working on my PhD research and was later formalised in 2017. It was soon after that that we applied for authorisation. The name was given in our first stakeholder meeting for KCN that was held on 5 March 2018 at Golden Apple Hotel in Kondoa.

Colin: What areas did you survey before you decided on Kondoa, and how far is this area from the closest large city?

Matogoro: Kondoa District is located around 140 km from Dodoma City; it is almost a two-hour car drive from Dodoma to Kondoa. Before selecting an area to host this project, a physical visit was made to three districts in Dodoma, namely, Bahi District, Chamwino District and Kondoa District. During the physical visit, it was found that Bahi and Chamwino Districts had better road and communication infrastructure compared to Kondoa District. The better infrastructure was partly because they are both located near Dodoma City, around 60 km and 20 km for Bahi and Chamwino respectively, but Kondoa District was very much isolated due to poor road infrastructure, which makes it unattractive for many businesses and hence led to a digital divide. It should be noted that more than three years ago, one would spend four to six hours in the car to drive from Dodoma City to Kondoa District.

Colin: Could you give a physical description of Kondoa District? Is it mountainous? Is it flat? And how many people live there?

Matogoro: It is a mountainous area and has around 269,704 people – 136,518 are male and

1 Matogoro, J., Mvungi, N. H., Justinian, A., Karandikar, A., & Singh, J. (2018). *Towards Affordable Broadband Communication: A Quantitative Assessment of TV White Space in Tanzania*. https://www.springerprofessional.de/en/towards-affordable-broadband-communication-a-quantitative-assess/15909208

2 An authorisation letter was required because the UHF spectrum band is licensed spectrum and therefore, for the university to transmit in this spectrum, we needed to have permission from the Tanzania Communications Regulatory Authority.

Kondoa Girls High School teachers, students and project team members.

133,186 are female.[3] KCN will connect three educational institutions to the internet. The first school was connected in May 2018. The school is a girls-only secondary school with around 810 students and 46 staff. Teachers and students are now connected to high-speed wireless internet delivered using TVWS technology. Kondoa Girls High School is located 4.6 km from the base station which has a backhaul connection. Teachers are very comfortable and are able to access materials online, undertake school management online tasks, and can access results online.

Colin: You have suggested some reasons – such as Kondoa's isolation – but can you say more about why you wanted to start the community network initiative?

Matogoro: A big motivation was proving that TVWS could work. We started the community network initiative because we believe it is among the feasible solutions to connect the unconnected. KCN is piloting the use of TVWS for community networks in rural Tanzania and addressing the current internet gap. A number of initiatives have been undertaken by various stakeholders in Tanzania to connect the unconnected using the traditional approaches. However, only 23 million users[4] have access to the internet and half of Tanzania's population remains unconnected. We believe that a community network initiative and the use of unused UHF spectrum will be a feasible solution to address internet connectivity issues in Tanzania.

Colin: Where did you get the idea of using the community network model? Did you maybe learn from other people also using the community network model or get advice from someone? Or did it just seem intuitively like the right way to go about connecting?

Matogoro: I came across the idea when I was working on a paper that was reviewing different approaches to connect the unconnected. It is from conducting a literature review that I found that a community network initiative could also be used as an alternative to connect the unconnected. In that direction we also tried to find out a feasible technology to connect the unconnected in rural Tanzania. We initially thought about using fibre, but it is very expensive and may not be feasible in places like Kondoa which is a mountainous area; hence laying fibre can be difficult and will be more expensive. Then we thought of Wi-Fi but we found that Wi-Fi can hardly cover an area of around 200 metres. In rural, mountainous areas like Kondoa, Wi-Fi could not be a feasible solution because you might need a lot of access points to cover a larger cell size. But with TVWS technology in the UHF spectrum band, you can penetrate mountains, trees and buildings with a good antenna gain. It is also a feasible solution in a rural area because no one is transmitting in this band. So that was the idea behind it, and we are very excited that our community network has managed to use TVWS to connect the unconnected in rural Tanzania.

3 United Republic of Tanzania. (2013). 2012 Population and Housing Census.

4 https://tcra.go.tz/images/documents/telecommunication/TelCom_Statistics_March_2018.pdf

FIGURE 1.

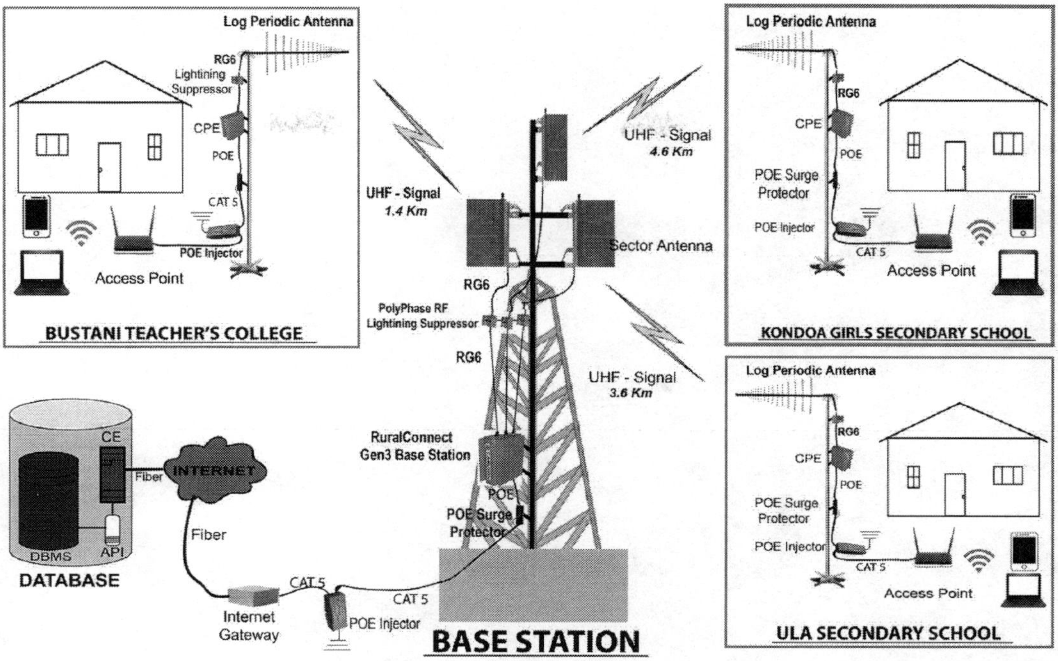

The network has four main parts: the transmitter (base station); the receiver (CPE installed at school); the geo-location spectrum database (hosted at the National Internet Data Center in Dar es Salaam); and the communication channel (UHF signal) and 2.4 GHz for end-users to access the internet using their laptop and/or smartphone.

Colin: Did you encounter any challenges in using TVWS?

Matogoro: In piloting the use of TVWS for the community network, KCN connected three educational institutions, namely Kondoa Girls High School – which I mentioned earlier – Ula Secondary School and Bustani Teachers College, which are located 4.6, 3.6 and 1.4 km from the transmitter. In the literature it is reported that UHF spectrum has good propagation characteristics to penetrate walls and trees, but in reality for that to happen one needs to have an antenna with good antenna gain. A 4 dBi antenna gain failed to establish a connection to Bustani Teachers College, which is just about 1.4 km from the transmitter. This failure is partly because between Bustani Teachers College and the transmitter there is a heavy forest which is also the source of water for Kondoa District. The team is working to replace this antenna with an 8 dBi antenna gain. So we can see that sometimes if there is a heavy forest, one may face challenges in establishing the link between the transmitter and receiver. However, TVWS remains a feasible and affordable technology in rural areas similar to Kondoa.

Jane: Is it a TVWS solution with a mesh network?

Matogoro: Yes, it is TVWS with a mesh network.

Colin: It definitely seems like TVWS is the good technology to use in the terrain you've described. Could you talk about the people that were initially involved in starting the community network, and maybe the motivations that brought you and others together to start the network?

Matogoro: The University of Dodoma[5] is leading the project from the technology side – but actually it is a project with a number of expert international team members. The project has the following team members: Jabhera Matogoro from the Department of Computer Science; Prof. Justinian Anatory from the Department of Telecommunication Engineering and Dean of the School of Informatics; Prof. Nerey Mvungi from the Department of Electrical Engineering at the University of Dar es Salaam;[6] Prof. Ermanno Pietrosemoli, a senior project researcher from the International Centre for Theoretical Physics

5 https://www.udom.ac.tz/home
6 https://www.udsm.ac.tz

Technical team members from KCN install equipment at Bustani Teachers College in Kondoa.

(ICTP)[7] in Italy; Dr. Marco Zennaro, also a senior project researcher from ICTP; Abibu Ntahigiye, the chief executive officer of the Tanzania Network Information Centre (tzNIC)[8] and ISOC Tanzania Chapter chairman; Nazarius Kirama, ISOC Tanzania Chapter secretary; and Rebecca Ryakitimbo from TechChix and an ISOC Tanzania Chapter member. Community members are represented in this project by the steering committee, which represents various stakeholders in Kondoa. I have worked with Prof. Anatory and Prof. Mvungi as research advisor at the University of Dodoma; I met Prof. Ermanno and Dr. Marco in a wireless tutorial class in Lusaka, Zambia during an African Network Operators Group (AfNOG)[9] workshop in 2013; I have also worked with Abibu, Nazar and Rebecca when I served as ISOC Tanzania Chapter secretary from 2015 to 2017. It is a team of dedicated people who are happy to lead the project.

Colin: Could you say more about what efforts you are making to discuss the community network and get involvement from community members, and maybe community members that don't have the same level of expertise as the researchers you are working with?

Matogoro: Yes, it is true that community members may not have same level of expertise, but we have found that most community members even in rural areas know what they expect from the internet. We have also found that in communities there are people who are naturally interested in some of these technical issues, and they are well known by most community members because of their involvement in undertaking similar technical activities. For instance, in Kondoa, we found that there are a number of youth who are able to install DSTV antennas. This made it easy to receive recommendations from community members forming the steering committee on who could help us. We managed to engage these youth in installing TVWS antennas. In short, we are trying to learn from community members in Kondoa and identify the local skills available and we build on this to train them on relevant skills to support the installed network. We have also found that the best approach to achieve technical

7 https://www.ictp.it
8 https://www.tznic.or.tz
9 https://www.afnog.org

sustainability in these communities is to adopt a learning-by-doing approach. This becomes interesting to them and they are happy to take the initiative forward.

Colin: Could you discuss the structure of the steering committee, who is on it, and how decisions are made?

Matogoro: The steering committee has members representing government, religious leaders, political leaders, educational institutions, youth and women around Kondoa District. The decisions are based on consensus among the steering committee members.

Colin: How do you think the approach to building this network differs from when a private company decides to deploy some kind of telecommunication infrastructure in a region?

Matogoro: The only difference and uniqueness of this approach is that this is a kind of bottom-up approach. You know, for the private company, deploying a network, in most cases, it's top down. They set the price and then people pay for it. But for the community network, the members of the community network have the power to set the price.

Jane: What advice would you give to someone who would like to start up a project like this?

Matogoro: One piece of advice is that the community network is the best approach. And having TVWS as the technology for the middle and last mile makes it a more feasible alternative to connect the unconnected, especially in hard-to-reach environments and rural areas with mountainous terrain similar to Kondoa. I know that there is still a challenge on the technical know-how, but let us work together to make sure that more communities are impacted and contribute to the digital economy. I am aware that most countries – especially developing countries – do not have policies supporting the operation of both community networks and TVWS, but that should not be an obstacle in exploring the benefit from these approaches. However, policy makers and government should create a favourable environment for community members to explore the potentials of these technologies. We need to have policies and regulations that favour this kind of technology and to develop capacities among community members. Generally, recommendations for the technologies to be used might be diverse depending on the community, but for the person who is working in a community like Kondoa, the community network is the best approach. But it is very important to make sure that the community members own the project.

Acknowledgments

Jabhera Matogoro would like to extend his thanks to the University of Dodoma and the Internet Society for supporting this project. The project was also supported by various stakeholders in different capacities, including the Tanzania Communications Regulatory Authority, KCN Steering Committee Members, faculty members from the College of Informatics and Virtual Education, Dr. Luzango Mfupe and his team from the Council for Scientific and Industrial Research,[10] Rajabu T. Kitunda and his team from Selcom Broadband Limited (UhuruOne),[11] Felix Alex who was engaged in the development of a geo-location spectrum database, and KCN Technical Team members (John Wambura, Phabian Misanya, Abdinasir Mohamed and Sidney Mbele).

10 https://www.csir.co.za

11 www.uhuruone.co.tz

THAILAND

BUILDING LAST-METRE COMMUNITY NETWORKS IN THAILAND

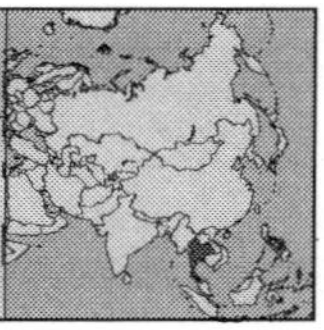

Internet Education and Research Laboratory (intERLab) – Asian Institute of Technology (AIT), Net2Home Social Enterprise
Kanchana Kanchanasut, Adisorn Lertsinsubtavee, Apinun Tunpan, Nisarat Tansakul, Preechai Mekbungwan, Nunthaphat Weshsuwannarugs and Parkpoom Tripatana
https://interlab.ait.ac.th

Introduction

Providing connectivity to rural communities has been one of the most challenging tasks worldwide, and Thailand is no exception. In this country, fibre optic connectivity to homes is still confined to provincial capitals and big towns. Although cellular coverage is extensive, data packages are expensive. More recently, fibre optic networks have been pushed by the government under the One Access Per Village Project, with 24,700 villages connected to the internet by the end of 2017 with one fibre end per village. But to access the net, villagers have to gather around the dropping points similar to those telecentre projects in the early days.

TakNet is one of the most successful community network projects in Thailand, bringing significant impact and improvement to people's lives in the province of Tak in the northwest of Thailand since 2013. Currently 15 communities have set up networks in their villages, with more than 1,000 residents using community networks on a daily basis. TakNet remains an experimental project that aims to provide internet connectivity to rural homes at affordable rates and acceptable quality. This makes it different from the telecentre approach. It was also planned for community members to be jointly responsible for the network.

At the time of writing, TakNet is the only community network operating in Thailand. This report discusses key success factors in setting up the network in response to a lack of connectivity at the local level.

Regulatory environment

Even though there is no specific policy in favour of running community networks in Thailand, local regulations allow us to use unlicensed 2.4 and 5 GHz bands in which the transmission range is limited. Nevertheless, this means less freedom to choose other wireless frequencies that many overseas community networks have. The frequency restriction limits our network size and coverage. The use of sub-1 GHz frequency bands such as TV white space (TVWS) is not allowed in our country. The Thai regulator, the National Broadcasting and Telecommunications Commission, is reluctant to ease regulations such as using high transmission for point-to-point communication and deploying rural broadband via TVWS. It is our policy to abide by existing rules and regulations in our network design and operation, limiting ourselves to the technology choices that are permissible. We are currently operating with an internet service provider (ISP) licence that has been issued to our Net2Home social enterprise company (see below).

Network set-up

The idea behind TakNet originated from our earlier work in post-disaster communication networks using mobile network technology. We realised that familiarity with technology is vital for successful deployment under stressful post-disaster scenarios; our work on community networks was built on this experience.

In general, a community network is a form of self-configuring and "self-healing" network where routers use dynamic routing protocols to form an ad hoc network. A community network allows individual users to join the network and share the connectivity by setting up their own relay routers at a village level up to city scale.

Initially, TakNet was a technical experiment connecting villagers using wireless mesh networks which could be easily transformed to become an emergency communication network in times of natural disasters. Now we use small mobile routers with firmware called DUMBO,[1] designed for post-disaster communication, to connect houses together using a limited 3G/ADSL gateway to the internet. The DUMBO firmware includes the Optimized Link State Routing (OLSR)[2] protocol to form the mesh

1 dumbo-technology.interlab.ait.asia
2 https://tools.ietf.org/html/rfc3626

network, SIP phones[3] and a few other applications like short message applications that users can deploy. The simplicity of DUMBO firmware enables the community to build or extend the TakNet network in an ad hoc manner to meet their own needs. This means that in case of natural disasters they would be able to repurpose their routers, currently fixed to their walls, to form a post-disaster communication network almost immediately.

Key success factors of TakNet

The key success factor of our project is its strong collaboration between three main players: the research and development team led by the Internet Education and Research Laboratory (intERLab)[4] of the Asian Institute of Technology, who develop and apply the DUMBO firmware for the community network; the corporate social responsibility (CSR) programme of the Thai Network Information Center Foundation (THNICF);[5] and the local community's participation. In addition, as mentioned, our Net2-Home social enterprise licence allows us to function as an ISP.

The intERLab team plays the key role in terms of software development by customising the DUMBO firmware to form the wireless mesh network. The intERLab team also concentrates on research and development to improve the network infrastructure using new technologies (e.g. TVWS, LTE small cell) as well as proposing new services for TakNet such as VoIP, instant messaging, video-on-demand and distributed ledger (or shared database) to support the day-to-day activities of community members. While doing research and development for TakNet, intERLab uses TakNet as its research test bed for other research projects, such as using the internet of things (IoT) for monitoring air quality.

In line with its objective of promoting the internet as part of infrastructure development, the THNICF cooperates with the intERLab team to expand the community network into nearby areas by using local technicians. Our experience shows that the key to success is to have simple technology where we can transfer the operation to local technicians who usually are without proper vocational education and have trouble reading English instructions. Involving local technicians and having them gradually take over running the network is our key strength. They monitor the network and act as our first-tier support. They are currently capable of troubleshooting and fixing most network issues. They are trained by the intERLab team and are the key to expanding the community network to nearby areas. They earn more income from installation fees and commissions if they get more members.

After five years, with our limited resources, we have expanded to 15 communities and have over 1,000 users. We are introducing different wireless technologies so that we can connect hard-to-reach areas.

A few years after our first community network village was connected, ISPs started to move in and offered services to the village. We viewed this as a positive benefit to the community – they would have better internet access and more choices. We worked with ISPs in reaching out to those villagers that could not afford the ISP's standard service price.

Today, despite the government's fibre-to-village project, and the penetration of the networks by the big ISPs, access is still quite limited. If the villagers opt for the government network, they have to visit a telecentre or go to other places where fibre optic lines get dropped. This can be quite far from their homes and cause concerns for parents. If the villagers opt for a commercial ISP, the monthly fees are still unaffordable for most of them. While last-mile fibre infrastructures are now receiving attention from the government, the regulator and big ISPs, we are convinced that the last metres in remote villages can be served very well by community networks.

Pathway to sustainability

Initially, we relied on young volunteers who were university students or new graduates attending the annual Thailand Networking Group (THNG) Camp[6] supported by the THNICF to help us with field deployments. These young volunteers went out to visit rural villagers and convinced them to get routers installed on the external walls of their homes. This provides a good start for building a reliable and resilient community network. DUMBO routers and other equipment were donated by various entities through the THNG Camp and other THNICF fundraising activities. After the network set-up was completed, the intERLab team and the THNG Camp members trained the villagers in how to use our network. When the villagers started to use our network, we asked for village volunteers to act as the local technical support team. Then our researchers taught the technicians how to maintain the network.

3 https://en.wikipedia.org/wiki/Session_Initiation_Protocol

4 https://interlab.ait.ac.th

5 https://www.thnic.or.th/en/home

6 www.thng.in.th/#thng-camps

It was not successful in the beginning. We had to persevere and teach them slowly. Now they know how to maintain, expand and deploy new sites.

During the first three years, TakNet was maintained on a volunteer basis where the local technicians cooperated with our technical staff to do some basic troubleshooting and collect users' fees. The fees entailed sharing the cost of the internet subscription fees with some electricity charges added. This amounted to about USD 2.5 per month per household. However, the fees collected were not sufficient for expansion or recovery in case of a major breakdown such as a router malfunction and replacement. Such a low fee paying model is not sustainable for the long term due to a lack of funding and human resource support. A solution was needed to develop TakNet further so that it could be attractive to more users who are also willing to pay more to sustain the network's operation and growth.

To implement this plan, we started a social enterprise called Net2Home in 2016 to fully manage the services and network deployment of TakNet. In this new model, the monthly fee for each participant is increased to USD 8 per month to cover the cost of network equipment installation, maintenance, internet connectivity and the use of local services (e.g. distributed ledger application, VoIP, video streaming, chat applications). However, this new subscription fee is still three times cheaper than those of the commercial ISP. After operating TakNet under the Net2Home company, the number of deployments increased from one to five communities per year.

We believe that a community network offers more value to members than just a connection to the internet. As mentioned, to ensure economic sustainability, TakNet aims to attract more members by introducing incentive applications like distributed ledger and chat applications. And with the IoT, community networks can be extended to provide services to agriculture and local village manufacturing activities, as well as offering waste or pollution management solutions inside a community. These economic and health value-added services could be very helpful in making the community network more sustainable.

Women at TakNet

TakNet provides local, national and international opportunities for women in a variety of activities, such as software/hardware development, network deployment in several rural villages in the province of Tak, and the chance to present scientific research at international conferences. At present, almost 50% of our team members at intERLab and Net2Home are women, while at the community level we also found that most of the community leaders working with us are women.

Conclusions

Internet subscriptions in Thailand are rather expensive even for urban residents with average wages higher than those in rural areas. The digital divide is a common social issue. While our experience may not impact on internet access for extremely remote villages, a digital divide still exists in villages with just one or two links to the internet, and we helped expand these one or two drop points to cover all houses within the village.

From our five years' experience, we have faced several challenges that have required concrete solutions to support the network's growth and long-term sustainability. Specifically, it will be very important to incorporate a broader, ambitious vision with core values when designing the next developmental phases of our rural community wireless mesh networks. Key issues like enabling faster digital transformation, improving agriculture yields, creating cleaner and greener manufacturing systems, and providing tighter socioeconomic integration of rural communities can serve as some important goals and milestones. To realise our vision, we need to re-think and prioritise from the perspective of villagers' painful experiences, to create solutions to tackle the issues they face, and to implement a business model that is integrated into these solutions. Modern developmental approaches such as the "lean and agile" methodology could prove to be highly valuable. We believe that technological innovations come through research, observation, participation and collaboration. Because of this, activities and funding in relevant research areas should be made available and managed effectively.

Action steps

The decreasing cost of cellular network services (e.g. 3G/4G) will definitely impact on the cost advantage of our community networks. Currently, there are low-cost 1 Mbps or 4 Mbps unlimited internet access plans offered by some mobile ISPs. These could entice our users to switch, but the ability to offer a higher speed (e.g. 10 Mbps to 30 Mbps during non-peak hours) at similar fees still remains our key advantage.

Newer and more cost-effective technologies, both in hardware and network frequencies, are being studied and researched. TVWS is a promising technology that is being studied and planned for testing. This year, the intERLab team and colleagues from other departments of the Asian Institute of Technology are going to run a series of experiments using TVWS/LTE on TakNet with a special approval for research purposes from the regulator. We have found that TakNet users are mostly from the younger generation, while the older generations do not have enough incentives to use the network. The THNICF is now supporting a project to build a distributed ledger application for TakNet users to take advantage of "last-metre" access more effectively as part of their day-to-day life. TakNet is and will be for all members of the community from all ages and gender groups; residents will appreciate TakNet for more than just buying internet access for their children.

TUNISIA

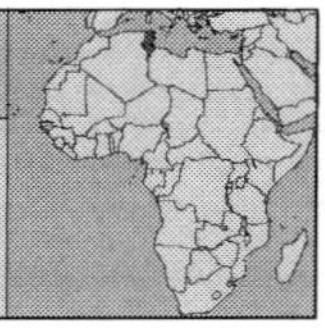

SAYADA INAUGURATES THE FIRST DECENTRALISED, FREE WI-FI COMMUNITY NETWORK IN TUNISIA

iGmena
Hamza Ben Mehrez
www.iGmena.org

Introduction

The Tunisian revolution brought the hope for a better system of internet governance – one that is participatory, democratic and open.[1] Today, the town of Sayada is pioneering an open government initiative. Since 2013 – and with the support of the local municipal government – civic technologists inaugurated the first decentralised, free community Wi-Fi mesh network in the country.[2] This is part of a recent local government code that aims to decentralise the power of decision making and spending to the local level.[3] Now, with the sharing of the Sayada community network experience, the country is in a strong position to make an open, free and resilient internet ecosystem a reality.

Connecting the unconnected

The challenges that Tunisia faces in increasing internet access, especially in rural areas, are significant. The country is ranked 65th out of 201 countries in terms of internet usage.[4] The country has 5,472,618 internet users with a penetration rate of 48.1%. The number of people *not* connected to the internet is a little more than that, in the order of 5,902,602 people. Tunisia is ranked ahead of Algeria and Morocco, its nearest comparable neighbours, in terms of internet usage.[5]

Official statistics published by the Tunisian national telecommunications regulator in April 2016 indicate that there are 516,061 fixed data subscriptions and 7,280,197 mobile data subscriptions, with a penetration rate of 16.02% for fixed data and 64.5% for mobile data.[6]

Most Tunisians are using a 3G/4G connection with a bandwidth capacity of 180 Gbps. Connecting 100% of the population is still a challenging issue and the focus of the government is to reach the unconnected people in the rural areas and households below the average income. The main factors related to increasing connectivity in rural areas are expanding infrastructure, ensuring affordability, increasing usability and developing innovative state policies.[7]

In this context, the coastal town of Sayada, some 140 km from Tunis and in the administrative governorate of Monastir, inaugurated the first free Wi-Fi community network in Tunisia in December 2013. The community, represented by students, technology experts, civil society activists, policy makers and municipality officials, was helped by CLibre,[8] a local association that aims to promote the culture of a free and open internet. Today, the Sayada community network – called "MeshSayada" – which covers 70% of the town, is a success story of a community-designed wireless network system.[9]

The community network also serves as a platform for locally hosted content, such as Wikipedia and OpenStreetMap,[10] and will expand to include more locally created content. The network consists of 11 rooftop nodes (using 12 routers), including the cultural centre, town hall and nine residences. The network links major areas of the town and covers the main streets, the weekly market, the train station and high schools. The routers have been placed in different interconnected nodes to have optimal

1 Jelidi, R. (2015, 28 November). Online Freedom of Expression in Tunisia: Finding the Right Balance. *iGmena*. https://www.igmena.org/-Online-Freedom-of-Experession-in-Tunisia-Finding-The-Right-Balance

2 Mesh networking is a type of network topology in which a device (node) transmits its own data as well as serving as a relay for other nodes. Routers are used to provide the best and most efficient data path for effective communication. In the event of a hardware failure, many routes are available to continue the network communication process.

3 Jouini, A. (2015, 22 October). Freedoms on the Internet: Where does Tunisia Stand? *iGmena*. https://www.igmena.org/Freedom-on-the-Internet-Where-does-Tunisia-Stand-

4 https://www.internetworldstats.com/af/tn.htm

5 Marzouk, H. (2017, 28 June). Utilisateurs d'internet en 2016 : la Tunisie classée neuvième en Afrique. *L'Economiste Maghrébin*. https://www.leconomistemaghrebin.com/2017/06/28/internet-user-by-country-2016-tunisie-neuvieme-en-afrique

6 www.intt.tn/fr/index.php?rub=265&srub=363&art=445& status

7 Verdomski, M. (2017, 21 June). Secure Connectivity – Why It's Important and How to Implement It. *IoT For All*. https://www.iotforall.com/iot-security-secure-connectivity

8 https://www.facebook.com/CLibretn

9 Naffati, W. (2013, 16 December). La ville de Sayada inaugure le 1er réseau wifi gratuit communautaire en Tunisie. *Tunisie Haut Debit*. https://thd.tn/la-ville-de-sayada-inaugure-le-1er-reseau-wifi-gratuit-communautaire-en-tunisie

10 https://www.openstreetmap.org

coverage of the network, whether on the street or inside the houses.

MeshSayada: A model of grassroots community network development

More than 15 Wi-Fi nodes were installed throughout the town via two unidirectional transmitters that have a diffusion angle of 60 degrees. Their range can reach two or three kilometres in a free zone without obstacles. The municipality of Mestir provided bandwidth for synchronisation and proxy of sites to the community of local developers who synchronised the Sayada web portal with the Sayada Wikimedia site.

Multistakeholder collaboration is critical to the project. On 13 November, 20 computer scientists and engineers came together in the Sayada Cultural Centre with a group of 14 girls (10 to 14 years old) from the city of Mestir and Sousse, Sayada community members, people from nearby towns, and groups from across Tunisia to learn and contribute with different skills toward building the network.

The Ubuntu Tunisia Association arrived from the neighbouring city of Sousse on the second weekend to help set up the server and to create the network portal page. Local engineers, academics and technology experts focused on participatory network planning, site planning, and solving various technical challenges. Everyone put their skills to practice when installing two directional routers on the town hall and using the router interfaces to test the mesh link distances to connect with battery-powered routers set up in the street. Fifty children came to attend the workshop. They later went home and explained MeshSayada to their parents.

Several IT developers from around Tunisia volunteered to contribute by adding additional applications and content to the local network. The applications and content are installed in the network service. These include OpenStreetMap, Wikipedia in French and Arabic, a collection of 2,500 free books in French, an Etherpad application for collaborative document editing, and a Media Grid[11] application for secure chat and file sharing.[12]

To consolidate the idea of inclusiveness and the bottom-up approach, a young developer created a local portal that links to each of these services, and that allows people to use easy-to-remember names such as Sayada.mesh or Wikipedia.mesh to access the local applications.

The project cost the association and the town of Sayada no money. The equipment was provided entirely by the Open Technology Institute (OTI),[13] a non-governmental organisation that works on community network development to help strengthen the participation of residents through open government processes. Residents donated their time and effort to build the Sayada network.

In terms of the network quality, it offers a useful user experience when accessing web pages, text chat and web-based maps. The challenge came when needing more bandwidth for intensive tasks, such as file sharing and video streaming. This would require additional work by local network administrators to optimise links and increase throughput.

The bandwidth capacity performance declined between two-hop and three-hop connections. On average, the two-hop throughput was 2 Mbps, and the three-hop links averaged 1.8 Mbps. However, the links are of an acceptable quality given the nature of the content on the local server; for web pages, text chat and web-based maps, the throughput constraints over the network should not result in a degraded user experience.

Conclusion

Six years after the revolution, a significant contribution has been made towards local development. After the passing of the local government code on 27 April 2018, 31 municipalities will benefit from financial autonomy to manage local affairs and to ensure the principle of free administration. Old and newly formed municipalities will work to include residents in local internet development policies and projects.[14] This will help not only the people of Sayada but all Tunisian citizens to participate in the local community development of their cities.[15]

Newly elected municipal councils could help create mesh community networks in other regions by organising awareness sessions on open budgets, transparency and corruption, the usefulness of information and communication technology (ICT) and the importance of community networks for their towns. Tunisian people hope that the newly elected municipal councils will invest resources that have been transferred from the centre to the periphery in modernisation projects, such as the roll-out of

11 mediagrid.org

12 Gerety, R., Gunn, A., & Hawkins, W. (2014). *Case Study: Mesh Sayada*. https://commotionwireless.net/files/posts/041814-Case-Study-Sayada.pdf

13 https://www.newamerica.org/oti

14 Gerety, R., Gunn, A., & Hawkins, W. (2014). Op. cit.

15 Abderrahim, T. (2017, 19 May). Tunisia's Decentralization Process at a Crossroads. *Carnegie Endowment for International Peace*. https://carnegieendowment.org/sada/70035

mobile networks so that all end-users can take advantage of digital opportunities and are able to use the internet wherever they are.

The new local government code is expected to help local authorities to be more financially independent by diversifying their sources of financing and providing guidance on how to allocate money for the development of infrastructure. The new legal framework for decentralisation could consolidate the multistakeholder bottom-up community networks model and ensure the participation of local communities in local decision making.

Action steps

The followings steps are suggested for Tunisia:

- Expanding infrastructure such as the national fibre-optic network through public-private partnerships is crucial. Partnerships and cooperation are essential to strengthen the national infrastructure backbone in rural areas.
- The promotion and the continued deployment of internet exchange points (IXPs) is another priority. These will stimulate the further development of the local community network content and ecosystem. Currently, Tunisia has two IXPs, TunIXP Tunis and TunIXP Enfidha. Rural areas need more IXPs to be deployed in other regions of the country to bring critical infrastructure and services such as interent service providers (ISPs), banks and data centres to these areas. Tunisia should take the further step of deploying IPv6 in the new services.
- It is also important to implement new data compression and caching techniques that would make telecommunications networks operate more efficiently. This can be possible by deploying open source hardware or making more efficient use of spectrum. This would benefit end-users and the community networks themselves with cost-oriented approaches, providing services that are tailored to the unique needs of the community, empowering local people, and thereby encouraging involvement in other grassroots efforts and community affairs and creating new working opportunities. The result would be networks that are better able to operate and deliver services and expand internet access. A virtuous cycle would be promoted by improving both access to and creation of local content and services.
- Tunisia has set an ambitious national strategic plan called "Digital Tunisia 2018" where deploying infrastructure is central to major national projects such as e-education, e-health and e-tax. Community networks have a role to play in this. However, future community networks in the country need to build trust in the communities. Community programmes may compete for funding from the same donor agencies, hampering their ability to collaborate. The selection of the right donors who have previous experience working together, both personally and professionally, can positively affect their ability to collaborate in attaining the community network objectives.

UGANDA

CONNECTING ISOLATED COMMUNITIES IN UGANDA

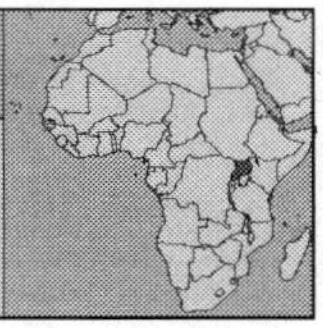

BOSCO Uganda
Tonny Okwonga
www.boscouganda.com

Introduction

Battery Operated Systems for Community Outreach (BOSCO) Uganda has been standing by isolated communities for the past 10 years, providing information and communications technology (ICT) connectivity and training in northern Uganda. One prominent observer once called this region the worst place in the world to be a child: remote, war-affected, culturally scoured, and deeply isolated over more than two decades of brutal insurgency. Being rooted in a tradition of solidarity with suffering people throughout these conflict years, there is a need to prevent technology from acting primarily as a conduit for unbridled social forces without concern for their impact on communities.

Technology in service of these forces carries the risk of simply amplifying the loudest social message, communicating to emerging global citizens that they are really just backward and poor Western consumers. We watch this threat unfold with the rise of institutions such as sports betting among the youth in northern Uganda, and the loss of cultural practices and norms.

We show that this does not have to be the case.

Background

BOSCO Uganda is a non-profit organisation under the trusteeship of the Archdiocese of Gulu. Back in 2006, many thought that the new technology that began to rock northern Uganda would deprive people of their traditional culture. Yet providing internet and voice over internet protocol (VoIP) telephony and solar-powered PCs to rural areas helped to both connect people and preserve culture.

Since 2006, BOSCO has supported many isolated, rural communities in gaining connectivity through ICTs and helped them become part of a broader networked community. Now, as one of Uganda's leading NGOs in the area of information and communications technology and development (ICT&D), it helps to carefully integrate ICTs into local community needs, leapfrogging missing technical infrastructure and working in areas indispensable to sustainable development, such as mentoring, e-learning for adolescent refugees, renewable energy, entrepreneurship, and research development. All these efforts attracted international attention and BOSCO received the inaugural Breaking Borders Award[1] in the technology category from Google and Global Voices.

BOSCO's historical competency lies in rural ICT connectivity and training, but our unique focus is using that technology for community building. BOSCO has provided a high-speed intranet to connect users with one another across regional communities, and then connected the network to a modest, shared internet connection. The network is powered by solar energy. The goal is to enable once-isolated peoples to leapfrog over not just missing technical infrastructure but, importantly, over the social infrastructure that is missing in war-affected rural villages and, thereby, build new foundations for emerging together into sustainable, globally participatory futures.

For instance, the small community networks, mushrooming amongst the brown ant hills in places like Pagak, Jengari, Unyama, and the Pabbo Parish and Catechist Training Centre, not only enabled the internet to bring news of events from elsewhere in Uganda and the world, but also fostered local pride. These communities have different abilities, yet travelling along different paths they can arrive at the goal of rural communities built around shared cultural values, such as shared land ownership and peace-building traditions. In short: connecting people, preserving culture.

The craft of building a cyber-catalysed community is a new manifestation of a lost art. It is pressed into contemporary expression both by the desire to overcome ever-widening disparities among peoples, resulting from the near-runaway pace of global change, coupled with unprecedented opportunities for collaboration created at the technology edge of that change. BOSCO's mission is precisely to leverage this frontier using ICTs for building international

1 BOSCO Uganda. (2010, 7 May). Archbishop Acceptance Speech – "Breaking Borders Award". boscouganda.com/2010/05/07/archbishop-acceptance-speech-breaking-borders-award

partnerships – both organisation-to-organisation and person-to-person – in support of the emergence of once-isolated communities into new forms of participatory global citizenship. But the fruit of these international collaborations must be bridging the traditional with the sustainable, along pathways that transform, rather than replace the old with the new.

BOSCO's integrated community approach: The case of CE3

BOSCO's immediate future is focused on developing connected, entrepreneurial and sustainable ecosystems with access to electricity in the communities we serve, to turn isolation and dependence into self-advocacy. Traditionally, many development projects follow a one-pronged approach, focusing time and resources on effecting change through singular pin-point interventions. For instance, take the case of education. Like many approaches, BOSCO includes computer education suited to the needs of the communities and the realities of global interconnectivity: a Web 2.0 collaborative approach with e-agriculture, educational outreach, e-government and others as well as computer-based entrepreneurial training. However, BOSCO's CE3 project integrates ICTs for the purpose of education with a new "eco-electrification" model. Inspired by Accenture's corporate citizenship values[2] on "Skills to Succeed" and "Environment", CE3 uses the collective impact of entrepreneurial education, energy, and connectivity to drive sustainable economic growth in rural communities.

The CE3 project adopts a community-focused and impact-driven approach to economic development, and was created after close discussions with rural unconnected communities and NGO partners operating in northern Uganda. Like many rural communities throughout rural Africa, our rural unconnected communities face limited economic opportunities daily. It is an environment where the youth mature in isolation without access to the internet, students complete their education and graduate with few local employment opportunities, culture discourages risk and entrepreneurship, and basic infrastructure such as electricity is unavailable – further limiting the development of education, ICT access and productive entrepreneurial ventures.

Over 50% of children in the developing world go to primary school without access to electricity – affecting over 291 million children worldwide. The lack of electricity at these schools limits students and teachers from being offered basic teaching and learning tools such as ICTs or e-learning solutions, as well as infrastructure such as lighting to study at night or electric water pumps for drinking and cooking. Often these schools are in close proximity to communities and business owners who also have a strong demand and willingness to pay for reliable energy, enhance their business and entrepreneurial skillsets to improve their current businesses or start new ones, and participate in the education and mentorship of students.

The CE3 project aims to equip communities with solar energy, Wi-Fi connectivity, entrepreneurial training and mentoring to create businesses and jobs that are more efficient, more diverse, and more lucrative – an economic ecosystem "in a box". To do so, the CE3 model comprises three interrelated pillars that build on one another.

The first pillar develops entrepreneurial skills using a blended learning approach: students work through a computer-based, self-paced, six-module course with assistance and guidance from a course facilitator. Designed for CE3, this interactive course walks students through the process of starting a business, and guides them through the development of a business plan, an output of course completion. After completing the course, entrepreneurs enter the mentorship programme, where they receive face-to-face, one-to-one mentoring from a local business professional and virtual one-to-one mentoring from an Accenture volunteer to help bring their business plan to fruition. Local mentors are comprised of community businessmen and women who have built successful businesses and offered to coach young programme graduates.

The skills acquired by programme participants under the first pillar enable them to leverage the energy and ICT components of the model to build businesses and grow the local economy.

The second pillar is energy. In each site location, the solar energy system is located in partnership with an institution that co-invests in the system setup, has built a secure facility to house the system and has agreed to pay for a certain percentage of the power.

Once power is available, each site is equipped with internet connectivity, enabling the site to participate in web-based activities such as entrepreneurship training. ICTs enable the programme to reach a significantly larger number of current and aspiring entrepreneurs and provide access to information that would not be available through traditional skills-building programmes.

2 Accenture is a partner on the project. See: https://www.accenture.com/us-en/company-corporate-citizenship

Building a community means more than just building ICT centres

Although our focus is still on implementing ICT and development centres, our activities go far beyond the provision of computer training. BOSCO engages in developing and providing long-term solutions for the social-economic development of the vulnerable population with whom we work.

Consider the example of expanding the scope of our work to reach out to the adolescent refugees and host communities in Adjumani and Arua, near the South Sudanese border. Between them, these regions host over half a million refugees from South Sudan and the Democratic Republic of Congo (DRC). BOSCO Uganda has built access points from locally available recycled oil drums (or in some cases metal plates), which makes them durable. These boxes (or "centres") are equipped with low-power-consuming netbooks and/or thin clients (both computer set-ups were used) that provide access to pre-loaded educational content called "KOLIBRI". However, to make these facilities relevant to the young people at the settlements, BOSCO staff members always interact and discuss the intervention with field-workers, such as case workers and caregivers, who are both refugees and nationals working in the child-friendly spaces and the early childhood development centres managed by Save the Children[3] and World Vision[4] and funded by UNICEF and other stakeholders. As a result, the learning content becomes integrated into the support offered to refugees in order to meet their needs in the most appropriate way. For example, a vulnerable teenage mother will, through content provided, learn skills such as catering, and with the support of her mentor, set up a small baking business.

Through the ICT centres, people living in the refugee settlements have the opportunity to connect to family and friends currently living in other settlements or still in the conflict zone. This could be achieved either through social media platforms or through software applications such as Skype, or similar applications pre-installed on each computer by us.

Although the primary beneficiaries of the intervention are youth living in the refugee settlements, the centre is also open to the general community living around the area. This encourages the possibility of stronger interaction between the Ugandan community and the South Sudanese refugees.

Expanding networks and infrastructures

Funded by AFRINIC under the FIRE Africa Grants[5] we launched a new project called "Expanding BOSCO-Uganda Internet/Intranet Network Access to the Rural Remote Communities in Northern Uganda". The project is focused on expanding the BOSCO solar power and internet infrastructure to remote areas by using long-distance wireless that connects rural areas to a central server station and to the world wide web at minimal running costs. The system is powered by solar energy, and offers VoIP telephony and an intranet which connects all stations through an internal high-speed network and a central server for easy information sharing amongst the users. The BOSCO expansion plan goes beyond borders: in the future we also want to promote ICTs on a non-profit basis outside Uganda and especially in the war-affected areas in South Sudan.

Action steps: Our vision for the future

Our vision for the future is based on our experiences and our achievements. The last 10 years have shown us that when you dream big, big things can happen. In this fast-changing world it is often challenging to define a single future for an organisation like BOSCO Uganda. We will continue to work in our three key areas of renewable energy, entrepreneurship and mentorship, and research and development.

Specifically:

- We will be strengthening our ICT backbone and entire infrastructure, for instance, by erecting towers.
- We need funds to support the acquisition of internet bandwidth to help scale up our network in remote areas.
- We need more capacity building in policy and advocacy to help the unconnected get connected in a way that they can afford.

3 https://uganda.savethechildren.net

4 https://www.wvi.org/uganda

5 https://www.fireafrica.org

UNITED STATES

BUILDING RESILIENCE WITH COMMUNITY TECHNOLOGY

Digital Equity Laboratory, The New School; Detroit Community Technology Project
Greta Byrum and Diana Nucera
https://www.digitalequitylab.org and https://detroitcommunitytech.org

Introduction

Our story begins almost a decade ago, when new technologies enabled resistance and sparked a wave of digital activism in Tunis, in Tahrir Square, in New York City's Zuccotti Park and Washington DC's K Street, and around the world. Many groups in the United States (US) recognised the potential of that moment: long-time advocates for media justice and literacy, public access media organisers, builders of community Wi-Fi and low-power FM radio, community organisers and civil rights leaders, open tech and data advocates, hackers and policy strategists. This is the story of a vision that emerged among people working together to create community resilience and digital justice between Detroit, New York City, Washington DC, and eventually in collaboration with international partners, by building community-owned internet infrastructure.

Community technology and digital stewardship

> Wireless internet can unlock the enormous potential in our local communities. These opportunities can only be sustained, however, if networked technology projects are led by people who are deeply invested in their community's welfare; that is, people with a deep understanding of – and a desire to maintain – the fabric that binds their community together.[1]
> – Diana Nucera

Community networks in the US have long struggled to grow in parallel with the large international community networks that have emerged, particularly in Europe. Some US standouts have achieved a sustainable operating scale and model, and have provided a critical long-time service for their communities. Tribal Digital Village Network (TDVNet) has been bringing free internet to community anchors in indigenous territories since 2001, currently with over 350 miles (over 560 kilometres) of point-to-point and point-to-multipoint links supporting 86 tribal buildings (and providing a net neutral service).[2] Monkeybrains in San Francisco is a nimble independent local wireless internet service provider (WISP) which uses a combined millimetre-wave, mesh and point-to-point system to serve 5,000 locations on a sliding-scale basis.[3] Sudo Mesh is a five-year-old community-owned and run local mesh network serving Oakland, California,[4] while Meta Mesh in Pittsburgh, Pennsylvania has built out to 65 live sites comprised of 109 devices.[5] And NYC Mesh has built out 178 nodes using volunteer labour and a decentralised governance model.[6]

But unlike guifi.net,[7] Freifunk[8] or Rhizomatica,[9] in the US we do not have a long history of expanding networks beyond discrete geographic areas or particular use cases. This has a lot to do with the consolidation and the political power of the incumbent telecommunications industry, which has taken many steps to place a stranglehold on local broadband[10] by creating state-level prohibitions on ownership of broadband facilities by city governments and by starving local networks of backhaul (bandwidth) by buying out or blocking competing wholesale bandwidth providers.[11] The capture of the US Federal Communications Commission at the national level by industry lobbyists has also had a

1 Nucera, D. (2014). Two-Way Streets: Forging the Paths Towards Participatory Civic Technology. *Civic Quarterly, 2*. https://civicquarterly.com/article/two-way-streets

2 https://sctdv.net/about-tdv

3 https://monkeybrains.net

4 https://sudoroom.org/wiki/Mesh/History

5 https://www.pittmesh.net

6 We have not listed many small-scale municipal broadband projects in the US (and one large one, in Chattanooga, TN), since this chapter is focused on community-led broadband. For information on municipal broadband in the US see Christopher Mitchell's work at https://muninetworks.org

7 https://guifi.net/en/node/38392

8 https://freifunk.net/en

9 https://www.rhizomatica.org

10 Koebler, J. (2015, 14 January). The 21 Laws States Use to Crush Broadband Competition. *Motherboard*. https://motherboard.vice.com/en_us/article/qkvn4x/the-21-laws-states-use-to-crush-broadband-competition

11 Breitbart, J. et al. (2007). *The Philadelphia Story: Learning from a Municipal Wireless Pioneer*. https://www.newamerica.org/oti/policy-papers/the-philadelphia-story

An access point tower constructed by Tribal Digital Village overlooks Pala, California. Photo: Tribal Digital Village

global effect by catalysing the recent repeal of net neutrality (the Open Internet Order of 2015).[12]

Another issue that has challenged scale for community networks in the US is that many do not find a subscriber base beyond loyal techie advocates and small communities. Without a broad base of users – including those who may not be able to afford the high cost of monopolistic broadband service, or who may not have (or want to have) the skills, time or patience for troubleshooting a do-it-yourself (DIY) system – community networks often rely upon one or two staff, or just volunteers. So, ironically, some community networks with a flat or decentralised governance approach end up serving already information- and technology-rich areas, since those areas are where volunteers live and work.

As journalist and community network documentarian Armin Medosch puts it, "far-sighted techies tend towards a linear extrapolation of technologies into the future without considering other factors, such as politics, the economy, the fundamental differences between people in class based societies."[13] Similarly, Alison Powell's research on community networks points out their tendency to reinforce "geek publics" rather than the "community publics" they purport to serve.[14] In the complex political and economic context of the urban US, the political and economic challenges around digital infrastructure, access and inclusion have kept many US community wireless networks from achieving or sustaining scale.

Commotion Wireless

Starting in around 2008, a group of media justice and community wireless activists had a different vision for community wireless. Led by developers and media activists who had built Champaign-Urbana, Illinois's Indymedia Center and its CUWiN network,[15] the Open Technology Institute (OTI)[16] began developing Commotion,[17] envisioned as an integrated plug-and-play OpenWrt-based mesh networking platform that communities could easily deploy, and which included secure encryption and a suite of locally hosted applications. With device-based peer-to-peer dynamic mesh routing, Commotion would be able to work with or without a connection

12 https://www.fcc.gov/document/fcc-releases-open-internet-order

13 Medosch, A. (2014). The Rise of the Network Commons, Chapter 1 (draft). *The Next Layer*. www.thenextlayer.org/node/1231.html

14 Powell, A. (2008). WiFi publics: producing community and technology. *Information, Communication & Society*, *11*(8), 1068-1088. eprints.lse.ac.uk/29545

15 https://web.archive.org/web/20041111094354

16 https://www.newamerica.org/oti

17 https://www.commotionwireless.net

to the global internet, and to route traffic around points of failure automatically. Based on the principles of security, resilience, and local control, Commotion would be an open-source local communication and media platform to be used from Tahrir Square to Detroit, in emergencies from Mubarak to Katrina.[18]

In 2011, OTI started testing beta versions of Commotion in the field in multiple locations, often working with groups who were familiar with local media and DIY radio "barnraisings" and interested in trying new technologies to advance their work – for example, the Media Mobilizing Project in Philadelphia,[19] the Allied Media Projects in Detroit,[20] and Occupy K Street in Washington DC.[21] Like many other attempts at building local wireless, these early tests showed that without enough local techies and engineers with dedicated time to spend fixing things, networks would break frequently, users would get frustrated, and user confidence and numbers would decline. Furthermore, stable electricity and bandwidth were a challenge in some locations, and depended on local relationships and governance – that is, network representatives to be in charge of different aspects of the networks, from physical hardware to communicating with users and node hosts.

The Detroit Digital Justice Coalition

Meanwhile, in the summer of 2009 at the Allied Media Conference in Detroit,[22] a group of leaders came together to investigate the role that local technology projects might play in restoring communities harmed by the US economic crisis,[23] by training people how to use the internet and technology to create local micro-economies. The resulting Detroit Digital Justice Coalition (DDJC)[24] was comprised of 13 member organisations and individuals including seniors, youth, environmental justice communities, welfare rights activists, hip hop community organisers, community gardeners, independent technologists and designers, each one believing that communication is a fundamental human right.[25] The DDJC had a plan to bring whole communities online, not just isolated individuals, so the internet would be a welcoming environment for new users.

For decades, Detroit has topped the list of "worst-connected cities" nationally, with 2013 data showing nearly 60% of its residents lacking in-home broadband subscriptions and 40% lacking any connection whatsoever;[26] 38% of Detroit residents live below the federal poverty level,[27] and since 2014, tens of thousands have faced water shut-offs and evictions due to tax foreclosure. Yet offline, Detroit's organisers knew that vibrant community leaders have been steadily transforming the city from the ground up with community gardens, land trusts, co-ops and a thousand other grassroots initiatives enabling local self-determination. As Allied Media Projects describes in their Media Literacy Guide, the DDJC's goal was:

> [T]o use digital technologies to strengthen these efforts, interconnect them, and make them more visible. This would shift the online narrative of the city while propelling communities to rewrite their offline reality – growing businesses, community programs, and community infrastructure through media-based organizing skills.[28]

The Coalition first started building their vision for digital justice offline by collaborating on a set of shared principles. To develop a common understanding of how to shape the role media and technology might play in communities, the Coalition asked members to answer the following questions:

- How are you currently using media and technology for organisation and economic development?
- What kind of support and collaboration would make your work stronger?
- What should digital justice in Detroit look like?

The Detroit Digital Justice Principles[29] were born through this process, presenting a unifying definition of what digital justice means to the community: access, participation, common ownership, and

18 Gall, C., & Glanz, J. (2014, 20 April). U.S. Promotes Network to Foil Digital Spying. *The New York Times*. https://www.nytimes.com/2014/04/21/us/us-promotes-network-to-foil-digital-spying.html?_r=0

19 https://mediamobilizing.org

20 https://www.alliedmedia.org

21 https://en.wikipedia.org/wiki/Occupy_D.C.

22 https://www.alliedmedia.org/ddjc/story

23 The US Subprime Mortgage Crisis of 2009 and resulting Great Recession hit Detroit hard, as the city was already suffering from the decline of the US auto industry. Foreclosures skyrocketed and the city went into bankruptcy. The federal government then appointed an Emergency Manager who had unilateral authority to alter or eliminate collective bargaining agreements, cut city services, and lay off public employees, overruling the democratically elected city government.

24 detroitdjc.org

25 At the same time, the Media Mobilizing Project (MMP) in Philadelphia responded to this opportunity by building a Digital Justice Coalition with groups focused on housing, workers' rights, youth, education, and public health. See Breitbart, J. (2014). A Victory for Digital Justice (Your Tax Dollars at Work). In D. Freedman et al. (Eds.), *Strategies for Media Reform: International Perspectives*. Fordham University Press.

26 Callahan, B. (2014, 3 November). America's worst-connected big cities. *Redistributing the future*. https://redistributingthefuture.blogspot.com/2014/11/americas-worst-connected-big-cities.html

27 https://www.census.gov/quickfacts/fact/table/detroitcitymichigan/PST045217

28 https://www.alliedmedia.org/files/dfm_final_web.pdf

29 https://detroitcommunitytech.wordpress.com/detroit-digital-justice-coalition-principles

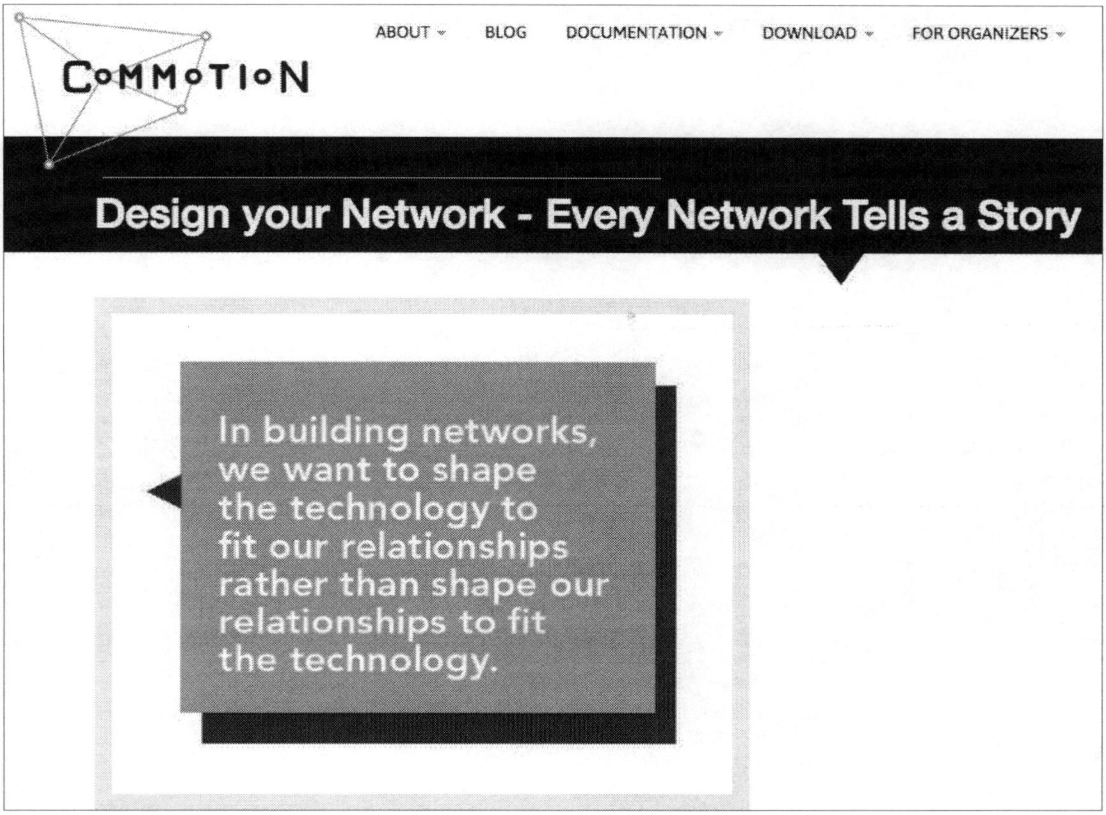

Commotion Construction Kit. Image: Commotion Wireless

healthy communities, each one describing a different aspect of digital justice.

With these principles and a proposal focused on a "community technology" approach to creating healthy technology ecosystems, the DDJC's member organisations got to work implementing their vision through the BTOP federal broadband stimulus programme.[30] The Detroit Future programmes built networks of teachers, youth and artists and trained them to use media production and web development for organising, teaching and helping small businesses. By 2011, the Detroit Future programmes had trained hundreds of Detroiters of all ages to use technology on their own terms to address a range of issues from housing to environmental degradation to water shortages, while also providing a platform for the city's growing creative entrepreneurship.

At the same time, OTI was working hard to solve the problem of how to maintain and expand Commotion community wireless networks locally, including in Detroit. In order to provide local technologists, builders and organisers with documentation, tools and resources, OTI together with Detroit-based social enterprise The Work Department[31] developed a prototype "Neighborhood Construction Kit", which included information modules on everything from how to make a flyer to promote your network to how to install a chimney mount on a rooftop. With additional technical information on how to configure Commotion firmware on standard routers (mostly Ubiquiti) and other devices like Android phones, this became the Commotion Construction Kit and moved online to live as documentation on the Commotion Wireless GitHub site.[32]

Still, even with resources available online, OTI found it difficult to build a stable foundation of network support among local organisations and advocates, especially when local organisations had so many

30 The Broadband Technology Opportunities Program (BTOP) was part of the American Recovery and Reinvestment Act (ARRA) of 2009, the Obama administration's response to the Subprime Mortgage Crisis and the Great Recession. Coalitions of local organisations could apply for Public Computer Center and Sustainable Broadband Adoption (SBA) grants under the BTOP; the DDJC and Allied Media Projects were awarded an SBA grant, which they used to create the Detroit Future Program, comprised of Detroit Future Schools, Detroit Future Youth and Detroit Future Media, all of which trained Detroiters on digital skills.

31 https://www.theworkdept.com

32 https://commotionwireless.net

International Community Technology Seed Grantees. Photo: Detroit Community Technology Project

competing demands on their time personally and in their community leadership work. It would take more organising, tapping into local movement building, dedicated expertise and resources, plus a whole new approach to building a new tech-supported economy and a method for teaching different kinds of learners about how to build and maintain networks. Detroit's organisers and community technologists once again held the key to bringing these efforts together.

The Detroit Community Technology Project

> By training local residents to be "digital stewards" of the networks, community organizers create employment opportunities and provide public Internet access while strengthening social networks within the community... At their most ambitious, these projects suggest a different way of thinking about work in the digital future: that we might manage our digital ecosystem with care and intention rather than constantly disrupt and respond to disruption. At minimum, these projects show the importance of localism and workforce development to maximize the economic benefits of new networks and produce technology that is attuned to a community's needs.[33] – Joshua Breitbart

In 2012, the DDJC and the Allied Media Projects' Detroit Future Media partnered with OTI to create the Digital Stewards Program,[34] which trains neighbourhood leaders in designing and deploying community wireless networks with a commitment to the Detroit Digital Justice Principles.

The Digital Stewardship training programme is based on the pedagogy of popular education, including the idea that everyone brings valuable knowledge and experiences into any learning space. Instructors take the role of facilitator, building peer-to-peer educational conditions through activities that work for all types of learners. This approach includes a process of envisioning all of the ways we can use a community network to strengthen neighbourhoods and solve local problems, beyond simply gaining access to the global internet. The Digital Stewards Program led to the formation of the Detroit Community Technology Project (DCTP)[35] in 2014. DCTP was developed to encompass broader community technology education, to organise work and to share best practices.

In 2014, OTI and DCTP received funding to work internationally to implement an international

33 Breitbart, J. (2015). *Telecommunications Policy and the Future of Work.* New America. https://www.newamerica.org/oti/policy-papers/telecommunications-policy-and-the-future-of-work

34 Allied Media Projects. (2012, 22 October). AMP partners with the Open Technology Institute to launch Digital Stewards Program. https://www.alliedmedia.org/news/2012/10/22/amp-partners-open-technology-institute-launch-digital-stewards-program

35 https://detroitcommunitytech.org/?q=story

Community Technology Seed Grants programme,[36] supporting 11 community groups internationally to adopt and modify the training for their own contexts. In the process of implementation, OTI and DCTP found that many international groups would have difficulty obtaining hardware that would reliably run Commotion, or that they were trying to achieve different community goals from what the platform would support. So the project forked again, this time moving all of the organising and general wireless curriculum onto the Community Technology Field Guide[37] so that it could be more generally applicable for different kinds of community technology projects, including but not limited to Commotion.

Red Hook, Brooklyn and Hurricane Sandy

In 2012, shortly after Digital Stewards launched in Detroit, OTI also helped bring the Detroit Digital Stewardship curriculum to Red Hook Initiative (RHI)[38] in New York. RHI wanted a mesh network to create a local online youth-produced radio station, so with OTI's help they adapted the Digital Stewardship curriculum on basic community organising, wireless engineering and construction for their workforce training programme for young adults living in public housing, and added learning modules on video production, web design, and professional development. In partnership with Parsons student Alyx Baldwin, RHI held participatory design workshops with residents, and by the fall of 2012, RHI's Digital Stewards had built a small network serving some of the major public spaces and community anchors in the neighbourhood.

Although New York is a wealthier city than Detroit, many of its residents face similar challenges in accessing broadband service: 31% of New Yorkers currently do not have broadband service at home, including 32% of Black and 33% of Latinx New Yorkers. That is considerably more than the 21% and 23% for White and Asian residents. Geographically, service is also not equitably distributed. In some neighbourhoods, people would have to pay on average 5% of their income on cable service in the current market, and would have only one option for service.[39]

When Hurricane Sandy struck New York in October 2012, flood-prone Red Hook was devastated. Cell phone service was down and internet service went out in places. The neighbourhood was dark, with chest-deep water in the streets – but with its small mesh network, RHI was still able to connect to its staff and communities in parts of the neighbourhood that had no communications or power at all for weeks after the storm.[40] RHI organised volunteers using the mesh to help distribute supplies to elders and others unable to leave the public housing towers in the neighbourhood, and gave the community a voice online to broadcast what was happening. People all over the world following RHI's Twitter feed put together online shopping lists and shipped supplies to Red Hook.

Though the Red Hook WiFi[41] project was in the works before Hurricane Sandy struck, it gained prominence and media attention after the storm. The Federal Emergency Management Agency (FEMA) boosted RHI's broadband connection with a satellite uplink, so where the regular internet was unavailable, residents and government workers could log on to the mesh to quickly find out where to pick up supplies or find government officials. Neighbourhood building owners who had been wary about allowing RHI to install equipment on their rooftops joined the network, seeing its importance in areas of the community where power and communications were out. The City of New York opened up a funding opportunity for projects like Red Hook WiFi, which had provided critical community-led service in the aftermath of the disaster using innovative technology.

While RHI had led the Red Hook project, OTI had provided the link to the Detroit curriculum and helped to adapt and implement it. Building upon the success in Red Hook, OTI's umbrella organisation, New America, was awarded a contract with the City of New York's Economic Development Corporation (EDC) using a federal Sandy recovery grant to scale up the Digital Stewardship approach in New York City, and created its new Resilient Communities initiative. Resilient Communities started work in 2015 by seeding funding and resources among community leaders and community-based anchor organisations committed to building resilience and supporting affected communities in five low-income Sandy-impacted neighbourhoods throughout the city (East Harlem, South Bronx, Far Rockaway,

36 Gerety, R. (2015, 4 October). Meet the Community Technology International Seed Grantees. *Allied Media Projects*. https://www.alliedmedia.org/news/2015/10/04/meet-community-technology-international-seed-grantees

37 https://communitytechnology.github.io

38 https://rhicenter.org

39 Mayor's Office of the Chief Technology Officer. (2018). *Truth in Broadband: Access and Connectivity in New York City*. https://prodnet.www.neca.org/publicationsdocs/wwpdf/042018nyc.pdf

40 Cohen, N. (2014, 20 August). Red Hook's Cutting-Edge Wireless Network. *The New York Times*. https://www.nytimes.com/2014/08/24/nyregion/red-hooks-cutting-edge-wireless-network.html

41 https://redhookwifi.org

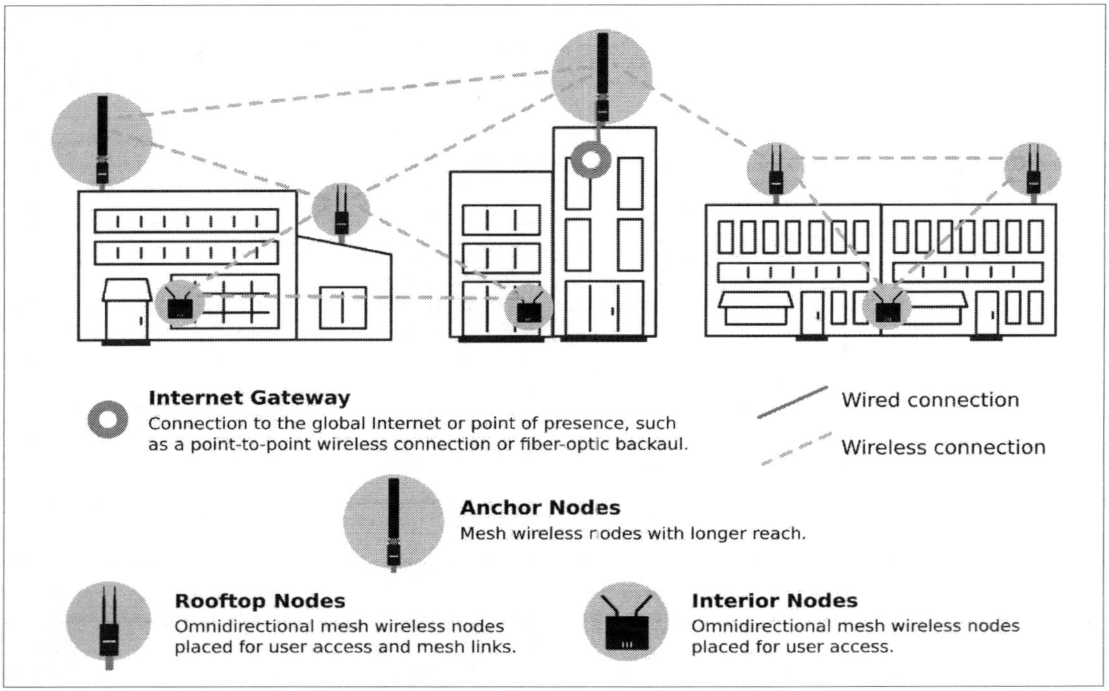

Community technology network distribution diagramme.

Sheepshead Bay and Gowanus). At the same time, the Red Hook Initiative also won a Sandy recovery contract from the City of New York's EDC to expand its network and add solar panels to increase resiliency. The City meanwhile embarked on an effort to build out free wireless systems in the Red Hook public housing development and two other New York public housing sites, Queensbridge and Mott Haven Houses, looking to weave together integrated systems of community- and City-led Wi-Fi throughout the city.[42]

2018: Community technology in Detroit and New York

To meet the need of scaling up the Digital Stewards Program for the five new communities, New America's Resilient Communities team once again teamed up with DCTP to expand upon the Digital Stewards training programmes, adding emergency management plans to the Digital Stewardship trainings based on an understanding of building resilience as a process of building trust and relationships, not just technological systems.

In the meantime, while the NYC Resilient Networks project launched in 2016, the Detroit Community Technology Project was also launching its new Equitable Internet Initiative (EII),[43] a programme training local residents on wireless broadband internet sharing in Detroit neighbourhoods, expanding the number of networks in the city built by Digital Stewards from around eight to 11. As we launched our work at both sites, Resilient Networks commissioned DCTP to develop the Community Technology Handbook,[44] a resource to share the approach and pedagogy of popular education for community technology to train Digital Stewardship trainers in both cities. Following the model of the Detroit Digital Justice Principles, project leaders also started the work by developing a set of shared principles to guide and ground the work in community needs.

EII's mission is to ensure that more Detroit residents have the ability to leverage online access and digital technology for social and economic development. Led by three community anchor

42 Lewis-Kraus, G. (2016, 3 November). Inside the Battle to Bring Broadband to New York City's Public Housing. *Wired*. https://www.wired.com/2016/11/bringing-internet-to-new-york-public-housing

43 Detroit Community Technology Project. (2017, 10 April). DCTP Launches the Equitable Internet Initiative Digital Stewards Training Program. *Allied Media Projects*. https://alliedmedia.org/news/2017/04/10/dctp-launches-equitable-internet-initiative-digital-stewards-training-program

44 Detroit Community Technology Project. (2016, 16 November). Introducing the "Teaching Community Technology Handbook". *Allied Media Projects*. https://www.alliedmedia.org/news/2016/11/16/introducing-%E2%80%9Cteaching-community-technology-handbook%E2%80%9D

Detroit Digital Stewards install networking equipment. Photos: EII

The Detroit Community Technology Project's Teaching Community Technology Handbook.

organisations, EII has now trained stewards across the city, and is already serving households in three neighbourhoods.[45] It has also successfully negotiated contracts for wholesale backhaul and transit with independent ISPs to provide low-cost wireless gigabit speeds to residents in its partner neighbourhoods, employing local Digital Stewards to perform maintenance and upkeep and continue to expand the networks. Forty young people throughout EII neighbourhoods received training and mentorship to build local apps to run on the networks, and EII's neighbourhoods are also designing resilience plans based on the networks.[46]

Resilient Networks NYC is designed to withstand shocks and stresses and provide community-maintained and cooperatively owned critical telecommunications infrastructure in flood-prone areas of New York City. But in the summer of 2017, even as Resilient Communities and partners had trained some 30 Digital Stewards citywide, and hurricanes were about to hit Houston, Miami and Puerto Rico, the Resilient Communities project had not been able to build a single node yet due to bureaucratic constraints (federal disaster recovery funding controlled by the city but regulated by federal officials has created multiple bureaucratic hurdles due to permitting, paperwork, environmental review, etc.).

Racing to develop a plan for that hurricane season, Resilient Communities adapted its planned network repair kits to develop the Portable Network Kit (PNK). The PNK is a collection of off-the-shelf consumer hardware that can be configured easily to make your own local online or offline Wi-Fi network for about USD 800 to USD 3,000, depending on the battery system. The PNK connects devices in a small area – anywhere from one building or public square

45 Mondry, A. (2017, 31 July). How a surprising partnership will result in internet for Detroiters lacking access. *Model D*. www.modeldmedia.com/features/equitable-internet-initiative-073117.aspx

46 https://detroitcommunitytech.org/eii/resiliency

Yamil Lora from The Point CDC in the South Bronx performs site surveys for a new resilient wireless network. Photo: Danny Peralta

to about a half square mile (1.3 square kilometres) if you add or "mesh" additional Wi-Fi devices to create a wider range. If you add additional kits, you can mesh them together to create an even wider range.

PNKs have found their way to Dominica and Puerto Rico post Hurricane Maria, and EII is now beginning to incorporate them into its network as well. Every participating partner in New York City will have PNKs to build out and expand their networks in emergencies or for community events and programmes.

Scaling up the community technology approach to resilience

> Revolutionary solidarity is what love looks like at scale. – Diana Nucera

As developed through the partnerships in Detroit and New York, community technology is a method of teaching and learning about technology with the goal of restoring relationships and healing neighbourhoods. Community technologists are those who have the desire to build, design and facilitate a healthy integration of technology into people's lives and communities, allowing them the fundamental human right to communicate.

We believe that increasing resilience means building and deepening relationships and developing creative solutions to strengthen communities in times of change and uncertainty. Our work starts from a set of core principles that focus on listening, participation and equity as a foundation for

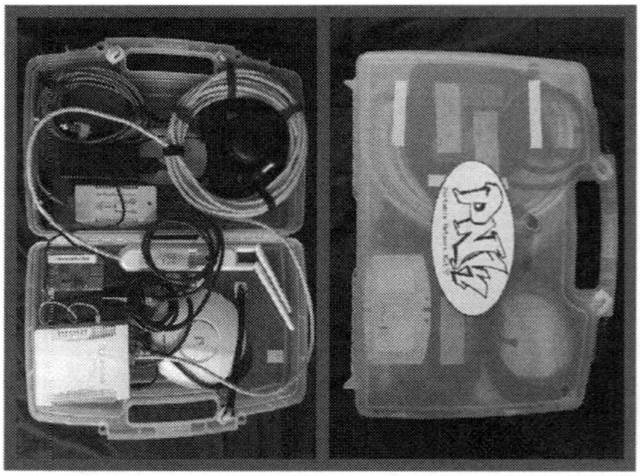

Portable Network Kit. Photo: Resilient Communities

building community resilience. We work with local leaders and groups to uplift and share creative, visionary and locally rooted solutions – and to make the systems and institutions they depend on more responsive.

With the current global emergency due to climate change, a rising tide of authoritarianism, and the nearly unchecked power of big tech, we see a critical opportunity now to develop future-ready solutions together with the communities most affected by these crises, particularly through the redesign and rebuilding of brittle physical and digital infrastructures.

Next up for DCTP/EII and Resilient Communities: DCTP will publish the Community Technology Workbook, which contains six years of learning modules and programming primarily designed by DCTP through the work and partnerships in Detroit and New York. Meanwhile, network build-out continues, as does the process of developing sustainable business models for all network anchor organisations in Detroit and New York City.

Our collaboration responds to all the tectonic political events and intertwined social, economic and natural disasters of the last decade: we in the US have learned about resilience from Maria, Katrina, Sandy, the Flint and Detroit water crises, PROMESA[47] in Puerto Rico and emergency management and the foreclosure crisis in Detroit, the US nation's ongoing oppression of people of colour and immigrants – and of course the rise of Donald Trump, the spread of surveillance, algorithmic and predictive criminalisation, and tech-enabled targeted deportation. We recognise that crises like these are happening around the world. Our hope is that, in the same way we have been able to scale our work in the US by creating an adaptable model of teaching, learning, and responding to local needs and circumstances, others around the world can adapt and scale up the community technology approach and tools.

Action steps

If you want to know more about us, the best way is to check out our websites:

- Detroit Community Technology Project – https://detroitcommunitytech.org/?q=story
- Resilient Communities – https://www.newamerica.org/resilient-communities
- Digital Equity Laboratory – https://www.digitalequitylab.org

We also have a GitHub full of community technology tools for those of you that want to get started planning and building at https://communitytechnology.github.io. Note that some of these resources may be out of date – stay tuned to the Detroit Community Technology Project's feed for updated material in our forthcoming Community Technology Workbook.

Below, we have crafted a few other ways in which you can be a part of the US community technology movement:

- Donate.[48] We can always use money. We can also use devices that are no more than two years old, or any Ubiquiti brand routers. If donating ethernet cables, we need to know the make and model in order to determine if they are capable.
- Hold critical conversations with your friends and family about technology and the future. Brainstorm ways in which we can re-imagine digital access and equity. Check out the chapter on facilitation in our Teaching Community Technology Handbook[49] for ideas on how to do this.
- Let us train you in community technology! If you feel your community is in need of a community wireless network, we may be able to train you. Please fill out the DCTP Community Technology Training Survey[50] if you want to plan a training in Detroit.
- Wherever you are, advocate for digital equity: net neutrality (no price discrimination for internet content; no preferential treatment for different service or content providers); privacy protections and accountable data stewardship for ordinary people; public and community ownership of infrastructure; and internet equity – digital opportunities, access, and quality of service.
- Remember: communication is a fundamental human right!

47 PROMESA is the Puerto Rico Oversight, Management, and Economic Stability Act, a 2016 US federal law which, as in Detroit, gives federally appointed emergency managers the authority to overrule local elected lawmakers, laws, and systems as an austerity measure.

48 https://www.alliedmedia.org/dctp/donate

49 https://detroitcommunitytech.org/teachcommtech

50 https://docs.google.com/forms/d/e/1FAIpQLSdUvgR7nGr-5vZhoDUYV3xyFg8AWNhombfoa3d23_xHFN-I6g/viewform

VENEZUELA

KEY CHALLENGES TO THE COMMUNITY NETWORK MODEL IN VENEZUELA

Maureen Hernández

Introduction

The digital gap undermines the development opportunities of communities profoundly. The asymmetric distribution and use of communication resources negatively affects a developing community, increases its vulnerability, and deepens the vicious cycle in which not having access to information inhibits the awareness of this information, and, therefore, the ability to claim it as a right.

While, as Ritu Srivastava says,[1] the internet has a democratising effect on society, this is not always the case in countries that are authoritarian states, or, such as Venezuela, have features in common with these states.[2]

Community networks, defined along clear principles such as non-discriminatory and open access, open participation and community ownership of infrastructure, do not exist in Venezuela – or, if they do, are not widely known to the internet community in the country.

The reason for this, as I argue in this report, is that community networks conceptualised in this way require certain structural conditions to be in place so that they can flourish. As I point out, at least three areas in Venezuela – the economy, policies on access, and laws impacting on freedom of expression – are particularly unsupportive of the open community network model found in many other parts of the world.

Economic factors

Hyperinflation and purchasing power

Venezuela has a rigid currency exchange control system, and to understand the economic problem and how to acquire any equipment not produced in the country, this problem must be understood first. While acquiring foreign currency is an extremely cumbersome process, the system has also changed a lot over the years, which makes it difficult to understand properly. However, what is not difficult to see is that due to the exchange control system, the import of any hardware that could be used in a community network is cost prohibitive.

According to calculations by Reuters, the last update of the currency auction system led to a devaluation of the Venezuelan bolívar fuerte (VEF) of more than 100% compared to last year. At the time of writing, the official exchange rate – known as the DICOM rate[3] – was fixed at 201,363.84 VEF per euro (EUR).

There is no official United States dollar (USD) rate, so for this we will apply a simple rule of three. At the time of writing, the exchange rate was at EUR 0.86 per USD 1:

- USD calculated rate = EUR rate in VEF x EUR per USD rate
- USD calculated rate = 201,363.84 x 0.86 = 173,172.90 VEF

The equivalent DICOM rate in USD contrasts with the more than VEF 3,524,330.15 that 1 USD was worth on the black market at the time of writing,[4] the latter which is the *de facto* exchange rate for a large part of the economy of the country that does not have access to the auction offered by the government.

This hyperinflation translates into a simple consequence: the purchasing power of the average Venezuelan is almost nil.[5]

The new minimum monthly salary[6] in the country is VEF 4,196,000, calculated based on 30 working days for social benefits with five days a week as workdays. According to the calculated DICOM rate (VEF 173,172.90/1 USD), the minimum

1 Srivastava, R. (2016). A Network by the Community and for the Community. In L. Belli, *Community Connectivity: Building the Internet from Scratch. Annual Report of the UN IGF Dynamic Coalition on Community Connectivity*. https://bibliotecadigital.fgv.br/dspace/handle/10438/17528

2 Urribarrí, R. (2013). Venezuela: De la Sociedad del Conocimiento al Socialismo del Siglo XXI. Paper presented at IV Congreso Venezolano de Investigadores de la Comunicación, Barquisimeto, Venezuela, 29-30 November. www.congresoinvecom.org/index.php/invecom2013/INVECOM2013/paper/viewFile/338/353

3 DICOM is now the country's only official exchange rate, although US dollars trade on the black market. www.dicom.gob.ve

4 On 29 July 2018 at https://dolartoday.com

5 Chinea, E., & Ramirez, M. (2017, 26 December). Venezuelans scramble to survive as merchants demand dollars. *Reuters*. https://www.reuters.com/article/us-venezuela-economy/venezuelans-scramble-to-survive-as-merchants-demand-dollars-idUSKBN1EK0XK

6 dctos.finanzasdigital.com/Gaceta-Oficial-6383-Ajuste-Sueldo-Minimo.pdf

TABLE 1.

Costs of wireless network equipment

Device	Mercado Libre Venezuela (VEF)	Amazon.com (USD)	Work time (days)*
Xiaomi MiWiFi R3	187,880,513[7]	39.99	1084.76
TP-Link WDR3600	200,000,000[8]	49.99	1154.73
Ubiquiti PicoStation M2	368,000,000[9]	81.49	2124.71

Cost of items on 28 July 2018.

wage represents only USD 30, and only USD 1.47 on the black market.

Now let us see how "low-cost" the acquisition of equipment for a wireless network can be in Venezuela. We will compare the prices found on local and international markets for some equipment needed for the installation of a LibreMesh[10] network (see Table 1).

For local reference we are using MercadoLibre, the Argentine e-commerce platform which is dominant in the region and by far the most popular portal in Venezuela. This platform is very often the only way to buy computer hardware locally, due to the fact that formal merchants are subject to economic regulations and often price-imposing policies by the government, frequently leading to their closure. As a reference for global markets, prices on Amazon.com are included. The work time is calculated dividing the monthly wage by 30 days.

Based on the figures in Table 1, for a community of 10 families with two economic providers per family, for a total of 20 monthly minimum wages, it would take them almost two whole months of salaries to acquire a Xiaomi MiWiFi R3.

Conclusion: Obviously, what is considered low-cost for many scenarios is difficult (or impossible) for a working-class community in Venezuela, in the same way the fees to cover costs associated with the maintenance of the network or to replace equipment would be too high for the community.

If this is the situation, why not resort to funding?

It is clear that the inability to pay for hardware is not a problem that we Venezuelans face alone, and there are several ways to secure financial support from various organisations. However, these too come with challenges.

When we talk about grants or subsidies that we can apply for in Venezuela, it is important to understand that the currency received by the organisation must be entered into the country legally, that is, at the DICOM rate. This results in a significant weakening of the purchasing power of the funding received.

To clarify this relationship let's look at an example. A grant of USD 20,000 at the DICOM rate (VEF 173,172.90) would end up having a net value in bolivars of 3,463,458,040 Bs.

- Funding in bolívars = Amount (USD) x DICOM rate (VEF/USD)
- Funding in bolívars = USD 20,000 x 173,172.90 = VEF 3,463,458,040

We can, further, use this equation:

- Funding in bolívars/Local cost of the item = Number of units.

In the case of a Ubiquiti PicoStation M2, using the prices in Table 1, VEF 3,463,458,040/368,000,000 = 9.41 units

This means that the entire amount of our initial subsidy would end up transforming into the equivalent of paying for 9.41 Ubiquiti PicoStation units at local prices. This can be compared to 245 units that the same money can buy in the United States with the initial amount.

Investment by the private or public sector

As of 22 March 2009, the internet was taken off the list of investment priorities for the nation and considered superfluous spending by presidential instruction, as gazetted in Official Gazette No. 39,146. This means that for the national plan or in the national government's agenda, access to the internet is considered a luxury. As a result, currency exchange petitions from this sector have been put at the bottom of the pile.

Technological factors

Current state of connectivity

Connectivity in Venezuela is going through a critical period. Internet speeds in the capital of the country can reach 20 Mbps with some providers (to which

7 https://articulo.mercadolibre.com.ve/MLV-503105794-enrutador-modem-conmutador-inalambrico-xiaomi-mi-fi-gestion-_JM

8 https://articulo.mercadolibre.com.ve/MLV-515144909-router-gigabit-inalambrico-de-banda-dual-n600-tl-wdr3600-_JM

9 https://articulo.mercadolibre.com.ve/MLV-464069979-antena-ubiquiti-nanostation-m2-24-ghz-11dbi-_JM

10 https://libremesh.org

FIGURE 1.

Average broadband internet speed in Venezuela

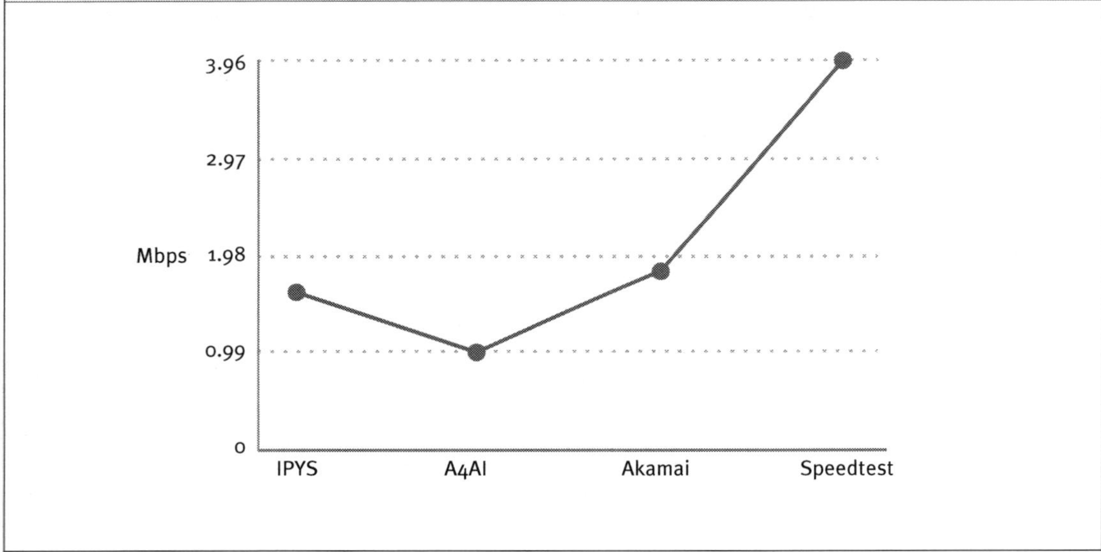

only a few have access), but when leaving the federal district, the chances of getting a quality service provider are very few. Venezuela occupies one of the last positions on global quality of service indexes[11] and the last position in the region.[12]

International phone calls can no longer be made because the fixed exchange control policy does not allow operators to cope with the devaluation of the currency.[13] Because of this, citizens turn to the internet to communicate with the rest of the world – even though the state of connectivity is as precarious as the economic situation.

Figure 1 compares average internet speeds in Venezuela. We draw on four major reports for measurements. Firstly, we use the report from IPYS Venezuela,[14] which is from our point of view the most valuable source of internet-related and freedom of communication insights in the country. Besides this report, we draw on data from the Alliance for Affordable Internet (A4AI),[15] Akamai[16] and Ookla (Speedtest).[17] As they suggest, the average broadband speed does not surpass 4 Mbps, and is reported to be lower than 2 Mbps by three of the reports.

Spectrum access

Venezuela has set aside the 324 MHz band of spectrum for mobile services, and the national regulator Conatel has identified portions of spectrum available in the 700 MHz, 800 MHz, 900 MHz, 1700 MHz, 1900 MHz, 2 GHz and 3.6 GHz bands, as disclosed in an administrative ruling published in the Official Gazette back in 2016. This has been confirmed on several occasions by measurements performed by other authors.[18]

However, these concessions have been granted for only two categories: community television and

11 www.speedtest.net/global-index#mobilewww.speedtest.net/global-index

12 According to the UN Economic Commission for Latin America and the Caribbean (ECLAC), Venezuela is one of the countries with the slowest high-speed connections, with about 0.2% connections with a speed of over 10 Mbps. Economic Commission for Latin America and the Caribbean. (2018). *State of broadband in Latin America and the Caribbean.* https://repositorio.cepal.org/bitstream/handle/11362/43670/S1800532_en.pdf?sequence=1&isAllowed=y

13 Associated Press. (2016, 11 April). Venezuela's cellphone providers suspend international calls. *Daily Mail.* www.dailymail.co.uk/wires/ap/article-3534144/Venezuelas-cellphone-providers-suspend-international-calls.html

14 Balbi, M., et al. (2018, 8 May). Internet surfing at its minimum. Situation of internet in Venezuela. *IPYS.* https://ipysvenezuela.org/2018/05/08/internet-surfing-at-its-minimum-situation-of-internet-in-venezuela

15 https://a4ai.org/affordability-report/data/?_year=2017&indicator=ITU_O&country=VEN

16 https://www.akamai.com/uk/en/about/our-thinking/state-of-the-internet-report/global-state-of-the-internet-connectivity-reports.jsp

17 www.speedtest.net/global-index/venezuela#fixed

18 Zennaro, M., & Arcia-Moret, A. (2013). Low cost spectrum measurements. In E. Pietrosemoli & M. Zennaro, *TV White Spaces: A Pragmatic Approach.* Abdus Salam International Centre for Theoretical Physics. wireless.ictp.it/tvws/book/tvws.pdf; Hernandez, M. (2016). Caracterizacion de los espacios en blanco del espectro radioelectrico: Caso de studio del Estado Merida. In L. Belli, *Community Connectivity: Building the Internet from Scratch. Annual Report of the UN IGF Dynamic Coalition on Community Connectivity.* https://bibliotecadigital.fgv.br/dspace/handle/10438/17528

radio stations. Currently there is no information about any other type of concession on the official portal of the regulator. This seems in line with the 180-degree about-turn taken by the current administration regarding open internet access.[19]

The threat of theft of infrastructure

The dire economic situation in the country makes this a key factor that needs to be considered when attempting to set up communications infrastructure.

In Venezuela, complaints about the massive theft of infrastructure are public knowledge and despite investment by private entities in security, the problem does not go away. The theft of equipment ranges from kilometres of fibre optic cable to repeaters, towers, batteries and antennas.

These forms of vandalism put pressure on any sort of bottom-up, collaborative governance structure that communities try to put in place. Like a recent Reuters article states,[20] local operators expect a worsening of service quality in the short term due to theft.

Legislation that works against freedoms

Discouraging policies

When a community network or a local project is set up by a community, the main thing is the people and the positive impact of this project on their lives. The purpose of providing a local connectivity tool is to improve the quality of the community and not, in any way, to put these people in danger or under the radar of any legislator. Unfortunately, in Venezuela there are recent laws that pose a real threat to freedom of expression, especially in digital media.

The first one is the Law against Hate, Intolerance and for Peaceful Coexistence, passed by the National Constituent Assembly in 2017.[21] According to the law, any speech could be described as "hate speech", and the law could be arbitrarily applied to imprison someone, which threatens freedom of expression and intimidates critics of the government.[22] At the same time, the Venezuelan constitution does not allow people to express opinions anonymously. This enables witch-hunting and threatens people's freedom of expression.

On the other hand, there is the creation of the Strategic Centre for Security and Protection of the Fatherland (CESSPA) in 2013.[23] One of the objectives of CESPPA is to serve as a central repository for information on security, defence and intelligence. The regulation proposes the creation of an entity whose functions will be the surveillance and monitoring of internet communications.

This, together with the new definition of "hate speech", begs the question: how do you establish a community network without allowing the users freedom of speech and having to censor certain conversations? Community networks base their principle of neutrality, according to Navarro et al., on "a commons oriented framework for community networks" where the network can be used for any participant for any purpose. Therefore, a network where some types of communications are prohibited expressions (while they are not illegal) is not a neutral network or not a community network.[24]

Action steps

While it is incredibly difficult to develop community networks in Venezuela, it is not fair to say it is impossible. The truth lies in trying to understand the sum of the difficulties and how to find allies to circumvent them.

Some of these difficulties are out of the control of the community: the exchange rate, the cost of equipment, laws that censor free speech, and even the likely theft of equipment, despite measures a community might take to safeguard their networks. These point to areas where civil society can engage legislators and policy makers in search of solutions.

19 Urribarrí, R. (2013). Op. cit.

20 Oré, D. (2016, 15 December). Crime wave worsens Venezuela's already shaky telecoms service. *Reuters*. https://www.reuters.com/article/us-venezuela-telecoms/crime-wave-worsens-venezuelas-already-shaky-telecoms-service-idUSKBN1441TH

21 Urribarrí, R. (2013). Op. cit.

22 espaciopublico.ong/ley-odio-monopolio-estatal-etica/#.WuZuMdNuZQK

23 www.el-nacional.com/noticias/politica/oficializan-creacion-del-centro-estrategico-seguridad-proteccion-patria_151643

24 Navarro, L., et al. (2016). A Commons-oriented Framework for Community Networks. In L. Belli, *Community Connectivity: Building the Internet from Scratch. Annual Report of the UN IGF Dynamic Coalition on Community Connectivity*. https://bibliotecadigital.fgv.br/dspace/handle/10438/17528

ZAMBIA

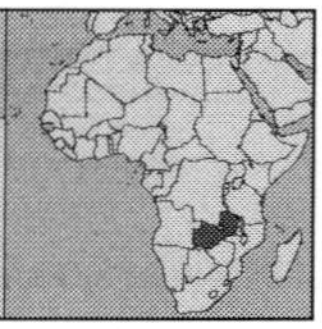

COMMUNITY ENGAGEMENT IN COMMUNITY NETWORKS IN RURAL ZAMBIA: THE CASE OF MACHA WORKS

Macha Works
Fred Mweetwa and Gertjan Van Stam
www.machaworks.org

Introduction

Macha is a typical village in the rural areas of Zambia's Southern Province. Some 135,000 villagers live in homesteads scattered over a 35-km radius. Its economy depends on seasonal agriculture and trade. The average income per villager is USD 1 a day, with subsistence farming of maize the main livelihood activity. The nearest town is Choma, some 70 km away. Macha has a community radio, and a mission station run by the Brethren in Christ church, which also operates a hospital, schools, and medical research institute. Its focus is on providing medical and educational services to Macha and four other neighbouring chiefdoms. Chief Macha, who is resident in the village, is the outspoken chief presiding over the Macha chiefdom.

Catalysed by the needs of the medical research institute, the Macha community has been experimenting with various technologies since 2003 in order to solve the challenge of communicating with the rest of the world. Apart from a VHF two-way radio connection with Choma city, the internet was the only way to communicate with the outside world until the arrival of the mobile network in late 2006.

After setting up a VSAT satellite link to the internet and building a local Wi-Fi network in 2004, a non-profit cooperative called Macha Works was established so that the community could take ownership of the network themselves and oversee its operations, maintenance and expansion. This report discusses the evolution of Macha Works: from a single satellite connection in a village shipping container, to a project that has resulted in community networks being launched in nearly all the provinces of the country.[1]

1 Macha Works is part of a network – Worksgroup – with separate local organisations in different countries. For instance, in Zambia there are, among others, Mukinge Works and Kalene Works, and in Zimbabwe there is Murambinda Works. Each organisation has its own governance structure and is in various stages of development.

Inspiring local communities

Macha Works' vision is to inspire people in rural areas to reach their collective and individual potential. Towards that goal, Macha Works has activities in many areas, among which are education (schools, vocational training in health and ICTs), financial services, energy, transport, and research and development. It operates inclusively, encouraging the participation of community members in its various initiatives on an equal basis. What we call "local talent" actively engage with Macha Works and push its projects forward. Macha Works also reaches out to other, equally remote communities. Those communities delegate their "local talent" to spend considerable time at Macha to learn "the tricks of the trade". When they go back to their villages, they set up community networks. In this way, nine successful community networks have been set up in total: in Macha, Kalene, Mukinge, Minga, Chitokoloki, Chilonga, Minga, Chikanta and Lusaka West, covering almost all provinces in Zambia.

The Macha Works model is anchored in the universal African principle of Ubuntu. Ubuntu emphasises communal love and the sharing of one's resources: the better-off share with the less-well-off. When resources are shared, costs are reduced and operations become more efficient, making services affordable for communities.

Working with the community the African way

The Macha Works experience shows that connecting rural communities in Africa is quite a challenge. Villages lack basic infrastructure, including electricity. Connecting through satellite links is extremely expensive. The erection of mobile towers means that mobile internet is proving to be an alternative. However, telecoms services do not allow communities to become empowered through setting up inclusive governance structures, and lack the collective vision offered by community networks.

The African experiences in rural areas are highly diverse. Each of the nine community networks has its own story to tell. Each has found a focus; but whether it is education (Macha) or helping the local citrus business (Kalene), all focus on the preservation of African culture and lifting the community "to the next level" by providing internet connectivity, among other services.

In the process, many social and technical hurdles are encountered. The community recognises that most challenges they face, such as poverty and a sense of not valuing their local cultures, are rooted in colonialism, and involve narratives drawn from imperialism and orientalism. Because of this, technical and social challenges have to be mediated in a way that takes a long-term view of change, requires stamina, and a constant rehearsal of local African values and practices. As mentioned, these practices are embedded in Ubuntu or "communal love". They also involve specific concepts that have been developed as a result of the project in Macha: *oratio*[2] or communicating embodied knowledge; *relatio* or "relational resource allocation"; *animatio*, the continuous present moment; and *domino*, the strive for maturity.

New models of communication for change

Academic literature – when accessible from Macha using the internet – did not provide useful guidance on how to go about building and maintaining the kind of community-led networks we wanted to. Because of this, Macha Works approached national universities and other academic partners to understand issues and potential ways of mediating better. In the process, we found that the ICT-for-development models applicable elsewhere in the world bore little relevance to the utterly complex socio-technical contexts and lived experiences in communities like Macha. Therefore, after acquiring authoritative guidance in collaboration with Chief Chikanta, in his former role as the Vice Chairman of the House of Chiefs, and after the reporting on the practical experiences we encountered in the Zambian House of Chiefs, we developed a methodology that involved community members in a way that was meaningful to them. As a result of this process, we have written numerous peer-reviewed and publicly accessible papers[3] about our approach and work and presented them in various academic settings.

Phases in the roll-out of the Macha Works community network model

Macha Works recognises three distinct phases in its community intervention and three distinct phases in the roll-out of its network.

Three phases of community intervention

In the process of responding to local needs, and in order to encourage local agency, Macha Works draws on the specific communication skills and ways of problem solving, mediation, and collaboration embedded in the community. This is important to ensure the long-term sustainability of the intervention. In particular, it identified three social processes in this regard: 1) community engagement, 2) workforce development, and 3) local thought leadership.

In the process of engaging the community, the organisation exercises sensitivity to local contextual frameworks and understandings, for instance, regarding time and space, affecting both the practice of human interaction and the assessment of realities. Social improvisation integrates the interests of stakeholders through observation and conversations in reiterative processes of interaction. Interactions follow local cultural behaviour patterns and involve consultation with the entire community. This process dissuades individualistic action, mediates any power imbalances experienced in relation to local structures, and engenders an environment with shared values, a common purpose, and sufficient levels of skill sharing to be able to integrate technologies in a way that is relevant to the community. This form of community engagement recognises the importance of human agency and sets up safety nets for the inevitable change in life conditions that the introduction of technologies brings.

By being practical and following the principle of "seeing-is-believing", local engineers working on the networks can implement appropriate technological solutions, overcoming challenges in a way that supports the sustainable progress of the community at large.

Developing a workforce conversant in ICTs and local meaning-making creates the agency for interventions. This needs to be an expression aligned with the local context and culture. It involves a revolution in education, aligning the capacity building in both content and format with local needs and learning processes.

Access to ICTs plays a crucial part in developing a workforce in the community and allows the community to access and share knowledge. The internet fuels local people with drive and vision, nurtures leadership and builds the community's technical know-how.

Lastly, thought leadership is a cross-cutting concern and involves a holistic progression through five phases, namely:

- Careful positioning, attaining the explicit right to influence others.

2 See, for example: Mawere, M., & van Stam, G. (2017). Oratio: A Framing of Knowledge in the Context of Technology and Academia. In M. Mawere & T. R. Mubaya (Eds.), *African Studies in the Academy: The Cornucopia of Theory, Praxis and Transformation in Africa?* Bamenda: Langaa RPCIG.

3 See: www.vanstam.net/gertjan-van-stam

- Expressed permission, grounded in lasting relationships.
- Tangible outcomes, with sustainable achievements through commendable actions.
- Capacity development, building abilities in individuals and organisations.
- Honourable representation, through recognition of wholesome and embodied knowledge.

The aim of thought leadership is to "give voice" to the local narrative: from the community, on activities in the community, on behalf of the community. These are the narratives of how ICTs, through the community network, have amplified the local, human intent.

Three phases in the network roll-out

Macha Works has also identified three phases in the implementation of community networks. These phases are: 1) sensitisation, 2) development and 3) fragmentation.

First, the sensitisation phase. This is when the community is not yet aware of the potential of the internet and has little experience accessing it. In these cases, LinkNet – the name that was given to the unit that oversees the technical operations and expansion of the network – would set up a work-station hub with connectivity to the internet in a modified shipping container. The set-up of each container is designed and tweaked according to the local realities. For example, solar power is used if the village is off the national electricity grid. The container is fitted with a satellite dish for internet connectivity. The container – which has been called the LinkNet Resource Container – acts as a socio-technical hub, a social business and a network operations centre. Its business model operates according to *relatio*, where the costs are shared by users and the community at large.

The container has been particularly successful in engaging the community right from the start, offering a secure, dedicated environment for "local talent" to meet. It also serves as a venue for ICT training. The value of this first phase has been recognised by the Zambian ICT regulator, which has financed the production and deployment of LinkNet Resource Containers in various rural communities.

In the second phase, from its base in the LinkNet Resource Container, a local/wide area network is developed. A multitude of set-ups, whether using mesh or point-to-point networks, have been implemented, all depending on the local context and the local expertise of the "local talent". At this stage, the community network has typically incorporated other community activities – whether offering training courses, business management support for local business, or even facilitating air transport to remote locations. Typically, after some years, the LinkNet Resource Container becomes too small to accommodate the various activities and the hub needs to be relocated. In Macha, Macha Works signed a 30-year lease and built extensive infrastructure for its activities, including training institutes.

In the third phase, commercial service providers start showing an interest in the community. As the demand for connectivity increases in the community, commercial service providers can start roaming the market. In some instances, there may be sufficient demand to justify laying a terrestrial broadband connection. In Macha, for instance, this has been done by Africonnect, a national internet service provider that was active in Choma.

At this point the market tends to fragment, with individual institutions in-sourcing their internet again, depending on individual donor involvement and the availability of well-skilled IT personnel in the community. At this time, the activities of Macha Works shift, with the LinkNet Resource Container becoming obsolete (it can be relocated to other communities). While Macha Works might continue facilitating training and capacity building on a number of topics, it now becomes involved in the broader issues faced by rural communities.

Conclusion

It is an indisputable fact that internet connectivity is a powerful tool to empower local communities and to guide their development. The internet can also help to preserve and share African cultural heritage, and, through the internet, rural Africa can become part of and enrich the global community.

But the mediation of tensions and conflict that are inevitable in the resource-deprived rural African environment needs committed leadership. The development of local skills is also essential. Socio-technical sustainability is dependent on the long-term engagement of local talent; local, national and international collaboration and alliances to withstand the tides and flows of super-colonial behaviour; community sensitisation on the benefit of accessing the internet; and a commitment to embodied knowledge, transparency and community ownership.

Macha Works has received the endorsement and support of many and various organisations and authorities in Zambia and abroad. They have recognised the uniqueness of the organisation's bottom-up approach, where the local community is responsible for its own development.

Action steps

The following steps are recommended to support community networks in Zambia:

- Enable connectivity and access through policy: Network constraints and access barriers suppress local voices, knowledge and inclusion.
- Involve trans-disciplinary engagement: Multiple and complementary understandings of society and technologies are necessary to reconcile an abstract international discourse – regimes of non-locally derived "truth" – with the challenging African realities and access constraints.
- Value the local: Many avenues are said to have been tried, but most appear to not take into account the local context, vocabulary, access and agency.
- Think local: Activating local meaning and relevance and the production of local systems and content are critical.
- Involve local: Iterative programmes involving local end-users and local talent engender embedded solutions and applications.
- Open development: Holistic, culturally aligned development involves the sharing of resources.
- Scaling is hard: Whether or not you can scale up depends on the local-level context and resource opportunities.

Made in the USA
Monee, IL
05 November 2026

67811024R10144